MODERN DEVELOPMENTS IN GAS DYNAMICS

MODERN DEVELOPMENTS IN GAS DYNAMICS

Based upon a course on Modern Developments in Fluid Mechanics and Heat Transfer, given at the University of California at Los Angeles

Edited by W. H. T. Loh

Manager, Science and Technology
Space Division, North American Rockwell Corporation
Downey, California

Ⴔ PLENUM PRESS • NEW YORK – LONDON • 1969

Library of Congress Catalog Card Number 69-14561

ISBN-13: 978-1-4615-8626-5   e-ISBN-13: 978-1-4615-8624-1
DOI: 10.1007/978-1-4615-8624-1

© 1969 Plenum Press, New York
Softcover reprint of the hardcover 1st edition 1969

A Division of Plenum Publishing Corporation
227 West 17th Street, New York, N. Y. 10011

United Kingdom edition published by Plenum Press, London
A Division of Plenum Publishing Company, Ltd.
Donington House, 30 Norfolk Street, London W. C. 2, England

To
Our Late Teacher Theodore von Kármán

PREFACE

During the last decade, the rapid growth of knowledge in the field of fluid mechanics and heat transfer has resulted in many significant advances of interest to students, engineers, and scientists. Accordingly, a course entitled "Modern Developments in Fluid Mechanics and Heat Transfer" was given at the University of California to present significant recent theoretical and experimental work. The course consisted of seven parts: I—Introduction; II—Hydraulic Analogy for Gas Dynamics; III—Turbulence and Unsteady Gas Dynamics; IV—Rarefied and Radiation Gas Dynamics; V—Biological Fluid Mechanics; VI—Hypersonic and Plasma Gas Dynamics; and VII—Heat Transfer in Hypersonic Flows.

The material, presented by the undersigned as course instructor and by various guest lecturers, could easily be adapted by other universities for use as a text for a one-semester senior or graduate course on the subject.

Due to the extensive notes developed during the University of California course, it was decided to publish the material in three volumes, of which the present is the first. The succeeding volumes will be entitled "Selected Topics in Fluid and Bio-Fluid Mechanics" and "Introduction to Steady and Unsteady Gas Dynamics."

Finally, I must express a word of appreciation to my wife Irene and to my children, Wellington Jr. and Victoria, who made it possible for me to write and edit this book in the very quiet atmosphere of our home.

W. H. T. Loh

March 1968
University of California
Los Angeles, California

CONTENTS

Chapter 2

Combined Heat and Mass Transfer Processes   63
by E. R. G. Eckert

Chapter 3

Hypersonic Viscous Flows   83
by Arthur Henderson, Jr.

Chapter 4

Hypersonic Gas Dynamics of Slender Bodies 131

by H. K. Cheng

Chapter 9

Plasma Dynamics 333

by S. I. Pai

Chapter 1

THEORY OF THE HYDRAULIC ANALOGY FOR STEADY AND UNSTEADY GAS DYNAMICS

W. H. T. Loh

Manager, Science and Technology
Space Division
North American Rockwell Corp.
Downey, California

INTRODUCTION *

Let us consider a supersonic parallel flow of gas. If a small cylindrical obstacle is situated in such a supersonic flow, the disturbance wave produced by the obstacle is propagated with respect to the moving gas with the local sound velocity. The waves are circular-cylindrical in shape. At each point, such waves arise continuously. All of them have as their common envelope two straight rays, the Mach rays, which form the Mach angle α with the direction of flow. If the obstacle is small, the intensity of the circular waves is small, and the disturbance is propagated only along the Mach rays. Now, instead of a parallel flow, we shall consider a general supersonic flow. The flow velocity and the sound velocity vary from point to point. For each sufficiently small partial region of flow, the same considerations as above are valid, the direction and Mach angle varying only from point to point. The disturbance arising from a small obstacle is now propagated along curved (broken Mach) lines.

It is possible with liquid fluids (water) to produce flows that show a far-reaching analogy to the two-dimensional flows of a compressible gas. A flow of this kind is obtained if water is allowed to flow over a horizontal bottom under the effect of gravity. The surface of the water is assumed to be free. At the sides, it must be bounded by vertical walls, or it must flow into water of a definite depth at rest. The fixed vertical walls correspond to

---

* The symbols used throughout most of this chapter are defined on p. 8 and pp. 59–60.

the boundaries of the gas flow. A channel with horizontal bottom and rectangular cross section with variable width is an example of this type of boundary. The water flowing into water at rest corresponds to a free gas jet.

1. TWO-DIMENSIONAL STEADY FLOW ANALOGY ([1])

1.1. Energy Equation

It will be assumed that the flow of the water is frictionless, so that conversion of energy into heat is excluded. The energy equation* then simply states that the sum of the potential and kinetic energies of a water particle is constant during its motion (Bernoulli's equation):

$$p + \tfrac{1}{2}\varrho V^2 + \varrho g z = p_1 + \tfrac{1}{2}\varrho V_1{}^2 + \varrho g z_1 \qquad (1)$$

On the surface of the water, p is constant and equal to the atmospheric pressure. In what follows, we may, without error, set this equal to zero, since only pressure differences are of physical significance in the case of incompressible flows. The magnitudes denoted with the subscript 1 refer to an arbitrary but fixed point of the flow. The magnitudes without subscript refer to a variable point. If the water flows out from an infinitely wide basin, then the velocity in the basin is $V_0 = 0$. In addition, the curvature of the free surface is zero. The corresponding water depth is denoted by h_0, and is at the same time the maximum depth which occurs.

Using the above point as reference, the Bernoulli equation reads:

$$p + \tfrac{1}{2}\varrho V^2 + \varrho g z = p_0 + \varrho g z_0$$

from which

$$V^2 = 2g(z_0 - z) + [2(p_0 - p)/\varrho] \qquad (2)$$

We now make a simplifying assumption, that the vertical acceleration of the water is negligible compared with the acceleration of gravity. Under this assumption, the static pressure at a point of the flow field depends linearly on the vertical distance under the free surface at that position,

$$p_0 = \varrho g(h_0 - z_0), \qquad p = \varrho g(h - z) \qquad (2')$$

This, substituted in (2), gives, finally,

$$V^2 = 2g(h_0 - h) = 2g\,\varDelta h \qquad (3)$$

---

* See Table I for notation used here and in the following section.

Since Eq. (3) does not contain the coordinate z, the velocity V is constant over the entire depth and is given only by the difference in height Δh between the total head and the free level, Δh being, at most, equal to h_0. The maximum attainable velocity therefore is $V_{max} = (2gh_0)^{1/2}$. The energy equation may thus be written

$$(V/V_{max})^2 = V^2/2gh_0 = \Delta h/h_0 \tag{3'}$$

In a gas, the energy equation is

$$(V^2/2g) + H = H_0 \tag{4}$$

and the maximum velocity is $V_{max} = (2gH_0)^{1/2}$; Eq. (4), corresponding to Eq. (3′), becomes

$$\frac{V^2}{V_{max}^2} = \frac{V^2}{2gH_0} = \frac{\Delta H}{H_0} = \frac{\Delta T}{T_0} \tag{5}$$

From these two equations, it may be seen that the ratio of the velocity to the maximum velocity for the water and gas flows becomes

$$(h_0 - h)/h_0 = (T_0 - T)/T_0$$

This is the case for

$$h/h_0 = T/T_0$$

With respect to the velocity, there therefore exists an analogy between the two flows if the depth ratios h/h_0 are compared with the gas-temperature ratios T/T_0. The water depth corresponds to the gas temperature, and conversely.

1.2. Continuity Equation

We consider at x, y a small fluid prism of edges dx and dy and height h (Fig. 1). Let u and v be the horizontal components and w the vertical component of the velocity V in the direction of the coordinate axes x, y, and z.

Neglecting the vertical acceleration of the water in comparison with the acceleration of gravity, Eq. (2′) is valid. From it, we have

$$\frac{\partial p}{\partial x} = \varrho g \frac{\partial h}{\partial x} \quad \text{and} \quad \frac{\partial p}{\partial y} = \varrho g \frac{\partial h}{\partial y}$$

The right sides of the above relations are independent of z, so that the horizontal accelerations for all points along a vertical also are independent of z. The horizontal velocity components u and v are thus constant over the entire depth because they were so in the initial state (of rest).

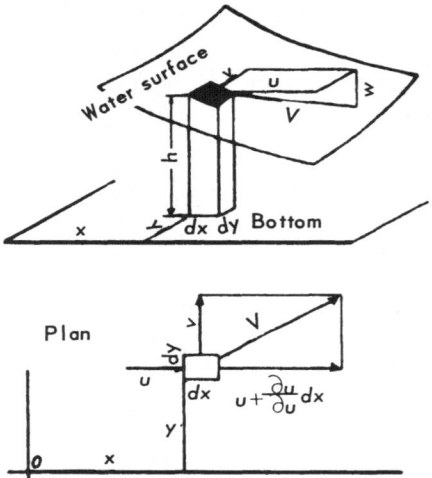

Fig. 1. Scheme for the derivation of the
continuity equation.

The continuity equation for the stationary flow simply expresses the fact that the quantity of fluid flowing into the prism (Fig. 1) per unit time is equal to the mass flowing out. Since the density of the water is constant, the same holds true for the fluid volume flowing in and for the volume flowing out. The inflow is $uh\,dy + vh\,dx$. The outflow is

$$\left(u + \frac{\partial u}{\partial x}\,dx\right)\left(h + \frac{\partial h}{\partial x}\,dx\right)dy + \left(v + \frac{\partial v}{\partial y}\,dy\right)\left(h + \frac{\partial h}{\partial y}\,dy\right)dx$$

This continuity condition written out and divided by $dx\,dy$ gives the equation of continuity

$$\frac{\partial(hu)}{\partial x} + \frac{\partial(hv)}{\partial y} = 0 \tag{6}$$

The continuity equation for a two-dimensional compressible gas flow is

$$\frac{\partial(\varrho u)}{\partial x} + \frac{\partial(\varrho v)}{\partial y} = 0 \tag{7}$$

Comparison of the two equations (6) and (7) shows that, just as with the energy equations, the continuity equations for the two flows have the same form. From this we may derive a further condition for the analogy, that the specific mass ϱ of the gas flow corresponds to the water depth h. It may be clearly seen now why the incompressible flow of the water may bear a relationship to the flow of a compressible gas. As a consequence of the compressibility in a two-dimensional gas flow, the gas mass per unit of bottom area is not a constant, but varies from point to point of the flow plane. Since the water depth in the flow with free surface varies, the mass per unit bottom area for this flow is also a variable.

From the continuity equation, we derived the result that the water depth h corresponds to the specific mass ϱ. By comparison of the energy equations of the two flows, it followed, however, that the water depth h was simultaneously the analogous magnitude for the temperature T. This is possible without contradiction only if a very definite assumption is also made as regards the nature of the comparison gas. For the gas flow, ϱ depends upon T, the relation between the two being the adiabatic equation

$$\varrho/\varrho_0 = (T/T_0)^{1/(k-1)} \tag{8}$$

Now $\varrho/\varrho_0 = h/h_0$ and simultaneously $T/T_0 = h/h_0$, and substituting in (8), we have the equation $h/h_0 = (h/h_0)^{1/(k-1)}$, which obviously is satisfied only for

$$k = 2 \tag{9}$$

Thus we have the result that the flow of the water is comparable to the flow of a gas having a ratio $k = c_p/c_v = 2$. Such gases are not found in nature. There are, however, many phenomena which do not depend strongly on the value of k, so that the analogy also has significance for actual gases.

1.3. Irrotational Motion

Before introducing the condition of absence of vorticity, we make a slight transformation of the continuity equation (6) taking account of the energy equation (3). The latter, solved for h, reads

$$h = h_0 - (V^2/2g)$$

Hence

$$\frac{\partial h}{\partial x} = -\frac{1}{2g}\frac{\partial(V^2)}{\partial x}$$

and using the fact that $V^2 = u^2 + v^2$,* this gives

$$\frac{\partial h}{\partial x} = -\frac{1}{g}\left(u\frac{\partial u}{\partial x} + v\frac{\partial v}{\partial x}\right) \tag{10}$$

---

* Since u and v are constant on a vertical, and since from (3), V also is constant, $w = [V^2 - (u^2 + v^2)]^{1/2}$ is also constant, and since w vanishes at the bottom, it may be neglected in comparison with the components u and v.

Similarly,

$$\frac{\partial h}{\partial y} = - \frac{1}{g} \left(u \frac{\partial u}{\partial y} + v \frac{\partial v}{\partial y} \right) \tag{11}$$

The continuity equation (6) may also be written in the form

$$\frac{\partial u}{\partial x} h + \frac{\partial h}{\partial x} u + \frac{\partial v}{\partial y} h + \frac{\partial h}{\partial y} v = 0$$

Substituting in the above the expressions (10) and (11), we obtain

$$\frac{\partial u}{\partial x} h - \frac{u}{g} \left(u \frac{\partial u}{\partial x} + v \frac{\partial v}{\partial x} \right) + \frac{\partial v}{\partial y} h - \frac{v}{g} \left(u \frac{\partial u}{\partial y} + v \frac{\partial v}{\partial y} \right) = 0$$

The above equation divided by h and rearranged gives

$$\frac{\partial u}{\partial x} \left(1 - \frac{u^2}{gh} \right) + \frac{\partial v}{\partial y} \left(1 - \frac{v^2}{gh} \right) - \left(\frac{\partial u}{\partial y} + \frac{\partial v}{\partial x} \right) \frac{uv}{gh} = 0 \tag{12}$$

We now introduce the condition for absence of vorticity. This will be true if $(\partial v/\partial x) - (\partial u/\partial y) = 0$. In this case, there exists a function $\Phi(x, y)$ such that $u = \partial\Phi/\partial x$, $v = \partial\Phi/\partial y$. Substituting $\Phi(x, y)$ into Eq. (12), the latter may be written:*

$$\Phi_{xx} \left(1 - \frac{\Phi_x{}^2}{gh} \right) + \Phi_{yy} \left(1 - \frac{\Phi_y{}^2}{gh} \right) - 2\Phi_{xy} \frac{\Phi_x \Phi_y}{gh} = 0 \tag{13}$$

This is the differential equation for the velocity potential of the ideal free-surface water flow over a horizontal bottom. The equation is partial of the second order and linear in the second derivatives. The coefficients depend on the derivatives of the first order and on these only. It is to be observed that gh is not a constant, but, according to the energy equation, is

$$gh = gh_0 - \tfrac{1}{2}V^2 = gh_0 - \tfrac{1}{2}(\Phi_x{}^2 + \Phi_y{}^2)$$

The well-known equation corresponding to (13) for the velocity potential of a two-dimensional compressible flow is

$$\Phi_{xx} \left(1 - \frac{\Phi_x{}^2}{a^2} \right) + \Phi_{yy} \left(1 - \frac{\Phi_y{}^2}{a^2} \right) - 2\Phi_{xy} \frac{\Phi_x \Phi_y}{a^2} = 0 \tag{14}$$

---

* Instead of $\partial\Phi/\partial x$, we write what follows in the usual notation: Φ_x; $\partial^2\Phi/\partial x^2 \equiv \Phi_{xx}$; $\partial^2/\partial x\,\partial y \equiv \Phi_{xy}$; etc.

The two equations (13) and (14) become identical if $(gh/2gh_0)$ is replaced by $(a^2/2gH_0)$. Here $(gh)^{1/2}$ is the basic wave velocity in shallow water, and corresponds to the velocity a in the gas flow.

1.4. Summary

We have yet to inquire what magnitude in the water flow is analogous to the gas pressure. Writing the equation of state $p = \varrho RT$ for an arbitrary state and for the state at rest, we obtain by division

$$p/p_0 = (\varrho/\varrho_0)(T/T_0)$$

Substituting for ϱ/ϱ_0 the corresponding value h/h_0, and for T/T_0 also h/h_0, we obtain the value corresponding to p/p_0:

$$p/p_0 = (h/h_0)^2 \tag{15}$$

This is also obtained directly from the adiabatic equation $(p/p_0) = (\varrho/\varrho_0)^k$ with $\varrho/\varrho_0 = h/h_0$ and $k = 2$. The pressure p_G on the bottom surface is proportional to the water depth h; with ϱ_w as the specific mass of the water, $p_G = \varrho_w gh$. This pressure has no analogy in the two-dimensional gas flow. In particular, it is not the magnitude corresponding to the gas pressure, since the magnitude corresponding to p is h^2, and not h. The force of the water flow per unit of length of the vertical wall is, because of the linear increase of the pressure with distance below the free surface, given by

$$P = \tfrac{1}{2}\varrho_w gh^2$$

For P, therefore, we have $P/P_0 = (h/h_0)^2$. Comparison with Eq. (15) shows that $p/p_0 = P/P_0$. The magnitude of the water flow corresponding to the gas pressure p is thus the force of the water on a unit strip of the side walls. The pressures in the two-dimensional compressible flow are analogous to the forces in the water on the vertical walls.

From the differential equation for the velocity potential, we have derived the fact that the velocity of sound a corresponds to the wave velocity $(gh)^{1/2}$. The differential equation arose through the combination of the energy and continuity equations. Thus the result $a \sim (gh)^{1/2}$ is not something essentially new but is only a consequence of the results $\varrho \sim h$, $T \sim h$, and $k = 2$ of these two equations. We have $a^2 = gkRT = g(k-1)H$, and for $k = 2$ and $H \sim h$, this gives $a^2 \sim gh$.

Since the velocity corresponding to a is $(gh)^{1/2}$, there corresponds to the subsonic flow $V/a < 1$ the flow with $V/(gh)^{1/2} < 1$. The water in this

TABLE I
Flow Analogy*

	Two-dimensional gas flow	Liquid flow with free surface in gravity field
Nature of the flow medium	Hypothetical gas with $k = c_p/c_v = 2$	Incompressible fluid (e.g., water)
Side boundaries geometrically similar		Side boundary vertical Bottom horizontal
Analogous magnitudes	Velocity, V/V_{max}	Velocity V/V_{max}
	Temperature ratio, T/T_0	Water depth ratio, h/h_0
	Density ratio, ϱ/ϱ_0	Water depth ratio, h/h_0
	Pressure ratio, p/p_0	Square of water depth ratio, $(h/h_0)^2$
	Pressure on the side boundary walls, p/p_0	Force on the vertical walls, $P/P_0 = (h/h_0)^2$
	Sound velocity, a	Wave velocity, $\sqrt{gh}$
	Mach number, V/a	Mach number, $V/\sqrt{gh}$
	Subsonic flow	Streaming water
	Supersonic flow	Shooting water
	Compressive shock	Hydraulic jump

* Additional nomenclature: g, acceleration of gravity; R, gas constant; ϱ, density; p, pressure; T, absolute temperature; H, heat content; c_p, specific heat at constant pressure; c_v, specific heat at constant volume; $k = c_p/c_v$, adiabatic exponent; Φ, velocity potential; x, y, z, rectangular coordinates in the flow space; u, v, w, components of the velocity in the x, y, and z directions; V_{max}, maximum velocity; a, in gas: velocity of sound; a, in water: propagation wave velocity $\sqrt{gh}$; *, critical value; h, water depth; h_0, total head (water depth for $V = 0$). (These symbols are used in Sections 1 and 2.)

case is said to "stream," while the water flow corresponding to the supersonic flow is said to "shoot." The essential difference in character between the supersonic and subsonic flows also exists in the case of water between streaming and shooting flows.

The analogy considered in this section holds for flows with Mach numbers smaller and greater than 1, and is summarized in Table I.

2. HYDRAULIC JUMPS (SHOCKS)

It is known that in "shooting" water under certain conditions, the velocity may strongly decrease over short distances and the water depth suddenly increase. An unsteady motion of this type is known as a hydraulic jump (Fig. 2). In Fig. 2, the water flows from forward to rear. In the forward

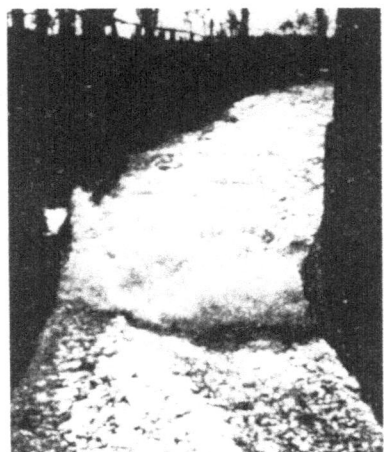

Fig. 2. Right (orthogonal) hydraulic jump.

part, the water "shoots." Over the entire width of the channel it jumps to a new water level and flows with considerably less velocity in the same direction toward the rear. The entire process is practically stationary.

Hydraulic jumps occur only in shooting water, i.e., in water whose velocity of flow is greater than the wave propagation velocity. In order to show this, let us imagine the forward water to be at rest, and that from behind there arrives the front of a water wave which arose from the opening of a large sluice. If the wave were very small, it would move forward with the basic wave velocity $(gh_1)^{1/2}$. Since, however, it has finite height $h_2 - h_1$, it moves, to a first approximation, with the velocity*

$$u_1 = [g(h_1 + h_2)/2]^{1/2}(h_2/h_1)^{1/2}$$

i.e., more rapidly than $(gh_1)^{1/2}$, and also than $(gh_2)^{1/2}$. In this coordinate system, moving with the shooting water, the wave is not stationary. The water may now be considered as moving with the velocity u_1 with respect to the wave. The latter then remains at rest in space. The water ahead, however, is not at rest, but has the flow velocity u_1, and this is greater than $(gh_1)^{1/2}$. It has thus been shown that such hydraulic jumps can be stationary only in shooting water. If the wave existed in streaming water, it would, because of its propagation velocity, which in this case is larger than the flow velocity, travel upstream. There would be the usual outflow from upper

---

* This formula is obtained from a simple application of the continuity and momentum equations; for $h_2 \to h_1$, it naturally passes over into $u_1 = (gh_1)^{1/2}$.

Fig. 3. Slant hydraulic jump.

to lower level without shock.* A shock (or hydraulic jump) in which the wavefront is normal to the flow direction is called a right hydraulic jump. It naturally has the property that the propagation velocity of the shock wave relative to the water is equal and opposite to the water velocity ahead of the jump.

More general than the right hydraulic jump is the less familiar slant hydraulic jump (Fig. 3). The water flows from left to right out of an open sluice. The water depth decreases and the velocity increases. The water flows from a constant upper water level into a basin with constant lower water level. Since the difference in head is greater than a third of the upper water depth, the water after escaping from the sluice receives a larger velocity than the basic wave propagation velocity, so that it shoots. It is thus possible for it to accelerate so rapidly that the water surface of the flow becomes lower than the lower water level. There is a portion of the

---

* The term "shock" will be used interchangeably with "hydraulic jump," and naturally has nothing to do with the compressibility of the water.

flow for which there is considerable pressure rise over a short distance. In this flow, however, the jump does not take place on a normal to the velocity, but instead along a line oblique to the flow direction, and we have a slant jump. On the meeting of the rear and forward jumps shown in the Fig. 3, there is a particularly strong pressure rise.

The slant jump, like the right jump, occurs only in shooting water. In order to be able to give a simple numerical treatment of the slant hydraulic jump, we make the assumption that the motion is entirely unsteady; i.e., that the water jumps suddenly along a line— the jump line—from the lower water level to the level after the jump. The simplest case of such a jump is obtained if a paral el flow is deflected by an angle β (Fig. 4). The shock in the supersonic flow of a compressible gas has been treated in standard texts. Here, however, for the shock of the shooting water, the analogy with a compressible gas flow for $k = 2$ no longer strictly holds. The previous considerations involved as assumption the validity of the Bernoulli equation, which is equivalent to the assumption that the flow was without losses. With shock, however, kinetic energy is converted into heat. In a gas flow, this again enters thermodynamically into the computation, whereas with the water flow, it is to be treated as lost energy.

2.1. Shock Polars

For the case of the deflection of a parallel flow by the angle β, the jump line is a straight line through the corner, making an angle γ (Fig. 4). For a very small deflection $\beta \to 0$, the two following limiting cases are possible: (1) A right jump; γ is then a right angle. (2) The flow goes through undisturbed. This is the limiting case of a jump whose effect approaches zero. The jump line passes over into the Mach line through the corner, γ is this case being the Mach angle.

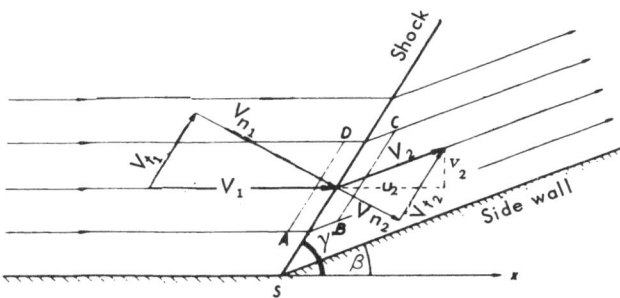

Fig. 4. Slant hydraulic jump (ground plan).

We shall take the x axis such that it has the direction of the velocity of approach V_1. The components in the x and y directions are thus $u_1 = V_1$ and $v_1 = 0$. Let the water depth before the jump be h_1, the velocity after the jump V_2, and its components in the x and y directions be u_2 and v_2. The water depth after the jump we shall denote by h_2, and by V_n and V_t, the components of the velocity normal and tangential, respectively, to the jump line. Here, too, we shall distinguish magnitudes before and after the jump by the subscripts 1 and 2, respectively. As control region for setting up the continuity and momentum equation, we choose the region $ABCDA$ (Fig. 4).

With the above notation, the continuity equation reads:

$$h_1 V_{n_1} = h_2 V_{n_2} \tag{16}$$

The momentum equation for the direction normal to the jump for the width $AD = b$ states that the decrease in outgoing momentum by that of the incoming momentum is equal to the force (area times pressure):

$$(\varrho V_{n_2} h_2 b) V_{n_2} - (\varrho V_{n_1} h_1 b) V_{n_1} = b h_1 (g \varrho h_1/2) - b h_2 (g \varrho h_2/2)$$

or, rearranging,

$$h_1 V_{n_1}^2 + (g h_1^2/2) = h_2 V_{n_2}^2 + (g h_2^2/2) \tag{17}$$

Finally, writing the momentum equation for the direction tangential to the jump,

$$(\varrho V_{n_1} h_1 b) V_{t_1} = (\varrho V_{n_2} h_2 b) V_{t_2}$$

we obtain, taking account of the continuity equation (16)

$$V_{t_1} = V_{t_2} \tag{18}$$

During the jump only the component of the velocity normal to the line is changed, the tangential component remaining unchanged.

As in the gas flow, it is also convenient for the treatment of the hydraulic jump to pass from the field of flow to the velocity plane. Taking account of Equation (18), we obtain the diagram shown in Fig. 5. The region of the flow before the jump is in the velocity diagram represented by T. After the jump, F is the point of the hodograph corresponding to the flow. The jump itself is represented by the transition from T to P. The direction of the jump line in the flow is given in the hodograph by the normal to the segment TP, since this has the direction of V_t.

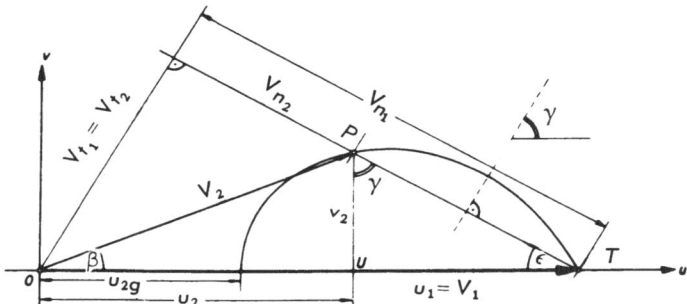

Fig. 5. Hydraulic jump in velocity diagram. Shock polar.

For a fixed velocity of approach V_1, i.e., for a fixed point of the hodograph T, we obtain for various deflection angles β, various end states P. The totality of all end states which correspond to a fixed initial state form a curve, the "shock polar" (Fig. 5). If the initial state T is changed, then to each point T there corresponds a shock polar. The entire family is the shock polar diagram (Fig. 6).

The equation of the shock polars $v_2 = f(u_2)$ will now be determined. We start from the continuity equation (16), the momentum equation (17), and the energy equation*

$$2gh_1 = 3a_1^{*^2} - u_1^2 \qquad (19)$$

The critical velocity a^* is given by the condition that the flow velocity is equal to the wave propagation velocity, $a = (gh)^{1/2}$, so that the Mach number $M = 1$. Thus if $V^2 = gh$, $a^* = V = (gh)^{1/2}$. Let us compute the water depth at the critical positions. From the energy equation, $V^2 = 2gh_0 - 2gh$, and this should be equal to $a^2 = gh$, i.e., $2gh_0 - 2gh = gh$, so that $h^* = \frac{2}{3}h_0$, and hence

$$V^{*^2} = a^{*^2} = \frac{2}{3}gh_0 \qquad (19')$$

The critical positions in a water flow without energy dissipation are located where the water depth is two-thirds of the total head. These positions in an accelerated flow are the transition points from "streaming" to "shooting" water, and conversely for decelerated flow.

Substituting the critical velocity a_1^* before the jump, we have

$$2gh_1 = 3a_1^{*^2} - u_1^2$$

---

* From the energy equation (3), we have: $2gh_1 = 2gh_0 - V_1^2 = 2gh_0 - u_1^2$ (since $v_1 = 0$).

We also need the two geometrical relations:*

$$\frac{V_{n_1}}{V_{n_2}} = \frac{u_1(u_1 - u_2)}{(u_1 - u_2)u_2 - v_2^2} \tag{20}$$

$$V_{n_1}(V_{n_1} - V_{n_2}) = u_1(u_1 - u_2) \tag{21}$$

In the five equations (16), (17), (19)–(21), there occur the variables V_{n_1}, V_{n_2}, h_1, h_2, u_1, u_2, and v_2. If we eliminate the first four, we will obtain the equation of the shock polar. In order to carry out this elimination, we first substitute in Eq. (20) the continuity equation (16):

$$\frac{h_2}{h_1} = \frac{u_1(u_1 - u_2)}{(u_1 - u_2)u_2 - v_2^2} \tag{22}$$

Substituting the continuity equation (16) into the momentum equation (17), we obtain

$$h_1 V_{n_1}^2 + (gh_1^2/2) = h_1 V_{n_1} V_{n_2} + (gh_2^2/2)$$

We thus have

$$2gh_2^2 = 2gh_1^2 + 4h_1 V_{n_1}(V_{n_1} - V_{n_2})$$

whence

$$(h_2/h_1)^2 = 1 + [4h_1 V_{n_1}(V_{n_1} - V_{n_2})/2gh_1^2] = [2gh_1 + 4V_{n_1}(V_{n_1} - V_{n_2})]/2gh_1$$

Substituting (19) and (21) in the above relations, we obtain

$$(h_2/h_1)^2 = (3a_1^{*2} - u_1^2 + 4u_1^2 - 4u_1u_2)/(3a_1^{*2} - u_1^2)$$

---

*From Figure 5, we may read off directly the two equations:

$$u_2^2 + v_2^2 - V_t^2 = V_{n_2}^2$$

and

$$u_1^2 \qquad\quad - V_t^2 = V_{n_1}^2$$

Their difference is

$$u_1^2 - u_2^2 - v_2^2 = V_{n_1}^2 - V_{n_2}^2 \tag{a}$$

From Fig. 5, it may also be seen that

$$(u_1 - u_2)^2 + v_2^2 = (V_{n_1} - V_{n_2})^2 \tag{b}$$

Dividing equation (a) by (b) and solving for the quotient V_{n_2}/V_{n_1}, we obtain relation (20). This relation must naturally exist since the three magnitudes u_1, u_2, and v_2 completely determine Fig. 5. For Eq. (21), we read off directly from Fig. 5

$$V_{n_1}(V_{n_1} - V_{n_2}) = (u_1 \cos \varepsilon)(V_{n_1} - V_{n_2})$$
$$= u_1[\cos \varepsilon \, (V_{n_1} - V_{n_2})] = u_1(u_1 - u_2)$$

Only the positive root applies, since the water depths h_1 and h_2, and hence also their ratio, are naturally positive. We obtain

$$\frac{h_2}{h_1} = + \left[\frac{3a_1^{*2} - 4u_1u_2 + 3u_1^2}{3a_1^{*2} - u_1^2} \right]^{1/2} \tag{23}$$

Setting finally the two right sides of Eqs. (22) and (23) equal to each other (elimination of h_2/h_1) and solving the relation thus obtained for v_2^2, the required equation of the polars is finally òbtained as

$$v_2^2 = [u_1 - u_2]\{u_2 - u_1[(3a_1^{*2} - u_1^2)/(3a_1^{*2} - 4u_1u_2 + 3u_1^2)]^{1/2}\} \tag{24}$$

Substituting $\bar{v}_2 = v_2/a_1^*$, $\bar{u}_2 = u_2/a_1^*$, and $u_1 = u_1/a_1^*$ in the above as dimensionless velocities referred to a_1^*, the equation of the polars becomes

$$\bar{v}_2^2 = [\bar{u}_1 - \bar{u}_2]\{\bar{u}_2 - \bar{u}_1[(3 - \bar{u}_1^2)/(3 - 4\bar{u}_1\bar{u}_2 + 3\bar{u}_1^2)]^{1/2}\} \tag{24a}$$

These are the curves $f(\bar{u}_2, \bar{v}_2, \bar{u}_1) = 0$ with $\bar{u}_1$ as parameter drawn in Fig. 6. They are similar to the shock polars of an ideal gas, but show a characteristic difference. Whereas for the maximum velocity, the shock polars in both cases become circles, the latter pass through the origin for water, while for a gas, the origin is not attained. In the case of a right jump, $v_2 = 0$. If we denote the velocity after the jump by u_{2g} (Fig. 5), Eq. (24a) for the latter becomes

$$0 = (\bar{u}_1 - \bar{u}_{2g})\{(\bar{u}_{2g} - \bar{u}_1[(3 - \bar{u}_1^2)/(3 - 4\bar{u}_1\bar{u}_{2g} + 3\bar{u}_1^2)]^{1/2}\} \tag{24a'}$$

From this we obtain $\bar{u}_{2g} = \bar{u}_1[(3 - \bar{u}_1^2)/(3 - 4\bar{u}_1\bar{u}_{2g} + 3\bar{u}_1^2)]^{1/2}$. If this equation is squared, multiplied through by the denominator, and rearranged,

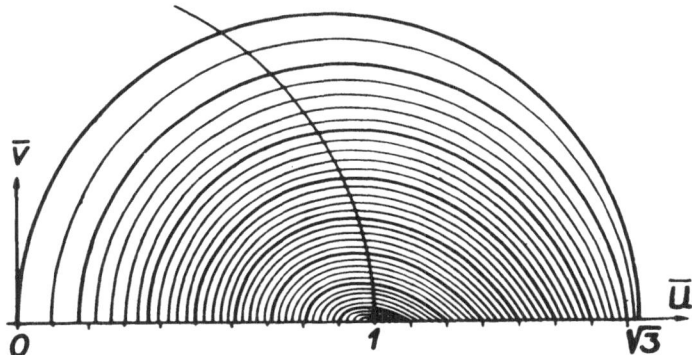

Fig. 6. Shock polar diagram for water.

the resulting expression may again be divided by $(\bar{u}_1 - \bar{u}_{2g})$, and we obtain a quadratic equation for $\bar{u}_1\bar{u}_{2g}$ whose positive solution is

$$\bar{u}_1\bar{u}_{2g} = \frac{3 - \bar{u}_1^2}{8}\left[1 + \left(1 + \frac{16\bar{u}_1^2}{3 - \bar{u}_1^2}\right)^{1/2}\right] \tag{24b}$$

The values computed from Eq. (24b) are collected in Table II. For an ideal gas the relation analogous to (24b) may be written in a very elegant manner:

$$\bar{u}_1\bar{u}_{2g} = 1 \tag{24c}$$

TABLE II

Normal Hydraulic Jump*

$\bar{u}_1$	M_1	h_1/h_0	u_{2g}^2	h_2/h_0	h_0'/h_0	$\bar{u}_1\bar{u}_{2g}$
1.0	1.0	0.667	1.0	0.667	1.0	1.000
1.02	1.032	0.653	0.980	0.679	1.000	1.000
1.04	1.063	0.639	0.960	0.692	0.999	0.999
1.06	1.093	0.625	0.941	0.704	0.999	0.997
1.08	1.127	0.611	0.922	0.715	0.998	0.995
1.10	1.161	0.597	0.903	0.726	0.998	0.993
1.12	1.198	0.582	0.884	0.737	0.997	0.990
1.14	1.237	0.567	0.865	0.747	0.996	0.986
1.16	1.276	0.552	0.846	0.756	0.994	0.981
1.18	1.318	0.536	0.827	0.765	0.993	0.976
1.20	1.359	0.520	0.808	0.772	0.990	0.970
1.22	1.404	0.504	0.789	0.779	0.987	0.963
1.24	1.448	0.487	0.770	0.785	0.983	0.955
1.26	1.498	0.472	0.751	0.791	0.979	0.946
1.28	1.550	0.454	0.731	0.795	0.973	0.936
1.30	1.608	0.437	0.712	0.798	0.967	0.925
1.32	1.667	0.419	0.692	0.799	0.959	0.914
1.34	1.730	0.401	0.672	0.800	0.951	0.901
1.36	1.796	0.383	0.652	0.799	0.941	0.887
1.38	1.867	0.366	0.632	0.798	0.931	0.872
1.40	1.942	0.347	0.611	0.795	0.919	0.855
1.45	2.164	0.300	0.557	0.778	0.881	0.808
1.50	2.45	0.250	0.500	0.751	0.834	0.750
1.55	2.84	0.199	0.438	0.706	0.770	0.679
1.60	3.41	0.147	0.368	0.640	0.685	0.589
1.65	4.43	0.092	0.285	0.542	0.566	0.470
1.70	7.25	0.036	0.174	0.36	0.369	0.296
$\sqrt{3}$	∞	0	0	0	0	0

* Before jump: h_0 is the total head; $\bar{u}_1$ the velocity referred to $a_1{}^*$, the critical velocity; M_1 the Mach number; and h_1 the water depth. After jump: h_0' is the total head, u_{2g} the velocity referred to $a_1{}^*$, and h_2 the water depth.

Only for the case $\bar{u}_1 = 1$ does (24b) accurately agree with (24c).* Otherwise, the right hydraulic jump leads to no simple relation like the normal shock of a gas. Within wide limits, however, Eq. (24c) may also be applied to water (see values of $\bar{u}_1 \bar{u}_{2g}$ in Table II).

We wish further to show that a very small jump has a jump line which in the limiting case is a Mach line. From the triangle TPU of Fig. 5, $\tan \gamma = (u_1 - u_2)/v_2$. From the equation of the shock polars (24a), we have

$$\left[\frac{\bar{u}_1 - \bar{u}_2}{v_2} \right]^2 = \frac{\bar{u}_1 - \bar{u}_2}{\bar{u}_2 - \bar{u}_1[(3 - \bar{u}_1{}^2)/(3 - 4\bar{u}_1\bar{u}_2 + 3\bar{u}_1{}^2)]^{1/2}} \tag{24d}$$

A small jump is obtained if $\bar{u}_2 \to \bar{u}_1$. The root in (24d) then approaches 1, and the entire expression becomes indeterminate. By differentiation of the numerator and denominator with respect to the critical variable $\bar{u}_2$, we obtain

$$\left[\frac{\bar{u}_1 - \bar{u}_2}{\bar{v}_2} \right]^2_{\bar{u}_2 = \bar{u}_1} = \frac{3 - \bar{u}_1{}^2}{3(\bar{u}_1{}^2 - 1)}$$

We then have

$$(\sin^2 \gamma)_{\bar{u}_2 = \bar{u}_1} = (\tan^2 \gamma)/(1 + \tan^2 \gamma) = (3 - \bar{u}_1{}^2)/2\bar{u}_1{}^2 \tag{25}$$

On the other hand, for the Mach angle α

$$\sin^2 \alpha = (a_1/u_1)^2 \tag{26}$$

For the wave velocity a_1, we have

$$a_1{}^2 = gh_1 \tag{27}$$

The energy equation (3) is

$$u_1{}^2 = 2h(h_0 - h_1) \tag{28}$$

As reference velocity, we choose $a_1{}^*$. For this, Eq. (19′) applies

$$\cdot \quad a_1^{*2} = \tfrac{2}{3}gh_0 \tag{29}$$

Eliminating from Eqs. (26)–(29) the magnitudes a_1, h_0, and h_1, we obtain for the sine of the Mach angle α the relation

$$\sin^2 \alpha = (3 - \bar{u}_1{}^2)/2\bar{u}_1{}^2 \tag{30}$$

---

* For $\bar{u}_1 = 1$, the shock polar shrinks into a point (Fig. 6). The only possible state after the jump is thus $\bar{u}_{2g} = 1$. For this case, we no longer have a finite jump, but then the gas flow also agrees with the water flow.

Comparison of (25) and (30) then shows that

$$(\sin \gamma)_{u_2=u_1} = \sin \alpha$$

2.2. Water Depths in Hydraulic Jump

Up to now, we have investigated how the velocity changes in the case of hydraulic jump. In this section, we shall treat the water depths in more detail.

For a flow without jump, the energy equation (3) holds between V and h

$$V^2 = 2g(h_0 - h)$$

where the total head h_0 is a constant. In the case of a jump, a portion of the kinetic energy of the water is converted into heat. For this reason, the total head after the impact—which, to distinguish from h_0, we shall denote by h_0'—is smaller than it was before. For the flow after the jump, the relation between the velocity and the water depth is given by the energy equation in the following form: $V^2 = 2g(h_0' - h)$. The new total head h_0' is constant along a streamline, but after the jump may vary from one streamline to another.

For gases, a clear picture of the pressures in the flow is obtained if the pressure is plotted as third coordinate over the velocity plane. For adiabatic flow, we thus obtained in the u, v, p space a surface of rotation whose meridian section represents p as a function of V:

$$V^2 = \frac{2k}{k-1} \frac{p_0}{\varrho_0} [1 - (p/p_0)^{(k-1)/k}] \tag{31}$$

For the two-dimensional flow of water with free surface, the magnitude V^2 corresponds to the pressure p in the gas flow. If we plot above the u, v plane, not the water depth, but the values $t = gh^2/2$, we shall find for the water in the u, v, t space the same relations that hold for a gas in the u, v, p space.

The representation in the u, v, $gh^2/2$ space is not very suitable for the practical computation of the jump. Nevertheless, we shall first consider the properties of this representation because it gives a very clear picture of the entire hydraulic jump process with regard to the velocity and the water depth simultaneously.

In the flow of water without energy dissipation, the water depth h and hence $gh^2/2$, depends not on u and v individually, but only on the absolute value of the velocity $V = (u^2 + v^2)^{1/2}$. Plotting t above the u, v plane, we

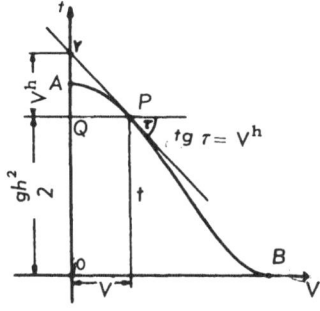

Fig. 7. A t–V curve.

obtain a surface of rotation. Let us consider its meridian section $t = f(V)$, abscissa V, ordinate t (Fig. 7). From the Bernoulli equation (3), we have

$$h = h_0 - (V^2/2g) \tag{32}$$

whence

$$t \equiv gh^2/2 = (1/8g)V^4 - (h_0/2)V^2 + (gh_0^2/2) \tag{33}$$

The characteristic shape of these curves of the fourth degree (Fig. 7), which in our problem have a physical sense only from A to B, may easily be understood from Fig. 8, which shows the parabola Eq. (32) and its "square," Eq. (33).

For each total head h_0', there is one such curve. The family of all these curves we shall denote as the t, V diagram (Fig. 9). As long as no jumps occur along a streamline, the relation between t and V, because of the constant total head, is given by a fixed curve of this family. As soon as jumps occur along the streamline, the t, V point on one curve "jumps" to another t, V curve.

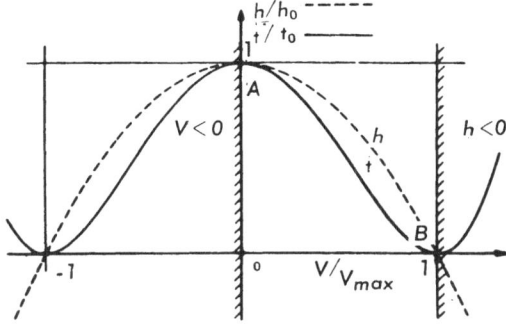

Fig. 8. Characteristic overall shape of the t–V curve. $V_{max}^2 = 2gh_0$, $t_0 = gh_0^2/2$.

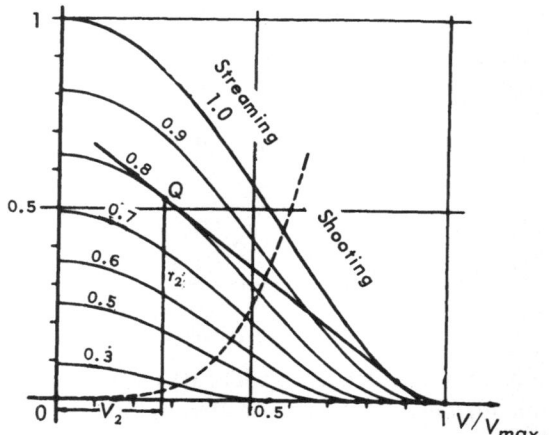

Fig. 9. Right hydraulic jump in the $t-V$ diagram.

Because the new total head for each jump is smaller than the previous one, we come each time to a curve lying closer to the origin, and not the reverse. To the curves of constant total head, $h_0' = \text{const}$, which give the relation between $gh^2/2$ and V for the zero-loss flow, there correspond the adiabatics in the gas flow, these being the lines of constant entropy, $s = \text{const}$. For the ideal gas, these are affine with respect to the V axis, but not for water.

The right hydraulic jump may be studied very simply in the t, V diagram. Let us compute first the slope of the tangent of the t, V curve to the axis of abscissas. From Eq. (33),

$$\frac{dt}{dV} = \frac{V^3}{2g} - Vh_0 = -V[h_0 - (V^2/2g)] \tag{34}$$

and with the energy equation (32) this slope becomes

$$dt/dV = -Vh \tag{35}$$

We shall, furthermore, compute the intercept of the tangent on the t axis, which is

$$QY = QP \tan \tau = V(Vh) = V^2h$$

Since $t \equiv gh^2/2$,

$$OY = QY + OQ = hV^2 + (gh^2/2) \tag{36}$$

Physically, both the slope of the tangent dt/dV and the intercept of the tangent on the t axis have a meaning. Through a vertical area in the flow normal to the streamlines, with width equal to the unit of length, there flows per unit time the volume Vh. The magnitude hV^2 in Eq. (36) represents, except for a constant factor ϱ, the momentum flowing through the same area per unit time, and the term $gh^2/2$ similarly, except for the constant factor ϱ, represents the pressure force on this surface.

For the right hydraulic jump, the continuity equation (16) is

$$h_1 V_1 = h_2 V_2 \tag{16'}$$

The momentum equation (17) becomes

$$h_1 V_1{}^2 + (gh_1{}^2/2) = h_2 V_2{}^2 + (gh_2{}^2/2) \tag{17'}$$

These two equations, compared with (35) and (36), indicate the following:

1. From (16') and (35): The tangent at the t, V curve at the point t_1, V_1 before the jump has the same slope as the tangent at the point t_2, V_2 at the t, V curve after the jump.

2. From (17') and (36): The t intercept of the tangent at t_1, V_1 is equal to the t intercept of the tangent at t_2, V_2.

This simply indicates that both tangents are one and the same straight line PQ (Fig. 9). If the magnitudes t and V are given before the jump, the right hydraulic jump is represented in the t, V diagram by a jump from $P(t_1, V_1)$ on the tangent to the t, V curve through this point to $Q(t_2, V_2)$, where this tangent touches another t, V curve.

Since as a result of a jump, we arrive at a t, V curve which lies nearer the origin than the t, V curve before the jump, it may be seen from Fig. 9 that the hydraulic jump is possible only for points P before the jump which lie on the curve to the right of its point of inflection. This is precisely the case for shooting water, since, according to (34).

$$d^2t/dV^2 = (3V^2/2g) - h_0$$

At the point of inflection, this must be equal to zero, so that

$$V^2 = 2gh_0/3$$

This is the limiting velocity for streaming and shooting water.

Let us consider the slant jump. This may no longer be drawn in the t, V plane; we require the u, v, t space. Plotting the values $gh^2/2$ perpendicularly above the u, v plane, we obtain, for the case of a flow without losses, the surface of rotation of a t, v curve. We shall denote such a surface as a

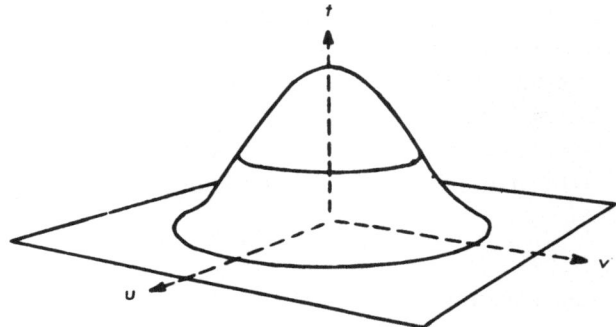

Fig. 10. The t-hill.

"t-hill" (Fig. 10). For each total head h_0', there is one such hill—each lying within the other. As long as no jumps occur in a flow, all possible corresponding values of u, v, and $gh^2/2$ are, because of the constant head h_0, given by a fixed t-hill. As soon as a jump occurs along a streamline, corresponding values of u, v, and t jump to a new, smaller t-hill, which corresponds to the new total head h_0'. After the jump, however, the relation is again given by a fixed new t-hill.

Let $P(V_1, 0, gh_1^2/2)$ denote the point in the u, v, t space before the jump, and $Q(u_2, v_2, gh_2^2/2)$ the point after the jump (Fig. 11). For the general slant jump, we obtain in the u, v, t space a clear representation similar to that for the right jump in the t, V diagram. This representation will include the right jump as a special case.

1. The slope of the tangential plane at the point P of the t-hill before the jump is, in the direction m_1, equal to zero, and in the direction r_1, equal to the slope of the meridian; i.e., equal to the slope of the t, V curve, which, according to Eq. (35), has the value $V_1 h_1$. The slope of this tangent

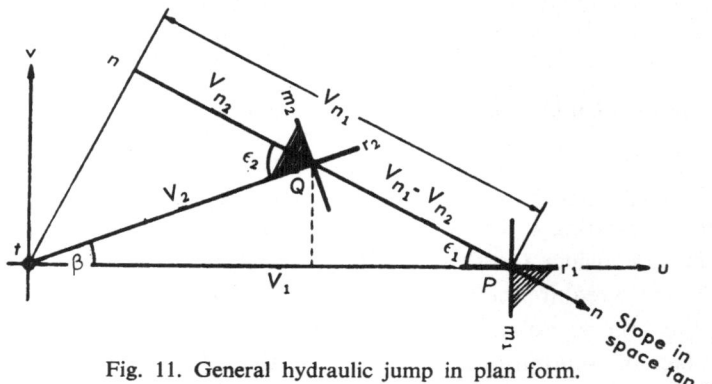

Fig. 11. General hydraulic jump in plan form.

plane at the point P in the direction $\overline{PQ}$ thus becomes:[*]

$$\tan \sigma_1 = V_1 h_1 \cos \varepsilon_1 = V_{n_1} h_1 \tag{37}$$

The point Q lies on a t'-hill. The tangent plane has a slope zero in the direction m_2, and in the direction r_2, according to (35), a slope $h_2 V_2$. The slope of the tangent plane at the point Q of the t'-hill in the direction $\overline{PQ}$ is thus

$$\tan \sigma_2 = V_2 h_2 \cos \varepsilon_2 = V_{n_2} h_2 \tag{38}$$

Comparing Eqs. (37) and (38) with the continuity equation for the slant jump (16), it is seen that the tangent plane at the t-hill at the point P before the jump in the direction $\overline{PQ}$ has the same slope as the tangent plane at the new hill at point Q after the jump in the same direction $\overline{PQ}$.

2. Let us now compute the slope of the segment $\overline{PQ}$ in space to the u, v plane. The height of the point Q is $t_2 = gh_2{}^2/2$; that of P is t_1. The slope of PQ then becomes:

$$\tan \sigma_3 = (t_2 - t_1)/(V_{n_1} - V_{n_2})$$

Since, however, P and Q are the points before and after the jump, the continuity equation (16) and the momentum equation (17) are applicable. Substituting these two equations in $\tan \sigma_3$, we obtain from (17)

$$t_2 - t_1 \equiv (gh_2{}^2/2) - (gh_1{}^2/2) = h_1 V_{n_1}^2 - h_2 V_{n_2}^2$$

and with (16), this becomes

$$t_2 - t_1 = h_1 V_{n_1}^2 - h_1 V_{n_1} c_{n_2} = h_1 V_{n_1}(V_{n_1} - V_{n_2})$$

Hence

$$\tan \sigma_3 = (t_2 - t_1)/(V_{n_1} - V_{n_2}) = h_1 V_{n_1} \tag{39}$$

Comparison with the result found above shows that the segment $\overline{PQ}$ has the same slope as the tangent plane at P and Q in the direction $\overline{PQ}$.

The segment $\overline{PQ}$ thus belongs to the two tangent planes, and, as a common line of these two planes, has the property that it is tangent both to the t-hill and the t'-hill. This result could also have been found by de-

---

[*] The slope of a plane in any given direction is equal to the slope of the plane in the direction of drop multiplied by the cosine of the angle between that direction and the given direction.

termining the line of intersection of the two tangent planes at P (given by m_1 and r_1) and at Q (given by m_2 and r_2). Then we would have obtained the straight line $\overline{PQ}$ as the line of intersection.

The general hydraulic jump is thus represented in the u, v, t space as follows: Let P be a point before the jump. Drawing through this point an arbitrary tangent at the t-hill (the only restriction on the choice of this tangent is that it must pass within the t-hill), the point indicating the state after the impact will be found where this tangent touches another t-hill of the family. To the degree of freedom of the tangent there corresponds the degree of freedom of the deflection angle β. The projection on the u, v plane of all possible points of contact Q of the tangent of a fixed point P is the already computed "shock polar" through P.

The right hydraulic jump is obtained for a direction PQ with the angle $\varepsilon = 0$. Figure 9 simply shows the vertical section with $\varepsilon = 0$ through the t-hill family.

By the intensity of a jump, we shall understand the ratio of the total head before the jump to the head after the jump, this ratio being a measure of the energy loss. The intensity is thus greater, the more nearly the angle between the shock wave front and the initial direction approaches a right angle, since ε then becomes smaller and the tangent PQ of the t-hill (total head h_0) at P touches t'-hills at Q (total head h_0') that lie more toward the interior. The right hydraulic jump has the maximum intensity. It may be remarked further that the point Q of an arbitrary slant or right jump, as the point of contact of a t'-hill, is that point of the straight line PQ for which the new total head h_0' is a minimum. Thus each jump is such that the energy loss becomes a maximum. For the ideal gas, the surfaces corresponding to the t-hill are surfaces of constant entropy, and the shock is such that the increase in entropy is a maximum.

The line joining all possible points of contact of the tangent at a fixed point P is the hydraulic-jump curve in the u, v, t space. It is a plane curve, its projection on the u, v plane being the already computed shock polar.

There is an entire family of shock curves in space (parameter point P). In their totality, they form a certain surface. This we shall denote as the shock surface in the u, v, t space. For the practical computation of the jump, the projections of the following three families of curves on the u, v plane are found convenient:

1. The curves of intersection of the shock surface with the tangent planes of all points $P(u, 0, t)$; these give the familiar shock polars (Fig. 6).

2. The curves of intersection of the shock surface with the planes

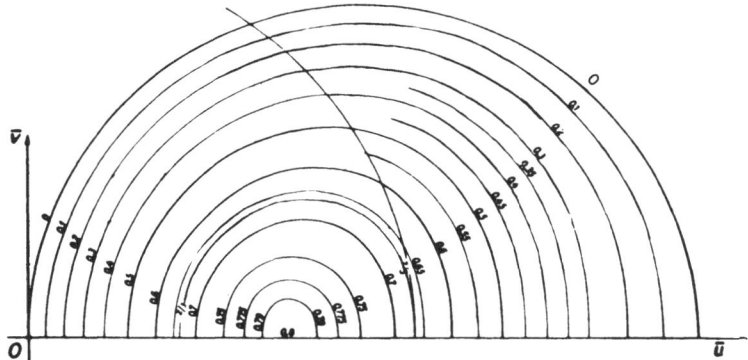

Fig. 12. Lines of constant water depth after the jump.

parallel to the u, v plane; i.e., contour curves. These are the lines of constant water depth h_2 (Fig. 12).

3. The curves of intersection of the shock surface with the family of t-hills. These give the lines of constant total depth after the jump; i.e. lines of constant energy loss (Fig. 13).

Let us consider the lines $h_2/h_0 =$ const. From the five equations (16), (17), (19)–(21) with the variables V_{n_1}, V_{n_2}, h_1, h_2, u_1, u_2, and v_2, we obtain an equation of the form $F(u_2, v_2, h_2) = 0$ if the four magnitudes V_{n_1}, V_{n_2}, h_1, and u_1 are eliminated. These are the curves of constant water depth after the jump. In order to obtain these curves, the elimination must be partly carried out graphically. The method used will be briefly explained below.

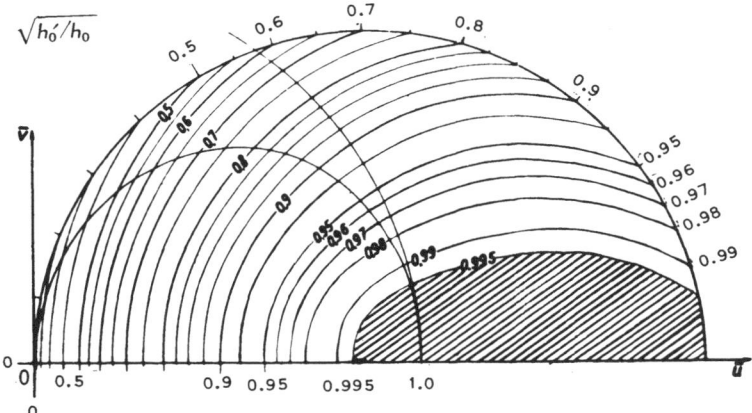

Fig. 13. Lines of constant total depth after the jump.

From (16) and (17) we obtain $gh_2{}^2/2 = (gh_1{}^2/2) + h_1 V_{n_1}(V_{n_1} - V_{n_2})$, which, with relation (21), gives $gh_2{}^2/2 = (gh_1{}^2/2) + h_1 u_1(u_1 - u_2)$, or

$$(h_2/h_0)^2 = (h_1/h_0)^2 + \tfrac{4}{3}(\tfrac{3}{2}gh_0)(h_1/h_0)u_1(u_1 - u_2)$$

Substituting the critical velocity equation (19′), $a_1^{*^2} = 2gh_0/3$, gives

$$(h_2/h_0)^2 = (h_1/h_0)^2 + \tfrac{4}{3}\bar{u}_1(\bar{u}_1 - \bar{u}_2)(h_1/h_0) \qquad (40)$$

Solving for $\bar{u}_2$,

$$\bar{u}_2 = \bar{u}_1 - \frac{(h_2/h_0)^2 - (h_1/h_0)^2}{\tfrac{4}{3}\bar{u}_1(h_1/h_0)} \qquad (40a)$$

We still need Eq. (19), which can be written $2gh_1/3a_1^{*^2} = 1 - \tfrac{1}{3}\bar{u}_1{}^2$. Substituting (19′), we obtain

$$h_1/h_0 = 1 - \tfrac{1}{3}\bar{u}_1{}^2 \qquad (41)$$

In order to draw the lines h_2/h_0, two methods were employed:

1. Assume a fixed value $h_2/h_0 = k$ and various values $\bar{u}_1$ for the variable. To each $\bar{u}_1$, there corresponds, by Eq. (41), a value h_1/h_0. With $\bar{u}_1$, the corresponding h_1/h_0, and the fixed h_2/h_0, there is obtained from (40a) the velocity component $\bar{u}_2$ after the jump. The point on the shock polar through $\bar{u}_1$ which has this abscissa $\bar{u}_2$ is a point of the curve $h_2/h_0 = k$. By varying $\bar{u}_1$, we obtain the complete curve $h_2/h_0 = k$.

2. Determine the values h_2/h_0 along an arbitrary straight line in the u, v plane (the straight line through $\bar{u} = 1$, $\bar{v} = 0$ was taken). Assume $\bar{u}_1$; measure $\bar{u}$ at the point of intersection of the straight line with the shock polar for $\bar{u}_1$; substituting $\bar{u}_1$, the corresponding value h_1/h_0 obtained from Eq. (41), and the above-determined value of $\bar{u}_2$ in Eq. (40) gives the value of h_2/h_0 at the point of intersection. By varying $\bar{u}_1$, we obtain h_2/h_0 along the entire straight line. From these values, we obtain by interpolation, points of the family of curves $h_2/h_0 = \text{const.}$

In particular, the values of h_2/h_0 may be computed for the right hydraulic jump. We have

$$(h_2/h_0)^2 = (h_1/h_0)^2 + \tfrac{4}{3}(h_1/h_0)\bar{u}_1(\bar{u}_1 - \bar{u}_{2g}) \qquad (40b)$$

Substituting the values for $\bar{u}_{2g}$ computed from Eq. (24b), in the above, we obtain the water-depth ratios h_2/h_0 given in Table II for the right hydraulic jump.

The maximum water depth in the state after the jump is obtained from equations (40b), (41), and (24b) for the jump which starts from the value $\bar{u}_1 = 3/\sqrt{5}$ (i.e., for $h_1/h_0 = 2/5$, $u_{2g} = 3/2\sqrt{5}$); the maximum is found to be $h_2/h_0 = 4/5$, and $h_0'/h_0 = 19/20$. We thus have the highest point of the shock surface described above in the u, v, t space.

2.3. Energy Loss during Hydraulic Jump

The energy loss during the jump bears a simple relation to the intensity of the jump, i.e., to the two total heads before and after the jump. In the flow over a horizontal bottom, the potential energy is a minimum if the water depth h is zero. If we set the potential energy equal to zero, then for a mass of water m at a depth h, the potential energy is $P = mgh/2$. Since the kinetic energy at points of rest is equal to zero, the energy loss ΔE which occurs in the hydraulic jump may be computed as the difference of the potential energy at a point of rest before and after the jump. For the mass of water m, this becomes $\Delta E = mg[(h_0/2) - (h_0'/2)]$. Dividing by the energy before the jump, $E = \tfrac{1}{2}mgh_0$, the relative energy loss is obtained as

$$\Delta e = \Delta E/E = 1 - (h_0'/h_0) \tag{42}$$

This is the relative energy converted into heat. For water, it is to be considered as "lost." In a gas, however, where the heat content is the magnitude that corresponds to the water depth, the total heat content remains the same before and after the shock. For the gas, the heat arising during the shock is not "lost" energy. The energy equation is the same before and after the shock: $V^2 = 2g(H_0' - H) = (H_0 - H)2g$.

We will now compute the curves of constant total depth h_0' after the jump. We start from the Bernoulli equation, which for the flow after the jump reads $V_2^2 = 2g(h_0' - h_2)$. This equation divided by $a_1^{*2} = 2gh_0/3$, gives

$$(V_2/a_1^*)^2 = 3[(h_0'/h_0) - (h_2/h_0)] \tag{43a}$$

Solving for h_0'/h_0,*

$$h_0'/h_0 = (h_2/h_0) + \tfrac{1}{3}\bar{V}_2^2 \tag{43b}$$

From the above formula, the curves $h_0'/h_0 = $ const (Fig. 13) were drawn similar to those for $h_2/h_0 = $ const. The following methods were employed:

* The values $\bar{V}_2$, $\bar{u}_2$, and $\bar{v}_2$ are velocities referred to a_1^*; e.g., $\bar{V}_2 = V_2/a_1^*$, $\bar{u}_{2g} = u_{2g}/a_1^*$.

1. The values h_0'/h_0 for the right hydraulic jump, i.e., along the u axis ($v_2 = 0$), are obtained by substituting the previously computed values h_2/h_0 and u_{2g} from Eqs. (24b), (40b), and (41) in Eq. (43b). These values are given in Table II.

2. Along the circle, $h_2/h_0 = 0$, $\bar{c}_2$ may be read off directly, and from Eq. (43b), $h_0'/h_0 = \frac{1}{3}\bar{V}_2^2$.

3. Along general arbitrary curves—in particular, along circles about the origin ($\bar{V}_2 = $ const), and along the curves given Fig. 12 ($h_2/h_0 = $ const) the values $\bar{V}_2$ and h_2/h_0 may be read off, and from (43b) we have h_0'/h_0 along these curves.

4. Points of fixed curves $h_0'/h_0 = k$ may also be computed directly. To each h_2/h_0, there corresponds with the assumed fixed ratio $h_0'/h_0 = k$, a value $\bar{V}_2$ from Eq. (43a). The intersection with the corresponding h_2/h_0 curve of the circle with this value of $\bar{V}_2$ as radius and the origin as center gives a point of the required curve $h_0'/h_0 = k$.

By means of the methods given above, the curves of constant energy shown as in Fig. 13 were drawn.

Since, in a gas, the heat content after the shock at points of rest remains the same, the critical velocity, which for an ideal gas is computed as $a^* = \{2g[(k-1)/(k+1)]H_0\}^{1/2}$, is a constant magnitude in the entire flow plane even when shocks occur. In a water flow, however, it is to be observed that the analogous critical velocity for water is not constant, Eq. (19') being valid: $a_1^{*2} = \frac{2}{3}gh_0$. This is constant only if the total head h_0 is constant, i.e., in a flow without hydraulic jumps. If these occur, however, we have seen that the total head is constant only between jumps, but for each discontinuity, "jumps" to a new value h_0', so that at the same time, there is a jump in the critical velocity—the latter after the jump assumes a new value a_2^*, which is smaller than a_1^*:

$$a_1^{*2} = \tfrac{2}{3}gh_0'$$

the ratio between the two being:

$$a_2^*/a_1^* = (h_0'/h_0)^{1/2} \tag{44}$$

The change of the critical velocity (the limiting velocity of streaming and shooting water) during hydraulic jump has the following important consequence: Let the critical velocity before the jump be a_1^*, and the flow velocity be V_1 (point P in Fig. 14). After the jump, let the velocity be V_2 (point Q). $\widehat{PQ}$ is a shock polar. As a result of the jump, the total head and

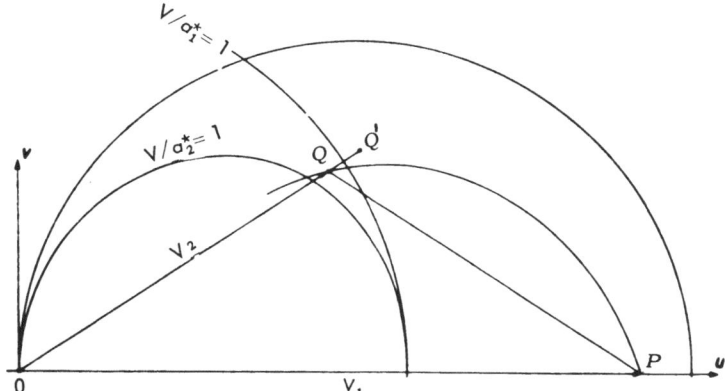

Fig. 14. Boundary line between streaming and shooting water after the jump.

hence a_2^* have become smaller than h_0 and a_1^*, respectively. It may then happen that in case V_2 is also smaller than a_1^*, V_2 nevertheless becomes larger than a_2^*. This means that the water continues to shoot after the jump, even if $V_2 < a_1^*$. There exists a curve $V_2/a_2^* = 1$ (Fig. 14). Depending on whether the point Q is without or within the area bounded by the curve and the u axis, the water, after the jump, is shooting or streaming. For a gas, this limiting curve, since $a_1^* = a_2^* = a^*$, is a circle about O.

The curve $V_2/a_2^* = 1$, which holds for water, is found in the following manner. Substituting in Eq. (43b) the relation (44), we have

$$h_0'/h_0 = (h_2/h_0) + (\tfrac{1}{3}h_0'/h_0)(V_2/a_2^*)^2$$

Putting $V_2/a_2^* = 1$, we obtain the equation:

$$h_0'/h_0 = \tfrac{3}{2}h_2/h_0$$

From the family of curves $h_0'/h_0 = $ const, and $h_2/h_0 = $ const, that curve along which this relation is satisfied is drawn. This is the required limiting curve.

Since hydraulic jumps occur in shooting water only, two cases are possible: (1) Shooting water goes over after the jump into streaming water and (2) the flow is also shooting after the jump. All right hydraulic jumps are followed by streaming water after the jump.

If the velocities are plotted in the characteristics and shock diagrams to an absolute velocity scale, then to each total head there would correspond its own diagram. All these would be similar to one another. If, however, we plot the dimensionless velocities (referred, e.g., to a_1^*), only a single diagram is required. It is to be observed, however, that in the shock diagram after the jump (point Q), we deal with the velocity V_2 referred to $a_1{}^*$. If, however, further changes in velocity are desired—whether of the characteristic diagram of a flow without losses, or of a new jump—the velocity V_2 must be referred to $a_2{}^*$, i.e., $V_2/a_2{}^*$. This is given in the hodograph by the point Q' (Fig. 14). It is obtained from $V_2/a_1{}^*$ by multiplication by $a_1{}^*/a_2{}^*$, i.e., from (44):

$$V_2/a_2{}^* = V_2/a_1{}^*(h_0/h_0')^{1/2}$$

For this reason, the curves of constant total head after the jump (Fig. 13) are denoted by $(h_0'/h_0)^{1/2}$ instead of by h_0'/h_0 as parameter.

In order to avoid having to pass from Q to Q' after the jump $\overset{\frown}{PQ}$, the shock polar could also have been defined as the geometric locus of all points Q' which correspond to a fixed point P. However one would have lost the property of the shock polars that the normals to their chords are parallel to the shock wave front in the flow.

2.4. Summary

We have seen that the flow of a compressible gas with $k = 2$ when shock is included is no longer analogous to the flow of water on a horizontal bottom. From Fig. 13, it may be seen, however, that the energy loss $\Delta e = 1 - (h_0'/h_0)$ is extremely slight over a large region. For shocks (hydraulic jumps), e.g., whose state after the shock is given by point Q lying in the hatched region, the relative loss is less than 1%. Because of this small shock loss, the analogy between the two types of flow is still satisfied to a first approximation for the case with shock.

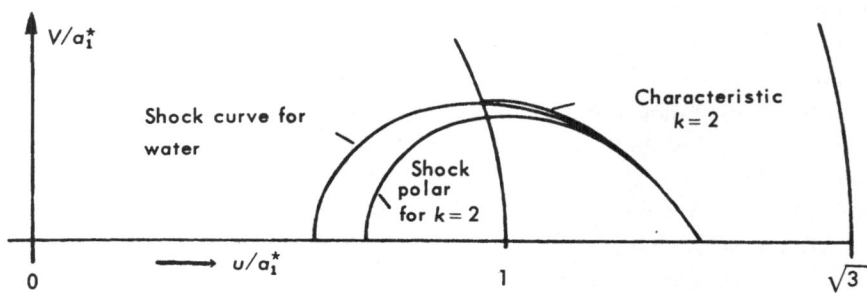

Fig. 15. Comparison of the shock polar for water, the shock polar for a gas with $k = 2$, and the ($k = 2$) characteristic.

In order to have a basis for comparison, Fig. 15 shows a shock polar for water and the corresponding shock polar for a gas ($k = 2$). Also given is the corresponding characteristic—the same curve for gas with $k = 2$ and water—in order to show that for continually decreasing shocks, the two shock polars approach one another and tend to coincide with the characteristic.

Preiswerk did the original work on the two-dimensional hydraulic analogy. The material presented here was derived from ([1]). For further details, the reader should refer to that work.

Following the discussion of the two-dimensional steady-flow hydraulic analogy, we may proceed to the one-dimensional unsteady-flow hydraulic analogy, first pioneered by Loh in 1944 at MIT [see ([5, 8, 12, 15, 16])] and later followed by other researchers ([6-11]). The following sections deal with the unsteady-flow hydraulic analogy and its applications.

3. ONE-DIMENSIONAL UNSTEADY GAS DYNAMICS BY HYDRAULIC ANALOGY

The hydraulic analogy arises from mathematical similarities in the basic equations. In two-dimensional steady flow, the equations of continuity, momentum, and energy were found in the preceding sections to be identical in mathematical form for (1) an irrotational isentropic ideal-gas flow with a specific heat ratio of 2, and (2) an incompressible frictionless water flow in an open horizontal channel of rectangular cross section. Pressure waves in the gas flow correspond to gravity waves in the water flow. The shock waves of gas dynamics correspond (although not rigidly) to the hydraulic jump ([1]) of hydraulics. This is the so-called "water table," which has been used frequently in the past to study, through the use of analogy, two-dimensional steady subsonic and supersonic flows.

The present section examines the problem of one-dimensional unsteady flow. It is shown here that the equations of continuity, momentum, and energy are identical in mathematical form for an isentropic ideal-gas flow of a specific heat ratio of any value, say k, on the one hand, and an incompressible frictionless water flow in an open horizontal channel of a cross sectional shape described by the equation $z = Cy^n$ on the other hand. Here, z is the local width and y is the local height of the cross section. It is also shown that the equations of waves and wave propagation are also identical in mathematical form for the two flows. In the case of one-dimensional unsteady flow, it is further found that the case of a specific heat ratio of 2 in a rectangular channel is a special case, with $n = 0$, of the general case.

3.1. Two-Dimensional Steady Flow

Let us examine the two-dimensional steady-flow case again. This case deals with the following two analogous flows: a flow of irrotational isentropic ideal gas of specific heat ratio equal to 2, and a flow of incompressible frictionless water in an open horizontal channel of a rectangular cross section. The equations of continuity, momentum, and energy for the two flows are:

Gas flow	Water flow

Continuity equation

$$u' \frac{\partial \varrho'}{\partial x'} + \varrho' \frac{\partial u'}{\partial x'} + v' \frac{\partial \varrho'}{\partial y'}$$

$$+ \varrho' \frac{\partial v'}{\partial y'} = 0$$

$$u' \frac{\partial h'}{\partial x'} + h' \frac{\partial u'}{\partial x'} + v' \frac{\partial h'}{\partial y'}$$

$$+ h' \frac{\partial v'}{\partial y'} = 0$$

Momentum equation

$$u' \frac{\partial u'}{\partial x'} + v' \frac{\partial v'}{\partial x'}$$

$$= -\left(\frac{1}{k-1}\right) \frac{\partial p'^{(k-1)/k}}{\partial x'}$$

$$u' \frac{\partial u'}{\partial y'} + v' \frac{\partial v'}{\partial y'}$$

$$= -\left(\frac{1}{k-1}\right) \frac{\partial \varrho'^{(k-1)/k}}{\partial y'}$$

$$u' \frac{\partial u'}{\partial x'} + v' \frac{\partial v'}{\partial x'} = - \frac{\partial h'}{\partial x'}$$

$$u' \frac{\partial u'}{\partial y'} + v' \frac{\partial v'}{\partial y'} = - \frac{\partial h'}{\partial y'}$$

Energy equation

$$\frac{u^2}{u_{\max}^2} = 1 - \left(\frac{T}{T_0}\right)$$

$$\frac{u^2}{u_{\max}^2} = 1 - \left(\frac{h}{h_0}\right)$$

Here the equations are shown in dimensionless form, so identical terms in corresponding equations may be put equal numerically. Comparison of the corresponding equations shows the following analogous terms:

Gas flow	Water flow

$$\varrho' \left(= \frac{\varrho}{\varrho_0}\right) \qquad\qquad h' \left(= \frac{h}{h_0}\right)$$

$$p'^{(k-1)/k} \left(= \frac{p^{(k-1)/k}}{p_0^{(k-1)/k}}\right) \qquad h' \left(= \frac{h}{h_0}\right)$$

$$T' \left(= \frac{T}{T_0}\right) \qquad\qquad h' \left(= \frac{h}{h_0}\right)$$

$$u' \left(= \frac{u}{[K(p_0/\varrho_0)]^{1/2}}\right) \qquad u' \left(= \frac{u}{[gh_0]^{1/2}}\right)$$

$$\left(\frac{1}{k-1}\right) \qquad = \qquad 1$$

The last condition specifies that the flow of water is comparable with the flow of a gas having a ratio of specific heats $k = c_p/c_v = 2$. Although no such gas with $k = 2$ is found in nature, the hydraulic analog has been found useful in a qualitative manner for the studies of two-dimensional steady subsonic and supersonic flows.

3.2. One-Dimensional Unsteady Flow

One-dimensional unsteady-flow equations were examined early in 1944 by Loh [see [5]] under the guidance of E. S. Taylor at MIT. It was found that these equations are identical in mathematical form for flow of an irrotational isentropic ideal gas (with a specific heat ratio of any value) and for the flow of incompressible frictionless water in an open horizontal channel with a cross section described by the equation $z = Cy^n$. Later, similar results were obtained independentaly at New York University by Probstein and Hudson [6]. Shapiro and Malavard [7] and Shapiro [8] of MIT essentially described Loh's results. Following Loh's early approach to one-dimensional unsteady flow, Szebehely and Whicker [9] of Virginia Polytechnic Institute generalized the hydraulic analogy to apply to the external steady-flow case. For internal unsteady flow, Schweitzer [10] and Schweitzer and Tsu [11] of Pennsylvania State University applied Loh's findings to a research study in exhaust dynamics. This type of analysis has been widely applied by several authors during the last few years [12]. It is the purpose of this section to present the analysis for the one-dimensional unsteady-flow hydraulic analogy [15]. The equations of continuity, momentum, and energy for the two analogous flows will be derived below.

3 2.1. *Basic Assumptions*

1. The fluid is frictionless, so that conservation of energy into heat is excluded both in gas and in water.

2. The flow is one dimensional and is in a duct (for gas) or channel (for water) of uniform cross section. This implies that the v and w components of fluid velocity are negligible compared with the u component of fluid velocity.

3. The vertical acceleration of the water is negligible compared with the acceleration of gravity. Under this assumption, the static pressure at a point of the flow field depends linearly on the vertical distance beneath the free surface at that position. In other words, $p = \varrho g(h - y)$. It is further assumed that the velocity is uniform and constant over any cross section perpendicular to the flow direction. The justification of this assumption for the case of one-dimensional unsteady flow is given in Appendix A.

4. In one-dimensional unsteady-gas flow, all parameters (pressure, temperature, velocity) are assumed to be uniform and constant across any section perpendicular to the direction of flow.

5. The viscosity and the surface tension of the water are neglected. In the wave analysis, the waves are considered to be long waves, so that the change of elevation of the water is considered to be small in comparison with h_0, the water depth at equilibrium position.

3.2.2. *The Continuity Equation*

a. Gas Flow. The continuity equation for one-dimensional unsteady flow has the following form:

$$\frac{\partial \varrho}{\partial t} + u \frac{\partial \varrho}{\partial x} + \varrho \frac{\partial u}{\partial x} = 0$$

Putting this into dimensionless form by letting $\varrho/\varrho_0 = \varrho'$, $u/a_0 = u'$, $x/l_a = x'$, and $t/t_a = t'$, it becomes

$$u' \frac{\partial \varrho'}{\partial x'} + \varrho' \frac{\partial u'}{\partial x'} + \frac{l_a}{a_0 t_a} \frac{\partial \varrho'}{\partial t'} = 0$$

b. Water Flow. The continuity equations say that the net mass rate of flow crossing the control surfaces must be equal to the net rate of change of mass inside the control volume. At a given instant, the mass rate of flow into the left boundary is $\varrho A u$ and the mass within the control volume is $\varrho A \, dx$. Hence we can write

$$-\frac{\partial(\varrho A u)}{\partial x} \, dx = \frac{\partial}{\partial t} (\varrho A \, dx)$$

Expanding this, and noting that ϱ is constant for water flow, we get

$$u \frac{\partial A}{\partial x} + A \frac{\partial u}{\partial x} + \frac{\partial A}{\partial t} = 0$$

Putting this into dimensionless form, noting that $A = \int_0^h y^n \, dy = h^{n+1}/(n+1)$, and using

$$h/h_0 = h', \qquad A/A_0 = h'^{n+1}$$

$$\frac{u}{[gh_0/(n+1)]^{1/2}} = u'$$

$$x/l_w = x', \qquad t/t_w = t'$$

it becomes

$$u' \frac{\partial h'^{n+1}}{\partial x'} + h'^{n+1} \frac{\partial u'}{\partial x'} + \frac{l_w}{[gh_0/(n+1)]^{1/2}t_w} \frac{\partial h'^{n+1}}{\partial t'} = 0$$

3.2.3. Momentum Equation

a. Gas Flow. The momentum equation for one-dimensional unsteady flow has the following form:

$$\frac{\partial u}{\partial t} + u \frac{\partial u}{\partial x} = -\frac{1}{\varrho} \frac{\partial p}{\partial x}$$

Putting this into dimensionless form as above and noting the isentropic relationship $p/\varrho^k = p_0/\varrho_0^k$, we get

$$\frac{\partial u'}{\partial t'} + \frac{a_0 t_a}{l_a} u' \frac{\partial u'}{\partial x'} = -\left(\frac{a_0 t_a}{l_a}\right) \frac{1}{k-1} \frac{\partial p'^{(k-1)/k}}{\partial x'}$$

b. Water Flow. Newton's law, as it applies to the fluid volume shown in Fig. 16, says that the net force to the right will be equal to the mass times the acceleration. The net force is the difference of pressure forces on a–a and b–b:

$$F_{a-a} - F_{b-b} = \int_0^h \varrho g(h-y)z \, dy - \int_0^{h+\partial h/\partial x \, dx} \varrho g(h-y)z \, dy$$

$$= -\frac{\varrho g}{(n+1)(n+2)} \left[h^{n+2} - \left(h + \frac{\partial h}{\partial x} dx\right)^{n+2} \right]$$

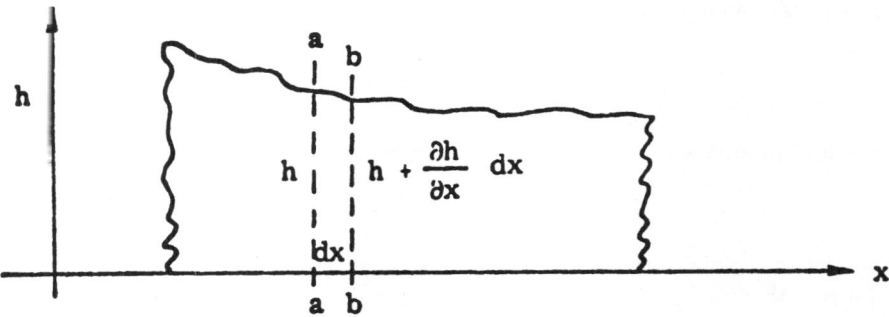

Fig. 16. Diagram for calculation of momentum equation.

The mass between a–a and b–b is (see Appendix B)

$$\varrho \frac{[h^{n+2} - (h + \partial h/\partial x \, dx)^{n+2}]}{(\partial h/\partial x)(n + 1)(n + 2)}$$

The acceleration is

$$\frac{Du}{Dt} = \frac{\partial u}{\partial t} + u\frac{\partial u}{\partial x}$$

By Newton's law of motion, we get, after simplification,

$$\frac{\partial u}{\partial t} + u\frac{\partial u}{\partial x} = -g\frac{\partial h}{\partial x}$$

Putting this into dimensionless form as done before, we get

$$\frac{\partial u'}{\partial t'} + \frac{[gh_0/(n+1)]^{1/2}t_w}{l_w}u'\frac{\partial u'}{\partial x} = -\left(\frac{[gh_0/(n+1)]^{1/2}t_w}{l_w}\right)(n + 1)\frac{\partial h'}{\partial x'}$$

3.2.4. Energy Equation

a. Gas Flow. The one-dimensional unsteady-flow energy equation has the following usual form:

$$\frac{\partial E}{\partial t} + u\frac{\partial E}{\partial x} + p\frac{\partial(1/\varrho)}{\partial t} + pu\frac{\partial(1/\varrho)}{\partial x} = 0$$

Putting this into dimensionless form by letting

$$E' = \frac{E}{E_0} = \frac{c_i T}{c_v T_0} = \left(\frac{T}{T_0}\right)$$

and using the other dimensionless quantities shown previously, it becomes

$$\frac{\partial E'}{\partial t'} + \frac{a_0 t_a}{l_a} u' \frac{\partial E'}{\partial x'} + (k-1)p' \frac{\partial(1/\varrho')}{\partial t'} + (k-1)\left(\frac{a_0 t_a}{l_a}\right)p'u' \frac{\partial(1/\varrho')}{\partial x'} = 0$$

b. Water Flow. The law of conservation of energy requires that the difference between the rate at which energy enters a control volume V and the rate at which energy leaves V must be equal to the net rate of increase of energy in V (Fig. 17). The rate of energy *entering* at a–a is given by

$$R_{a-a} = \int (\tfrac{1}{2}u^2)\varrho u \, dA + \int (gy)\varrho u \, dA + \int \varrho u \, dA$$

The rate of energy *leaving* at b–b is

$$R_{b-b} = [\int (\tfrac{1}{2}u^2)\varrho u \, dA + \int (gy)\varrho u \, dA + \int pu \, dA]$$
$$+ (\partial/\partial x)[\int (\tfrac{1}{2}u^2)\varrho u \, dA + \int (gy)\varrho u \, dA + \int pu \, dA] \, dx$$

The rate of energy increase in volume V is

$$R_{inc} = (\partial/\partial t)[\int (\tfrac{1}{2}u^2)\varrho \, dA \, dx + \int (gy)\varrho \, dA \, dx]$$

Therefore we may write the energy equation according to the law of conservation of energy as stated above:

$$- (\partial/\partial x)[\int (\tfrac{1}{2}u^2)\varrho u \, dA + \int (gy)\varrho u \, dA + \int pu \, dA] \, dx$$
$$= (\partial/\partial t)[\int (\tfrac{1}{2}u^2)\varrho \, dA + \int (gy)\varrho \, dA] \, dx$$

Expanding this relation by noting that $dA = y^n \, dy$ and $p = \varrho g(h - y)$, and further simplifying it by using the known continuity and momentum

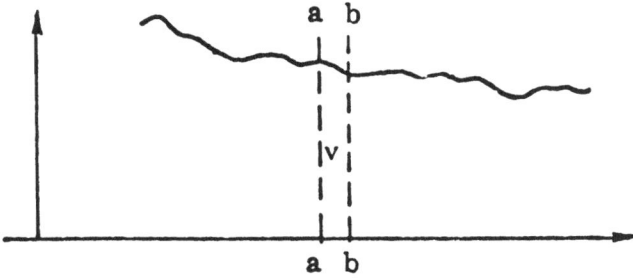

Fig. 17. Diagram for calculation of energy equation.

equations already derived, we get

$$\frac{\partial h}{\partial t} + u\frac{\partial h}{\partial x} + \frac{h^{n+2}}{n+1}\frac{\partial(h^{1/(n+1)})}{\partial t} + \frac{h^{n+2}}{n+1}u\frac{\partial(h^{1/(n+1)})}{\partial x} = 0$$

Putting this into dimensionless form, we get

$$\frac{\partial h'}{\partial t'} + \frac{[gh_0/(n+1)]^{1/2}t_w}{l_w}u'\frac{\partial h'}{\partial x'} + \frac{h'^{n+2}}{n+1}\frac{\partial(h'^{1/(n+1)})}{\partial t'}$$
$$+ \frac{h'^{n+2}}{n+1}\left(\frac{[gh_0/(n+1)]^{1/2}t_w}{l_w}\right)u'\frac{\partial(h'^{1/(n+1)})}{\partial x'} = 0$$

3.3. Equations of Standing Waves

In order to make the hydraulic analogy more clear for the present case of one-dimensional unsteady flow, it is best to check further the mathematical similarities of fundamental wave equations. Only the simple equations of wave formation produced by simple causes are considered here; otherwise the equations would be too complicated and difficult to operate. However, it is believed that the analogy of waves used in these simple equations will also exist with respect to complicated wave mechanisms. Based on this point of view, the following equations, derived in Section 4 are given here for comparison:

<table>
<tr><td align="center">Gas flow</td><td align="center">Water Flow</td></tr>
</table>

$$\frac{\partial^2\xi}{\partial t^2} = -\frac{1}{k}a_0^2\frac{\partial}{\partial x}\left(\frac{p}{p_0}-1\right)$$

$$\frac{\partial^2\xi}{\partial t^2} = -\frac{n+1}{n+2}\frac{gh_0}{n+1}\frac{\partial}{\partial x}\times\left[\left(\frac{h}{h_0}\right)^{n+2}-1\right]$$

$$\frac{\partial^2\xi}{\partial t^2} = a_0^2\frac{\partial^2\xi}{\partial x^2}$$

$$\frac{\partial^2\xi}{\partial t^2} = \frac{gh_0}{n+1}\frac{\partial^2\xi}{\partial x^2}$$

$$\frac{\partial^2}{\partial t^2}\left(\frac{p}{p_0}-1\right) = a_0^2\frac{\partial^2}{\partial x^2}\left(\frac{p}{p_0}-1\right)$$

$$\frac{\partial^2}{\partial t^2}\left[\left(\frac{h}{h_0}\right)^{n+2}-1\right] = \frac{gh_0}{n+1}\frac{\partial^2}{\partial x^2}\left[\left(\frac{h}{h_0}\right)^{n+2}-1\right]$$

$$\frac{\partial^2}{\partial t^2}\left(\frac{\varrho}{\varrho_0}-1\right) = a_0^2\frac{\partial^2}{\partial x^2}\left(\frac{\varrho}{\varrho_0}-1\right)$$

$$\frac{\partial^2}{\partial t^2}\left[\left(\frac{h}{h_0}\right)^{n+1}-1\right] = \frac{gh_0}{n+1}\frac{\partial^2}{\partial x^2}\left[\left(\frac{h}{h_0}\right)^{n+1}-1\right]$$

$$a = \left(\frac{dp}{d\varrho}\right)^{1/2} = \left(k\,\frac{p}{\varrho}\right)^{1/2} \qquad\qquad c = \left(\frac{gA}{b}\right)^{1/2} = \left(\frac{gh_0}{n+1}\right)^{1/2}$$
$$= (kRT)^{1/2}$$

Now we have derived the basic equations of continuity, momentum, and energy; we have also derived the basic equations of simple waves and wave propagation. Summarizing these equations and comparing them term by term, we will have the analogous terms in the two flows. These identical terms in corresponding equations may be put equal numerically, because they are already in dimensionless form.

3.4. Summary of Analogous Equations

Summarizing all the above equations, we have:

Gas flow   Water flow

Equation of Continuity

$$u'\,\frac{\partial\varrho'}{\partial x'} + \varrho'\,\frac{\partial u'}{\partial x'} + \frac{l_a}{a_0 t_a}\,\frac{\partial\varrho'}{\partial t'} = 0 \qquad u'\,\frac{\partial h'^{n+1}}{\partial x'} + h'^{n+1}\,\frac{\partial u'}{\partial x'}$$

$$+ \frac{l_w}{[gh_0/(n+1)]^{1/2}t_w}$$

$$\times \frac{\partial h'^{n+1}}{\partial t'} = 0$$

Momentum equation

$$\frac{\partial u'}{\partial t'} + \frac{a_0 t_a}{l_a}\,u'\,\frac{\partial u'}{\partial x'} \qquad\qquad \frac{\partial u'}{\partial t'} + \frac{[gh_0/(n+1)]^{1/2}t_w}{l_w}\,u'\,\frac{\partial u'}{\partial x'}$$

$$= -\frac{a_0 t_a}{l_a}\,\frac{1}{k-1}\,\frac{\partial p^{(k-1)/k}}{\partial x'} \qquad = -\frac{[gh_0/(n+1)]^{1/2}t_w}{l_w}$$

$$\times (n+1)\,\frac{\partial h'}{\partial x'}$$

Energy equation

$$\frac{\partial E'}{\partial t'} + \frac{a_0 t_a}{l_a}\,u'\,\frac{\partial E'}{\partial x'} \qquad\qquad \frac{\partial h'}{\partial t'} + \frac{[gh_0/(n+1)]^{1/2}t_w}{l_w}\,u'\,\frac{\partial h'}{\partial x'}$$

$$+ (k-1)p'\,\frac{\partial(1/\varrho')}{\partial t'} \qquad\qquad + \frac{1}{n+1}\,h'^{n+2}\,\frac{\partial(h'^{1/(n+1)})}{\partial t'}$$

$$+ (k-1)\frac{a_0 t_a}{l_a}\,p'u'\,\frac{\partial(1/\varrho')}{\partial x'} \qquad\quad + \frac{1}{n+1}\,\frac{[gh_0/(n+1)]^{1/2}t_w}{l_w}$$

$$= 0 \qquad\qquad\qquad\qquad\qquad \times h'^{n+2}u'\,\frac{\partial(h'^{1/(n+1)})}{\partial x'} = 0$$

Equations of standing waves

$$\frac{\partial^2 \xi}{\partial t^2} = -\frac{1}{k} a_0^2 \frac{\partial}{\partial x}\left(\frac{p}{p_0} - 1\right)$$

$$\frac{\partial^2 \xi}{\partial t^2} = -\frac{n+1}{n+2}\frac{g h_0}{n+1}\frac{\partial}{\partial x}$$
$$\times \left[\left(\frac{h}{h_0}\right)^{n+2} - 1\right]$$

$$\frac{\partial^2 \xi}{\partial t^2} = a_0^2 \frac{\partial^2 \xi}{\partial x^2}$$

$$\frac{\partial^2 \xi}{\partial t^2} = \frac{g h_0}{n+1}\frac{\partial^2 \xi}{\partial x^2}$$

$$\frac{\partial^2}{\partial t^2}\left(\frac{p}{p_0} - 1\right)$$
$$= a_0^2 \frac{\partial^2}{\partial x^2}\left(\frac{p}{p_0} - 1\right)$$

$$\frac{\partial^2}{\partial t^2}\left[\left(\frac{h}{h_0}\right)^{n+2} - 1\right]$$
$$= \left(\frac{g h_0}{n+1}\right)\frac{\partial^2}{\partial x^2}\left[\left(\frac{h}{h_0}\right)^{n+2} - 1\right]$$

$$\frac{\partial^2}{\partial t^2}\left(\frac{\varrho}{\varrho_0} - 1\right)$$
$$= a_0^2 \frac{\partial^2}{\partial x^2}\left(\frac{\varrho}{\varrho_0} - 1\right)$$

$$\frac{\partial^2}{\partial t^2}\left[\left(\frac{h}{h_0}\right)^{n+1} - 1\right]$$
$$= \left(\frac{g h_0}{n+1}\right)\frac{\partial^2}{\partial x^2}\left[\left(\frac{h}{h_0}\right)^{n+1} - 1\right]$$

$$a = \left(k\frac{p}{\varrho}\right)^{1/2} = (kRT)^{1/2}$$

$$c = \left(\frac{gA}{b}\right)^{1/2} = \left(\frac{gh}{n+1}\right)^{1/2}$$

$$a_0 = \left(k\frac{p_0}{\varrho_0}\right)^{1/2} = (kRT_0)^{1/2}$$

$$c_0 = \left(\frac{gA_0}{b_0}\right)^{1/2} = \left(\frac{g h_0}{n+1}\right)^{1/2}$$

A comparison of these basic equations shows that they are identical in mathematical form; therefore an analogy exists in one-dimensional unsteady flow between an isentropic ideal-gas flow having any value of specific heat ratio and an incompressible frictionless water flow in an open horizontal channel with a cross section described by the equation $z = Cy^n$. The analogous quantities are given in Table III. The last condition is the one the water-channel cross section must possess in order to represent a gas flow of desired specific heat ratio k.

3.5. Discussion

The relationship $1/(k-1) = n+1$ shows that change of the shape of the cross section changes the values of k. For ordinary gas problems, the values of k given in Table IV are of interest.

It is also interesting to note that if $n = 0$, the shape of the cross section is rectangular, and the present results are exactly reduced to those given

TABLE III

Summary of Analogous Quantities

Gas flow	Relationship	Water flow
Local speed of sound $a = (kRT)^{1/2}$	$\approx$	Local speed of wave $c = [gh/(n + 1)]^{1/2}$
Local Mach number $M = u/(kRT)^{1/2}$	$=$	Local Mach number $M = u[gh/(n + 1)]^{-1/2}$
Local density ratio ϱ/ϱ_0	$=$	Local water depth ratio $(h/h_0)^{n+1}$
Local pressure ratio p/p_0	$=$	Local water depth ratio $(h/h_0)^{n+2}$
Local temperature ratio T/T_0	$=$	Local water depth ratio h/h_0
Flow similarity number l_a/a_0t_a	$=$	Flow similarity number $l_w\{[gh_0/(n + 1)]^{1/2}t_w\}^{-1}$
Function of gas specific heat ratio $1/(k - 1)$	$=$	Function of channel cross section exponent $(n + 1)$

in the two-dimensional steady-flow case (classical results):

$$\frac{p}{p_0} = \left(\frac{h}{p_0}\right)^2, \qquad \frac{\varrho}{\varrho_0} = \frac{h}{h_0}, \qquad \frac{T}{T_0} = \frac{h}{h_0},$$

$$a = \left(K\frac{p}{\varrho}\right)^{1/2}, \qquad c = (gh)^{1/2}, \qquad \frac{1}{k - 1} = 1$$

Because the velocity in a water channel can be made as small as one thousandth or less of that occurring in a gas stream, according to

TABLE IV

Flows	k	n	Equation of channel cross section	Shape of channel cross section	Flow
$P = $ const	0	-2	$y^2z = $ const	Hyperbolic	Isopiestic
$V = $ const	∞	-1	$yz = $ const	Hyperbolic	Isometric
$T = $ const	1	∞	Indeterminate	Indeterminate	Isothermal
$p/\varrho^k = $ const	1.4	1.5	$z = cy^{1.5}$	Parabolic	Isentropic
$p/\varrho^{1.5} = $ const	1.5	1.0	$z = cy$	Triangular	Approximately isentropic
$p/\varrho^2 = $ const	2.0	0	$z = $ const	Rectangular	Classical result

the analogous relationships: sound velocity $= [k(p/\varrho)]^{1/2}$ and wave velocity $= [gh/(n + 1)]^{1/2}$, the time scale can be lengthened correspondingly. This makes the hydraulic analogy a relatively easy way for observing and studying transient effects in gas dynamics. The slow motion can even be seen unaided. This feature is particularly useful for studies such as those of unsteady flow in ducts and of traveling shocks, and of unsteady flow such as occurs in wave engines, and might also be useful for the study of transient phenomena such as occur during the starting of a diffuser, and of the stability of a diffuser during operation.

4. HYDRAULIC ANALOGY FOR LONGITUDINAL PLANE WAVES. DERIVATION OF EQUATIONS OF STANDING WAVES

The gas wave is considered to be set up in a duct, and the water wave is considered to be set up in an open horizontal channel with a cross section described by the equation $z = Cy^n$. Both the duct and the channel have a constant cross section. The waves considered are plane waves and are set up in the fluid medium, which is considered to be motionless except for small oscillations about the equilibrium positions. In both cases, the restoring force responsible for keeping the wave going is simply the opposition that the medium exhibits against being compressed.

4.1. Continuity Equation

4.1.1. Air Waves

Consider the air in a straight pipe of uniform cross section of area S as shown in Fig. 18. When a sound wave (or any elastic wave) passes through the pipe, the planes at different points along the pipe will be displaced from

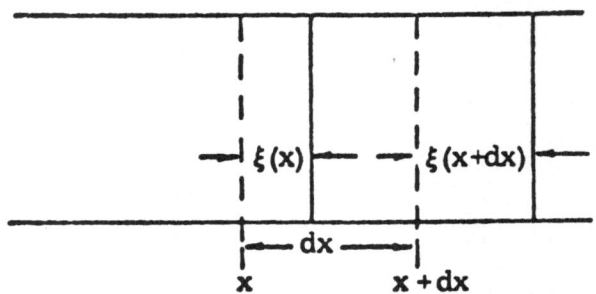

Fig. 18. Diagram for calculation of continuity equation for air waves.

their equilibrium positions back and forth along the pipe. This displacement depends on both t and x; the gas ahead of one plane will always be ahead of that plane, the gas between two planes will always be between these two planes; this means that gas particles already on the same plane will remain on the same plane which the sound wave penetrates. Now in Fig. 18, examine two planes which at equilibrium are at the distance x and $x + dx$. The gas between them has a density ϱ_0, so the mass is $\varrho_0 S\, dx$. When the planes are displaced to the dotted position, the mass between them will not be changed, but the volume is changed because the displacement of one plane is $\xi(x)$ and that of the other is $\xi(x + dx) = \xi(x) + \partial\xi/\partial x\, dx$, and hence the density changes to ϱ in order to keep mass constant. Therefore the continuity equation is

$$\varrho S[dx + \xi(x+dx) - \xi(x)] = \varrho[S\,dx + S\,\partial\xi/\partial x\,dx] = \varrho_0 S\,dx \qquad (45)$$

Let the relative change in density be $\delta(x, t)$, where

$$\delta = (\varrho - \varrho_0)/\varrho_0 \qquad \varrho = \varrho_0(1 + \delta) \qquad (46)$$

Substituting (45) into (44), one obtains

$$\varrho_0(1 + \delta)S\,dx\,(1 + \partial\xi/\partial x) = S\varrho_0\,dx$$
$$(1 + \delta)(1 + \partial\xi/\partial x) = 1 \qquad (47)$$

Since the change of density and displacement are small, the product of two small quantities $\delta\,\partial\xi/\partial x$ may be neglected,

$$1 + \delta + \partial\xi/\partial x = 1, \qquad \delta = -\partial\xi/\partial x \qquad (48)$$

4.1.2. Water Waves

Consider two planes perpendicular to the length of the channel, which move with the fluid, and therefore always contain the same particles, and which before the fluid was disturbed were at a distance dx part (Fig. 19); at time t, their distance apart will have become

$$dx + \xi(x+dx) - \xi(x) = dx + \xi(x) + (\partial\xi/\partial x)\,dx - \xi(x) = dx + (\partial\xi/\partial x)\,dx$$

but the quantity of fluid between them will be unaltered, and therefore, by the equation of continuity, one obtains

$$\frac{h_0^{n+1}}{n+1}\,dx = \left(dx + \frac{\partial\xi}{\partial x}\,dx\right)\frac{(h_0 + \eta)^{n+1}}{n+1}$$

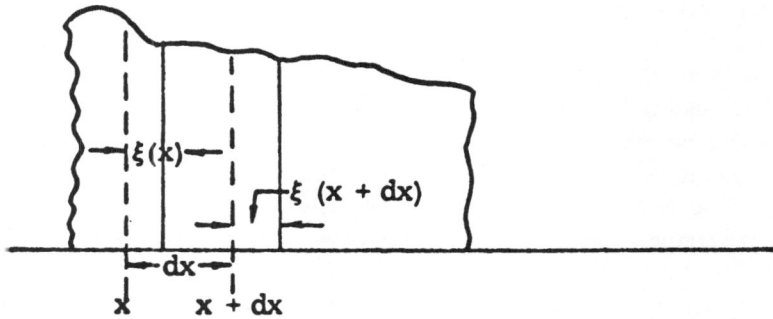

Fig. 19. Diagram for calculation of continuity equation for water waves.

where h_0 is the water height at undisturbed condition, and η is the change of elevation due to the disturbance with reference to h_0, and is equal to $h - h_0$, where h is the instantaneous water height after the disturbance.

Now,

$$\left(1 + \frac{\partial \xi}{\partial x}\right)\left(1 + \frac{\eta}{h_0}\right)^{n+1} = 1 \qquad (49)$$

Comparing (49) with (47), one obtains

$$(1 + \delta) = \left(1 + \frac{\eta}{h_0}\right)^{n+1}$$

4.2. Equation of Thermodynamics and Equation of Hydraulics

4.2.1. *Air Waves. Equation of Thermodynamics*

We have

$$\frac{dp}{p} + \frac{dv}{v} = \frac{dT}{T}$$

Now, if $v = (\text{const})\ dv = 0$, then $dp = (p/T)\ dT$,

$$dQ = \frac{\partial Q}{\partial p}\, dp = \frac{\partial Q}{\partial p}\, \frac{p}{T}\, dT, \qquad c_v = \left(\frac{dQ}{dT}\right)_v = \frac{\partial Q}{\partial p}\, \frac{p}{T}$$

Similarly,

$$\frac{\partial Q}{\partial p} = c_c \frac{T}{p}, \qquad \frac{\partial Q}{\partial v} = c_p \frac{T}{v}$$

For adiabatic change between volume and pressure, $dQ = 0$,

$$dQ = \frac{\partial Q}{\partial v} \, dv + \frac{\partial Q}{\partial p} \, dp = T\left(c_p \frac{dv}{v} + c_v \frac{dp}{p}\right) = 0$$

$$c_p \frac{dv}{v} = -c_v \frac{dp}{p}$$

(50)

Now, consider the gas in a pipe contained between the planes originally at x and $x + dx$; the volume of the gas is $S \, dx$ and the change in volume is $S(\partial \xi / \partial x) \, dx$. Since

$$S \, dx - S[dx + \xi(x + dx) - \xi(x)]$$
$$= S \, dx - S[dx + \xi(x) + (\partial \xi / \partial x) \, dx - \xi(x)] = -S(\partial \xi / \partial x) \, dx$$

the pressure is P_0, and the change in pressure is p (excess pressure over P_0). Applying Eq. (50), the equation of thermodynamics, one obtains

$$\frac{c_p S(\partial \xi / \partial x) \, dx}{S \, dx} = -c_v\left(\frac{p}{P_0}\right), \qquad p = -k P_0\left(\frac{\partial \xi}{\partial x}\right)$$

(51)

Substituting (48) into (51), one obtains

$$p = k P_0 \delta; \qquad \partial p / \partial x = k P_0 \, \partial \delta / \partial x$$

(52)

4.2.2. Water Waves. Hydraulics Equation

The hydraulics equation giving the relation between the change of hydrostatic force on the volume and change of water height of the volume is obtained from

$$F = \int_0^h \varrho g(h - y) z \, dy = \frac{\varrho g h^{n+2}}{(n+1)(n+2)} = \frac{\varrho g}{(n+1)(n+2)} (h_0 + \eta)$$

Therefore the hydraulics equation becomes

$$\frac{\partial F}{\partial x} = (n + 2) \frac{\varrho g}{(n+1)(n+2)} (h_0 + \eta)^{n+1} \frac{\partial \eta}{\partial x}$$

$$= \frac{\varrho g}{n + 1} (h_0 + \eta)^{n+1} \frac{\partial \eta}{\partial x}$$

(53)

4.3. Equation of Motion

4.3.1. Air Waves

The equation of motion can be written immediately (Fig. 20):

$$S[P_0 + p(x)] - S[P_0 + p(x) + (\partial p / \partial x) \, dx] = \varrho_0 S \, dx \, \partial^2 \xi / \partial t^2$$

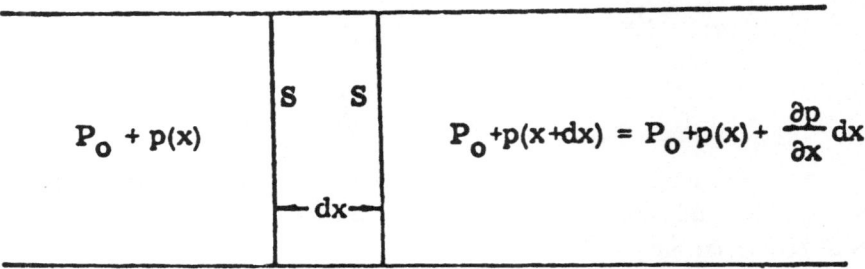

Fig. 20. Diagram for calculation of equation of motion.

where ξ is the displacement of the plane along the pipe;

$$\varrho_0 \frac{\partial^2 \xi}{\partial t^2} = -\frac{\partial p}{\partial x}$$

$$\frac{\partial^2 \xi}{\partial t^2} = -\frac{1}{k} a_0^2 \frac{\partial(p/P_0)}{\partial x} \tag{54}$$

Putting this in dimensionless form as before,

$$\frac{\partial^2 \xi'}{\partial t'^2} = -\frac{1}{k}\left(\frac{a_0 t_a}{l_a}\right)\frac{\partial p'}{\partial x'} \tag{55'}$$

4.3.2. Water Waves

$$\varrho \frac{h_0^{k+1}}{n+1} dx \frac{\partial^2 \xi}{\partial t^2} = -\frac{\partial F}{\partial x} dx \tag{56}$$

here $\partial^2 \xi/\partial t^2$ is the acceleration of the small volume between two sections dx apart in undisturbed state, and $\partial F/\partial x \, dx$ is the differential change of the hydrostatic force on the volume.

Substituting (53) into (55), one obtains

$$\begin{aligned}
\frac{\partial^2 \xi}{\partial t^2} &= -g\left(1 + \frac{\eta}{h_0}\right)^{n+1}\frac{\partial \eta}{\partial x} \\
&= -\frac{g h_0}{n+2}\frac{\partial}{\partial x}\left[\left(1 + \frac{\eta}{h_0}\right)^{n+2} - 1\right] \\
&= -\frac{g h_0}{n+1}\frac{n+1}{n+2}\frac{\partial}{\partial x}\left[\left(1 + \frac{\eta}{h_0}\right)^{n+2} - 1\right]
\end{aligned} \tag{56}$$

Putting this into dimensionless form as before,

$$\frac{\partial^2 \xi'}{\partial t'^2} = -\frac{[g h_0/(n+1)]^{1/2} t_w}{l_w}\frac{n+1}{n+2}\frac{\partial}{\partial x'}\left[\left(1 + \frac{\eta}{h_0}\right)^{n+2} - 1\right] \tag{56'}$$

Comparing (56) with (54), one obtains immediately

$$\left(\frac{a_0 t_a}{l_a}\right) = \frac{[gh_0/(n+1)]^{1/2} t_w}{l_w}, \qquad k = \frac{n+1}{n+2}$$

$$\frac{p}{P_0} = \left(1 + \frac{\eta}{h_0}\right)^{n+2} - 1$$

4.4. Wave Equations

4.4.1. Air Waves

Substituting (48) into (52), one obtains

$$\partial p/\partial x = -kP_0\, \partial^2 \xi/\partial x^2 \tag{57}$$

Substituting (57) into (54'),

$$\partial^2 \xi/\partial t^2 = a_0^2\, \partial^2 \xi/\partial x^2 \tag{58}$$

Similarily,

$$\partial^2 p/\partial t^2 = a_0^2\, \partial^2 p/\partial x^2 \tag{59}$$

$$\partial^2 \delta/\partial t^2 = a_0^2\, \partial^2 \delta/\partial x^2 \tag{60}$$

Equations (58)–(60) are the fundamental wave equations.

4.4.2. Water Waves

Between (49) and (56') either η or ξ can be eliminated; the result in terms of ξ is simpler, so partial differentiation of (49) with respect to x gives

$$\frac{-\partial^2 \xi/\partial x^2}{(1 + \partial \xi/\partial x)^2} = (n+1)\left(1 + \frac{\eta}{h_0}\right)\frac{1}{h_0}\frac{\partial \eta}{\partial x} \tag{61}$$

Substituting (61) into (56'), one obtains

$$\frac{\partial^2 \xi}{\partial t^2} = \left(\frac{gh_0}{n+1}\right)\left(1 + \frac{\eta}{h_0}\right)\frac{\partial^2 \xi/\partial x^2}{(1 + \partial \xi/\partial x)^2}$$

Substituting (49) into the above equation,

$$\frac{\partial^2 \xi}{\partial t^2} = \left(\frac{gh_0}{n+1}\right)\frac{\partial^2 \xi/\partial x^2}{(1 + \partial \xi/\partial x)(2n+3)/(n+1)}$$

From Eq. (49),

$$\left(1 + \frac{\partial \xi}{\partial x}\right) = \left(1 + \frac{\eta}{h_0}\right)^{n+1}$$

Therefore $\partial \xi / \partial x \approx (n+1)(\eta / h_0)$. Since η / h_0 is a small quantity, $\partial \xi / \partial x$ is also small in comparison to 1,

$$\left(1 + \frac{\partial \xi}{\partial x}\right)^{(2n+3)/(n+1)} \approx 1$$

$$\frac{\partial^2 \xi}{\partial t^2} = \left(\frac{gh_0}{n+1}\right) \frac{\partial^2 \xi}{\partial x^2}$$

(62)

Similarly,

$$\frac{\partial^2}{\partial t^2}\left[\left(1 + \frac{\eta}{h_0}\right)^{n+2} - 1\right] = \frac{gh_0}{n+1} \frac{\partial^2}{\partial x^2}\left[\left(1 + \frac{\eta}{h_0}\right)^{n+2} - 1\right] \quad (63)$$

$$\frac{\partial^2}{\partial t^2}\left[\left(1 + \frac{\eta}{h_0}\right)^{n+1} - 1\right] = \frac{gh_0}{n+1} \frac{\partial^2}{\partial x^2}\left[\left(1 + \frac{\eta}{h_0}\right)^{n+1} - 1\right] \quad (64)$$

Equations (63) and (64) can be proven, since the differential equations are satisfied by the functions of η found in the brackets of (63) and (64). The proof is as follows: Substituting (49) into (63), one obtains

$$\frac{\partial^2}{\partial t^2}\left[\left(1 + \frac{\partial \xi}{\partial x}\right)^{-(n+2)/(n+1)} - 1\right] = \frac{gh_0}{n+1} \frac{\partial^2}{\partial x^2}\left[\left(1 + \frac{\partial \xi}{\partial x}\right)^{-(n+2)/(n+1)} - 1\right]$$

$$\frac{\partial^2 \xi}{\partial t^2} = \left(\frac{gh_0}{n+1}\right) \frac{\partial^2 \xi}{\partial x^2}$$

The result is the same as Eq. (62), which has already been proved. Since Eq. (62) is satisfied by ξ, Eq. (63) is satisfied by η. Similarly, Eq. (64) may be proved. Comparing Eq. (62), (63), and (64) with Eqs. (58), (59), and (60), respectively, after putting them into dimensionless form, one again obtains

$$\frac{p}{P_0} = \left[\left(1 + \frac{\eta}{h_0}\right)^{n+2} - 1\right], \qquad \delta = \left[\left(1 + \frac{\eta}{h_0}\right)^{n+1} - 1\right]$$

4.5. Equation of Wave Propagation Velocity

4.5.1. *Air Waves*

The wave propagation velocity in air is well known as the sound velocity in air, which can be expressed as

$$a^2 = dp/d\varrho = k\,p/\varrho$$

so the wave propagation velocity in undisturbed stream is

$$a_0{}^2 = kp_0/\varrho_0$$

4.5.2. *Water Waves*

The wave propagation velocity of water in an open channel of a constant cross section of any shape with a straight horizontal bottom was derived mathematically by McCowan [14]:

$$c^2 = gA/b \tag{65}$$

where A is the area of the cross section and b the breadth at the free surface. Applying Eq. (65) to the cross section $z = y^n$,

$$c^2 = \frac{gh^{n+1}/(n+1)}{h^n} = \frac{gh}{n+1}$$

so the wave propagation velocity in the undisturbed stream is

$$c_0^2 = gh_0/(n+1)$$

4.6. Summary

In this section, the following equations have been derived:

4.6.1. *Equations for Gas*

1. Continuity equation:

$$(1+\delta)(1+\partial\xi/\partial x) = 1$$

2. Equation of thermodynamics:

$$\partial p/\partial x = kP_0 \,\partial\delta/\partial x$$

3. Equation of motion:

$$\frac{\partial^2 \xi}{\partial t'^2} = -\frac{1}{k}\left(\frac{a_0 t_a}{l_a}\right)^2 \frac{\partial p'}{\partial x'}$$

4. Wave equations:

$$\partial^2\xi/\partial t^2 = a_0^2\,\partial^2\xi/\partial x^2$$

$$\partial^2 p/\partial t^2 = a_0^2\,\partial^2 p/\partial x^2$$

$$\partial^2\delta/\partial t^2 = a_0^2\,\partial^2\delta/\partial x^2$$

5. Equation of wave propagation velocity:

$$a = [k(p/\varrho)]^{1/2}$$

4.6.2. *Equations for Water*

1. Continuity equation:

$$\left(1 + \frac{\eta}{h_0}\right)^{n+1}\left(1 + \frac{\partial \xi}{\partial x}\right) = 1$$

2. Hydraulics equation:

$$\frac{\partial F}{\partial x} = \frac{\varrho g}{n+1}(h_0 + \eta)^{n+1}\frac{\partial \eta}{\partial x}$$

3. Equation of motion:

$$\frac{\partial^2 \xi'}{\partial t'^2} = -\frac{n+1}{n+2}\left(\frac{[g h_0/(n+1)]^{1/2}t_w}{l_w}\right)^2\frac{\partial}{\partial x'}\left[\left(1 + \frac{\eta}{h_0}\right)^{n+2} - 1\right]$$

4. Wave equation

$$\frac{\partial^2 \xi}{\partial t^2} = \frac{g h_0}{n+1}\frac{\partial^2 \xi}{\partial x^2}$$

$$\frac{\partial^2}{\partial t^2}\left[\left(1 + \frac{\eta}{h_0}\right)^{n+2} - 1\right] = \frac{g h_0}{n+1}\frac{\partial^2}{\partial x^2}\left[\left(1 + \frac{\eta}{h_0}\right)^{n+2} - 1\right]$$

$$\frac{\partial^2}{\partial t^2}\left[\left(1 + \frac{\eta}{h_0}\right)^{n+1} - 1\right] = \frac{g h_0}{n+1}\frac{\partial^2}{\partial x^2}\left[\left(1 + \frac{\eta}{h_0}\right)^{n+1} - 1\right]$$

5. Equation of wave propagation velocity:

$$c = [gh/(n+1)]^{1/2}$$

Comparison of the corresponding equations in the two cases shows that they have the same forms. From these, we may conclude that the following analogous magnitudes must hold for the analogy between the two cases to hold:*

$$1 + \delta = \left(1 + \frac{\eta}{h_0}\right)^{n+1}$$

$$\frac{a_0 t_a}{l_a} = \frac{[g h_0/(n+1)]^{1/2}t_w}{l_w}$$

$$\frac{1}{k} = \frac{n+1}{n+2}, \qquad \frac{p}{P_0} = \left[\left(1 + \frac{\eta}{h_0}\right)^{n+2} - 1\right], \qquad \delta = \left[\left(1 + \frac{\eta}{h_0}\right)^{n+1} - 1\right]$$

* Note here the symbol p is the excess pressure.

5. EXPERIMENTAL VERIFICATION

In order to verify the theoretical analysis, an application was made to the induction system of a four-stroke reciprocations engine ([5]). The hydraulic model was built (Figs. 21 and 22) and experiments were run (Figs. 23 and 24). Results were compared to actual engine data obtained by Boden and Schecton (Figs. 25 and 26). Good agreement was generally obtained between the actual engine and the hydraulic model. These experimental results have soundly verified the theoretical analysis, in which the classical analysis is only a special case when $n = 0$.

The dynamic behavior (one-dimensional unsteady motion) of air inside a straight induction system of a single cylinder four stroke reciprocating engine, set up by the periodic opening and closing of the intake valve in conjunction with the piston motion, is a typical example which was studied by hydraulic analogy ([5]).

5.1. Equations for Experimental Model

5.1.1. Gas Flow

$$m\ddot{x} + F_3\dot{x} + k_a x = A_v[\sigma a_0^2 \varrho_0(1 + \sin 2\omega t) - P_k \cos(K_a\omega t - \Phi_k)]$$

$$x = \frac{\sigma v_0}{A_v}\left[(1 + \sin 2\omega t)\frac{P_k}{\varrho_0 a_0^2 \sigma}\cos(K_a\omega t - \Phi_k)\right]$$

$$\frac{\dot{x}}{a_0} = \frac{\pi N_a \sigma v_0}{30 A_v a_0}\left[\cos 2\omega t - \frac{K_a P_k}{2\varrho_0 a_0^2 \sigma}\sin(K_a\omega t - \Phi_k)\right]$$

$$\frac{Q_a}{v_0} = \sigma\left[1 + \sin 2\omega t - \frac{P_k}{\sigma \varrho_0 a_0^2}\cos(K_a\omega t - \Phi_k)\right]$$

$$e = \frac{Q_a}{2\sigma v_0} = \frac{1}{2}\left[1 + \sin 2\omega t - \frac{P_k}{\sigma \varrho_0 a_0^2}\cos(K_a\omega t - \Phi_k)\right]$$

5.1.2. Water Flow

$$m\ddot{x} + F_3\dot{x} + k_w x = A_v\left[\sigma\left(\frac{gh_0}{n+1}\right)\varrho_w(1 + \sin 2\omega t) - \varrho_w g H_k \cos(K_w\omega t - \Phi_k)\right]$$

$$x = \frac{\sigma v_0}{A_v}\left[(1 + \sin 2\omega t)\frac{g H_k}{\sigma\left(\dfrac{gh_0}{n+1}\right)}\cos(K_w\omega t - \Phi_k)\right]$$

$$\frac{\dot{x}}{\sqrt{\dfrac{gh_0}{n+1}}} = \frac{\pi N_w \sigma v_0}{30 A_v \sqrt{\dfrac{gh_0}{n+1}}}\left[\cos 2\omega t - \frac{K_w H_k g}{2\sigma\left(\dfrac{gh_0}{n+1}\right)}\sin(K_w\omega t - \Phi_k)\right]$$

Fig. 21. Photographic views of (above) inlet and exhaust system, and (below) piston and cylinder system.

Fig. 22. Photographic views of (above) recording system, and (below) valve and cam system.

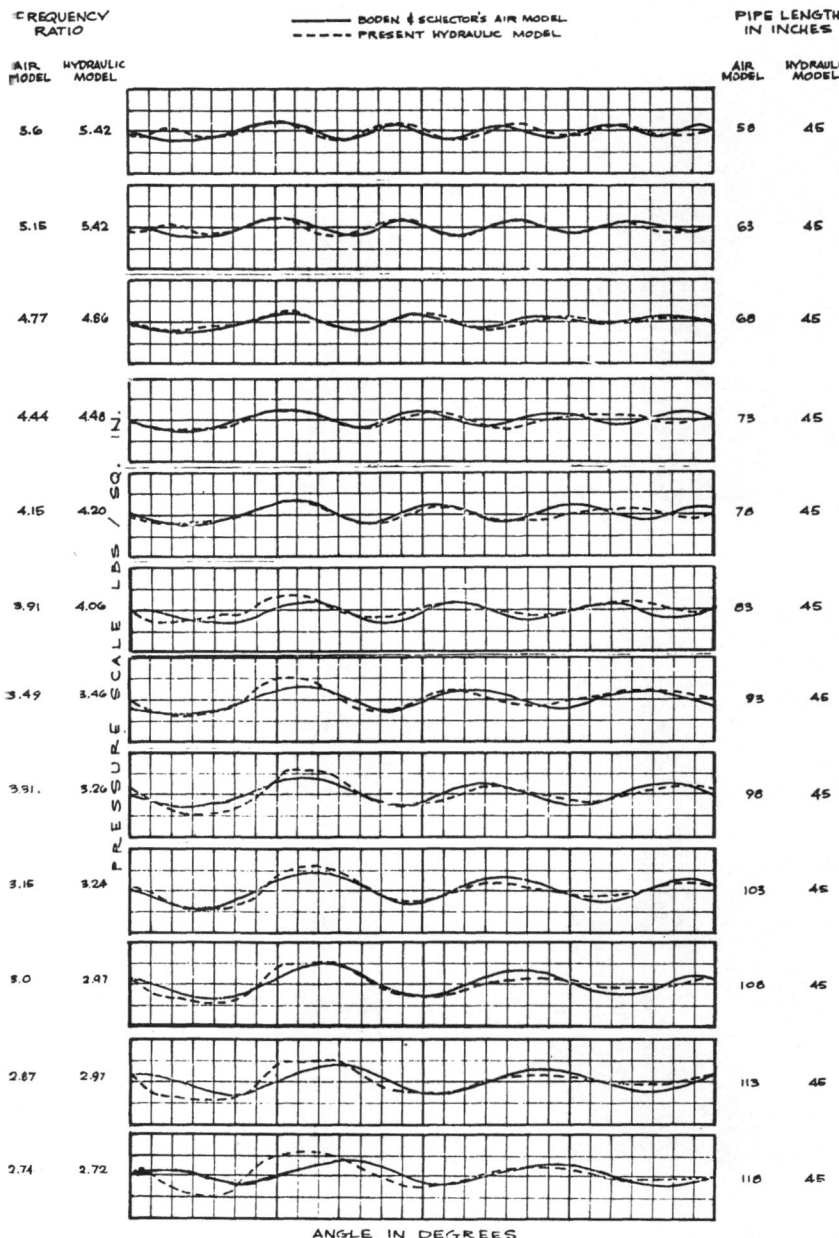

Fig. 23. Comparison of one-dimensional pressure waves between air model and hydraulic model.

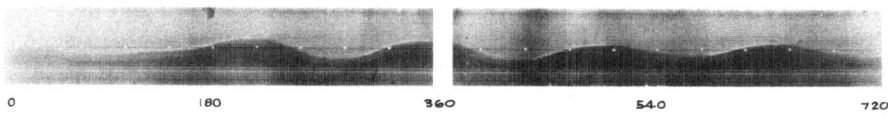

FREQUENCY RATIO = 5.55

Fig. 24. Typical photographic results.

PRESSURE VARIATIONS IN THE INLET SYSTEMS
(PRESSURE TAKEN NEAR THE INLET VALVE PORT)

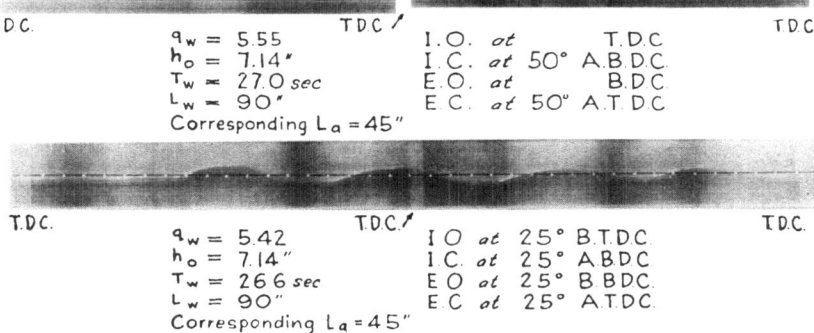

$q_w = 5.55$ I.O. *at* T.D.C
$h_o = 7.14''$ I.C. *at* 50° A.B.D.C.
$T_w = 27.0 sec$ E.O. *at* B.D.C
$L_w = 90''$ E.C. *at* 50° A.T.D.C
Corresponding $L_a = 45''$

$q_w = 5.42$ I O *at* 25° B.T.D.C.
$h_o = 7.14''$ I.C. *at* 25° A.B.D.C.
$T_w = 26.6 sec$ E O *at* 25° B.B.D.C.
$L_w = 90''$ E.C. *at* 25° A.T.D.C.
Corresponding $L_a = 45''$

Fig. 25. A photographic representation of the hydraulic results.

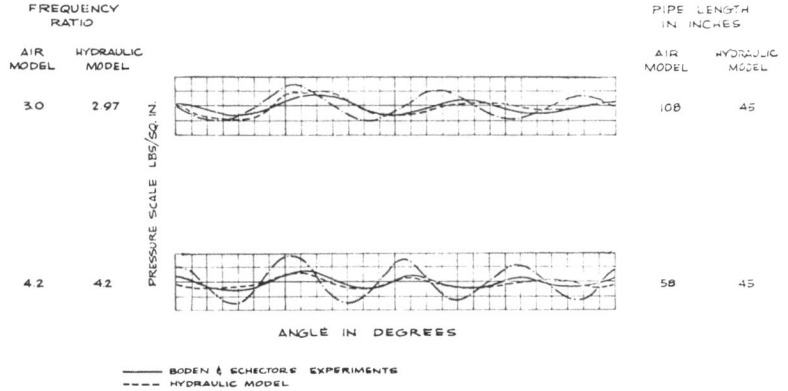

Fig. 26. A comparison of one-dimensional pressure waves between air model, hydraulic model, and acoustic model.

$$\frac{Q_w}{v_0} = \sigma\left[1 + \sin 2\omega t - \frac{H_k g}{\sigma\left(\dfrac{gh_0}{n+1}\right)} \cos(K_w \omega t - \Phi_k)\right]$$

$$e = \frac{Q_w}{2\sigma v_0} = \frac{1}{2}\left[1 + \sin 2\omega t - \frac{H_k g}{\sigma\left(\dfrac{gh_0}{n+1}\right)} \cos(K_w \omega t - \Phi_k)\right]$$

5.2. Analogous Quantities for Hydraulic Model Design

Gas flow		Water flow
$\dfrac{A_{ca}}{A_{va}}$	$=$	$\dfrac{A_{cw}}{A_{vw}}$
$\dfrac{P_k}{\varrho_0 a_0{}^2 \sigma_a}$	$=$	$\dfrac{H_k}{\dfrac{1}{g}\left(\dfrac{gh_0}{n+1}\right)\sigma_w}$
$\dfrac{N_a \sigma_a v_{0a}}{A_{va} a_0}$	$=$	$\dfrac{N_w \sigma_w v_{0w}}{A_{vw}\sqrt{\dfrac{gh_0}{n+1}}}$
K_a	$=$	K_w
$\therefore K_a \cong q_a = 30\dfrac{a_0}{N_a l_a}$	$=$	$30\dfrac{\sqrt{\dfrac{gh_0}{n+1}}}{N_w l_w} = q_w$
	$\therefore q_a$	$= q_w$ (Frequency ratio)

APPENDIX A. Discussion of the Assumption that the Velocity Is Constant Over the Whole Cross Section

Consider the surface s which encloses a volume v fixed in space. The law of conservation of energy requires that the difference in the rate of supply of energy to the volume v and the rate at which energy goes out through s must be the net rate of increase of energy in the volume v. Let U be the total energy per unit mass; then we have immediately

$$-\int_s U\varrho u_j n_j \, ds = \frac{\partial}{\partial t}\int_v U\varrho \, dv$$

By Green's theorem,

$$\int_s A n_j \, ds = \int_v \frac{\partial A}{\partial x_j} \, dv$$

to transform the surface integral into volume integral,

$$\int_V \left\{ -\frac{\partial}{\partial x_j}(U\varrho u_j) - \frac{\partial}{\partial t}(U\varrho) \right\} dv = 0$$

Since v, is arbitrarily chosen the integrand itself must be zero. Therefore

$$\frac{\partial}{\partial x_j}(U\varrho u_j) + \frac{\partial}{\partial t}(U\varrho) = 0$$

Expanding and substracting the continuity equation, we have

$$\frac{\partial}{\partial x_j}(U\varrho u_j) + \frac{\partial}{\partial t}(U\varrho) - \varrho \left[\frac{\partial U}{\partial t} + u_j \frac{\partial U}{\partial x_j} \right] = 0$$

In the present case of one-dimensional unsteady incompressible flow, this becomes

$$\frac{\partial}{\partial t}\left[\frac{\varrho}{2}u^2 + p + \varrho g y \right] + u\frac{\partial}{\partial x}\left[\frac{\varrho}{2}u^2 + p + \varrho g y \right] = 0$$

Now, if we assume that the vertical acceleration of the water is negligible compared with the acceleration of gravity, the pressure at a point of the flow field depends on the vertical distance beneath the free surface at that point. In other words,

$$p = \varrho g(h - y)$$

Substituting this into above equation,

$$\frac{\partial}{\partial t}\left[\varrho\frac{u^2}{2} + \varrho g(h - y) + \varrho g y \right] + u\frac{\partial}{\partial x}\left[\varrho\frac{u^2}{2} + \varrho g(h - y) + \varrho g y \right] = 0$$

Therefore

$$\frac{\partial}{\partial t}\left(\frac{u^2}{2} + gh \right) + u\frac{\partial}{\partial x}\left(\frac{u^2}{2} + gh \right) = 0$$

Since the equation does not contain the coordinate y, the velocity (the particle to be considered is an arbitrary distance y above the horizontal bottom) is constant over the entire depth y. In other words, the velocity is constant over the entire cross section and is a function of h only.

APPENDIX B. Mass–Volume Calculation

In Fig. 16, consider the volume between a–a and b–b, which are dx apart. In order to help the derivation of the volume integral between a–a and b–b, the enlarged sketch of Fig. 27 is included. Now, referring to Fig. 27,

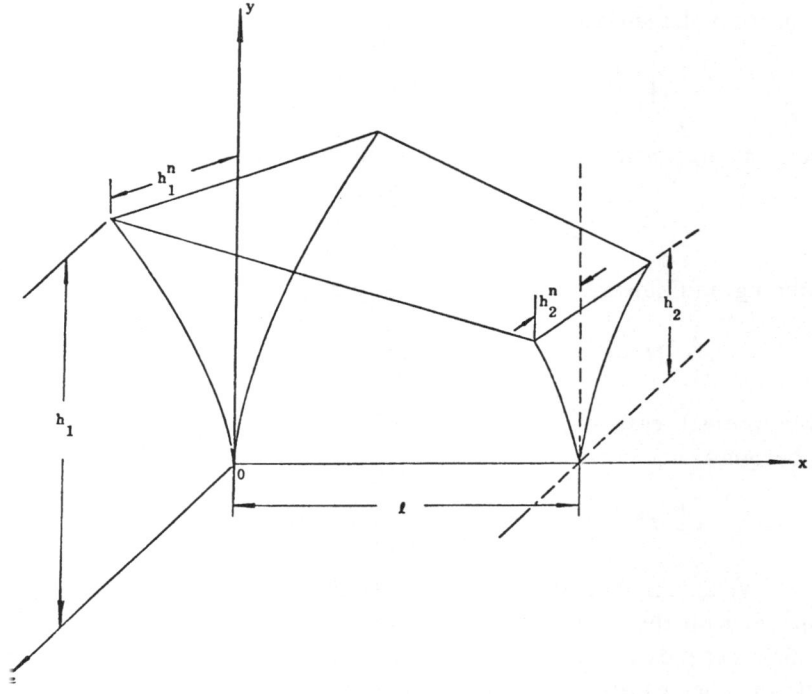

Fig. 27. Diagram for mass–volume calculation.

the volume is one which is generated by moving the plane section of area A_x perpendicular to Ox from $x = 0$ to $x = l$. Thus $V = \int_0^l A_x\,dx$, where A_x is a function of x. Assuming the surface change between sections a–a and b–b is straight, because they are only dx apart in this treatment, then from the Fig. 21,

$$y = \frac{h_2 - h_1}{l} x + h_1, \qquad dy = \frac{h_2 - h_1}{l}\,dx$$

$$V = \int_0^l y^n \, dy \, dx = \int_0^l \frac{y^{n+1}}{n+1}\,dx$$

$$= \int_0^l \frac{y^{n+1}}{n+1} \frac{l}{h_2 - h_1}\,dy$$

$$= \frac{l}{(h_2 - h_1)(n+1)(n+2)} [h_1^{n+2} - h_2^{n+2}]$$

Applying the result to the volume between a–a and b–b by using the notation shown in the instantaneous flow diagram (Fig. 27); we have

$$h_1 = h, \qquad h_2 = h + (\partial h/\partial x)\, dx, \qquad l = dx$$

$$V = \frac{dx\left[h^{n+2} - \left(h + \dfrac{\partial h}{\partial x}\, dx\right)^{n+2}\right]}{\left[h + \dfrac{\partial h}{\partial x}\, dx - h\right](n + 1)(n + 2)}$$

$$V = \frac{\left[h^{n+2} - \left(h + \dfrac{\partial h}{\partial x}\, dx\right)^{n+2}\right]}{\left(\dfrac{\partial h}{\partial x}\right)(n + 1)(n + 2)}$$

Therefore the mass between a–a and b–b is

$$\varrho V = \varrho\, \frac{\{h^{n+2} - [h + (\partial h/\partial x)\, dx]^{n+2}\}}{(\partial h/\partial x)(n + 1)(n + 2)}$$

NOTATION

A	cross sectional area of channel
a	local velocity of sound $= [k(p/\varrho)]^{1/2}$
C_p	specific heat at constant pressure
C_v	specific heat at constant volume
c	local wave propagation velocity $= [gh/(n + 1)]^{1/2}$
h	local water depth of flow
k	ratio of specific heats $= C_p/C_v$
l	some characteristic length in flow
n	exponent in relation $z = y^n$, with z the width of the channel, y the height of the channel
p	local pressure
T	local temperature in gas flow
t	time
t_a	some characteristic time in gas flow
t_w	some characteristic time in water flow
u	local velocity of flow in x direction

Greek Letters

δ relative change in density $= (\varrho - \varrho_0)/\varrho_0$

ϱ local density of fluid

η change of elevation of water due to disturbance with reference to h_0

ξ a function of x which represents the displacement of the plane disturbed by the wave motion

Subscripts

a air flow

w water flow

0   undisturbed condition or the initial equilibrium condition

REFERENCES

1. E. Preiswerk, "Application of the Methods of Gas Dynamics to Water Flows with Free Surfaces," NACA TM-934 and 935.
2. A. T. Ippan and R. T. Knapp, "Study of High Velocity Flow in Curved Channels of Rectangular Cross Section," *Trans. Am. Geophys. Union* (1936).
3. D. Riabouchinsky, "Sur l'Analogie Hydraulique des Mouvements d'un Fluide Compressible," *Compt. Rend.* **195**, 998–999 (1932).
4. T. von Kármán, "Eine praktische Anwendung der Analogie zwischen Überschall-Strömung in Gasen and überkritischer Strömung in ofenen Gerinnen," *Z.A.M.M.*, **18-1**, 49–56 (1938).
5. W. H. T. Loh, "A Study of the Dynamics of the Induction and Exhaust Systems of a Four Stroke Engine by a Hydraulic Analogy," ScD Thesis, MIT, 1946.
6. R. F. Probstein and G. E. Hudson, "A Water Analogue of the Isentropic Flow of Compressible Gases which Have Arbitrary Ratios of Specific Heats," Project SQUID Report No. NYU-20-R (August 1948).
7. A. H. Shapiro and L. Malavard, "Analogue Methods in Physical Measurements in Gas Dynamics and Combustion," in: *High Speed Aerodynamics and Jet Propulsion*, Vol. IX, Princeton University Press, Princeton, New Jersey (1954), pp. 309–321.
8. A. H. Shapiro, "An Appraisal of the Hydraulic Analogue to Gas Dynamics," Massachusetts Institute of Technology Meteor Report 34 (1949).
9. V. G. Szebehely and L. F. Whicker, "Generalization and Improvement of the Hydraulic Analogy," in: *Proc. 1st US Nat'l.Congress Applied Mech.*, 1951, Edwards Brothers, Ann Arbor, Michigan (1952), pp. 893–901.
10 P. H. Schweitzer, "Research in Exhaust Manifolds," *Trans. ASME* **74**, 517–528 (1952).

11. P. H. Schweitzer and T. C. Tsu, "Improving Engine Performance by Exhaust Pipe Tuning," ONR Report (1951).

12. W. H. T. Loh, "Hydraulic Analogy for Two Dimensional and One Dimensional Flows," *J. Aerospace Sci.* **26**, 389–390 (1959).

13. P. M. Morse, *Vibration and Sound*, McGraw-Hill Book Co., New York (1936).

14. J. McCowan, *Phil. Mag.* **33**, 258 (1892).

15. W. H. T. Loh, "Hydraulic Analogue for One Dimensional Unsteady Gas Dynamics," *J. Franklin Inst.* (January 1960).

16. W. H. T. Loh, "One Dimensional Unsteady Gas Dynamics by Hydraulic Analogy," in: *Proc. 3rd Symposium High Speed Aerodynamics and Structures* (1958).

Chapter 2

COMBINED HEAT AND
MASS TRANSFER PROCESSES

E. R. G. Eckert

Regents' Professor and Director, Thermodynamics and Heat Transfer
School of Mechanical and Aerospace Engineering
University of Minnesota

INTRODUCTION *

The purpose of this chapter is twofold: First, to discuss generally the basic processes involved in combined heat and mass transfer in a flowing fluid and to establish the conservation equations and the constitutive equations which describe these processes; second, to apply the above equations to boundary layer flow of gases at high temperatures and with chemical reactions, and to obtain some solutions through similarity considerations. The discussion will be restricted to such temperatures and pressures that the fluid involved can be considered a continuum and that local thermodynamic equilibrium for the various energy states is closely approached, so that the local condition in the field can be described by a single temperature.

1. CONSERVATION EQUATIONS

To describe local conditions in the fluid, the conservation equations have to be set up for an infinitely small volume element arbitrarily located in the fluid. We will use a Cartesian coordinate system and a volume element fixed in position relative to the coordinate system with the fluid moving through this element. Tensor notation will be used to bring the equations into a manageable form. The coordinates will be denoted by x_1, x_2, x_3, or, generally, by x_i or x_j, where i or j can assume values from 1 to 3. Figure 1 shows such a volume element. Its size is dV, and a fluid with n components flows through it. The masses of the various components inside the element

---

* Symbols used in this chapter are defined on pp. 80–82.

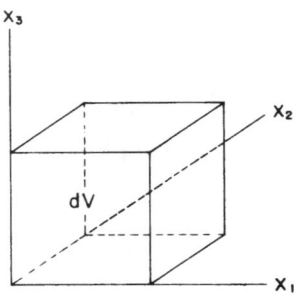

Fig. 1. Control volume.

are dm_1, dm_2, dm_3, ..., dm_m, ..., dm_n. The total mass is $dm = \Sigma_n \, dm_m$. Partial densities can then be defined as $\varrho_m = dm_m/dV$ and a total mixture density as

$$\varrho = \Sigma \, \varrho_m \tag{1}$$

The partial densities can be used to describe the composition of the mixture within the volume element. They are also referred to as concentrations. The conservation equations, however, take on a more convenient form when mass fractions defined as $w_m = dm_m/dm$ are introduced. The sum of all mass fractions adds up to 1:

$$\sum_n w_m = 1 \tag{2}$$

One of the difficulties in studying the literature on mass transfer is that various parameters in addition to concentration and mass fraction are used to describe the composition of the mixture; e.g., mole fraction and particle density. In the present chapter, the analysis will be based on mass fractions. The velocities of the various components at the location of the volume element will have different values as a consequence of the diffusion processes occurring.

The velocity of component m can, as a vector, be described by three components defined as $v_{m,i} = (1/\varrho_m) \, d\dot{M}_m/dA_i$, where $d\dot{M}_m$ indicates the mass flow of component m per unit time through the area element dA_i. The index i again refers to the directions of the three components; dA_i indicates the area element of the surface of the volume normal to the i direction. The velocity of the mixture is defined by $v_i = (1/\varrho) \, d\dot{M}/dA_i$, with $d\dot{M} = \Sigma \, d\dot{M}_m$. Correspondingly, the following relation exists between the mixture velocity and the component velocities:

$$\varrho v_i = \Sigma \, \varrho_m v_{m,i} \tag{3}$$

It is useful to introduce transport quantities relative to a coordinate system which moves with the fluid (for which $v_i = 0$). The mass flow of component m per unit time through a unit cross-sectional area may be denoted by $j_{m,i}$, and the three components of which are given by the equation

$$j_{m,i} = \varrho_m(v_{m,i} - v_i) \tag{4}$$

It can easily be checked that the following relation holds:

$$\Sigma j_{mi} = 0 \tag{5}$$

The mass fluxes $j_{m,i}$ are caused by the diffusion processes involved. There will also be an enthalpy transport by diffusion, according to the equation

$$\Sigma i_m j_{m,i} = \Sigma \varrho_m i_m(v_{m,i} - v_i) \tag{6}$$

where i_m denotes the enthalpy per unit mass of component m. This enthalpy transport will, in general, have a value different from zero. In addition, energy transport will be caused by conduction, by turbulence, and by similar processes which will be discussed later. The sum of all of these processes will be denoted by ε_i. The enthalpy transport ε_i refers again to a system moving with the velocity v_i. The diffusion of the various components will also generate a momentum transport, according to

$$\Sigma \varrho_m(v_{m,i} - v_i)(v_{m,j} - v_j) \approx 0 \tag{7}$$

The indices i and j in this equation can again assume values from 1 to 3 independently of each other. The existence of this momentum transport was first pointed out by von Kármán. Its value is, however, negligibly small for boundary layer situations ([1]), and will not be included in the considerations of this chapter.

A momentum transport which cannot be neglected, however, is connected with the mass flux of the mixture through the surface of the control volume as a consequence of the Maxwellian distribution of the molecular velocities. This momentum transport is introduced into the conservation equations in the form of stresses acting on the surface and denoted by p_{ij}. For $i \neq j$, these stresses are shear stresses. From a balance of the angular momentum of the forces on the control volume around its center of gravity, it follows that $p_{ij} = p_{ji}$. There also exist normal stresses describing the effect of the surrounding fluid on the fluid mass in the volume element. These are subdivided into two parts: normal viscous stresses p_{ii}, and a pressure p with the property that it is independent of direction. A more

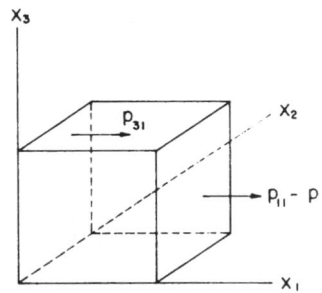

Fig. 2. Viscous stresses and pressure.

specific definition of the normal viscous stress depends on the nature of the fluid. For a Newtonian fluid, the normal viscous stress has to be introduced in such a way that it decreases to a value zero when the velocities decrease to zero. Figure 2 shows examples of the stresses acting on the surface of the volume element.

The following equations describe the continuity of mixture mass [Eq. (8)], of momentum [Eq. (9)], of energy [Eqs. (10) or (11)], and of the component masses [Eq. (12)] ([2]):

$$\frac{\partial p}{\partial \tau} + \frac{\partial (\varrho v_i)}{\partial x_i} = 0 \tag{8}$$

$$\varrho \frac{Dv_i}{d\tau} = -\frac{\partial p}{\partial x_i} + \frac{\partial p_{ji}}{\partial x_j} + \varrho g_i \tag{9}$$

$$\varrho \frac{Di^\circ}{\partial \tau} - \frac{\partial p}{\partial \tau} = -\frac{\partial \varepsilon_i}{\partial x_i} + \frac{\partial}{\partial x_j}(v_i p_{ji}) + \varrho v_i g_i \tag{10}$$

$$\varrho \frac{Di}{d\tau} - \frac{Dp}{d\tau} = -\frac{\partial \varepsilon_i}{\partial x_i} + p_{ji}\frac{\partial v_i}{\partial x_j} \tag{11}$$

$$\varrho \frac{Dw_m}{d\tau} = \frac{\partial j_{mi}}{\partial x_i} + \psi_m \tag{12}$$

The second term on the right in (11) is the dissipation function.

According to the rules of vector notation, terms in which an index i or j is repeated stand for the sum of terms obtained by replacing i with 1, 2 or 3; e.g.,

$$\frac{\partial (\varrho v_i)}{\partial x_i} = \frac{\partial (\varrho v_1)}{\partial x_1} + \frac{\partial (\varrho v_2)}{\partial x_2} + \frac{\partial (\varrho v_3)}{\partial x_3}$$

This, then, is a scalar expression corresponding to the divergence in vector notation. Equation (9) stands for three equations obtained by replacing i with 1, 2, or 3. The term $\partial p_{ji}/\partial x_j$ again stands for a sum; g_i stands for the three components of any body force. In Eq. (10), i° indicates the total enthalpy of the mixture composed of internal enthalpy i and kinetic energy, according to the equation

$$i^\circ = i + \tfrac{1}{2}v_i^2, \qquad v_i^2 = v_1^2 + v_2^2 + v_3^2$$

The enthalpy i is defined including chemical energy. In these equations, $D/d\tau$ is the substantial derivative:

$$\frac{D}{d\tau} = \frac{\partial}{\partial \tau} + v_i \frac{\partial}{\partial x_i}$$

A total of $n - 1$ equations of the form of Eq. (12) are required for an n-component mixture together with Eq. (2). The term ψ_m in Eq. (12) indicates a mass source term (per unit time and volume) for component m in the volume element. Such mass source terms are caused by chemical reactions. The terms ε_i and $j_{m,i}$ have been defined before.

These equations contain too many unknowns. Some of them are described by the thermodynamic equilibrium equations when local equilibrium is postulated; e.g., $\varrho = \varrho(p, T, w_1, w_2, \ldots, w_n)$ and $i = i(p, T, w_1, w_2, \ldots, w_n)$.

The terms ψ_m have to be described by chemical reaction rate equations or by the existence of some limiting conditions, e.g., local chemical equilibrium or chemically-frozen condition. The terms $j_{m,i}$, ε_i, and p_{ij} are described by the so-called constitutive equations, which will now be discussed.

2. CONSTITUTIVE EQUATIONS

The terms $j_{m,i}$, ε_i, and p_{ij} can be considered as fluxes—namely, as mass fluxes, energy fluxes, and momentum fluxes, respectively. It is found that they are caused by some driving potential (temperature gradients, concentration gradients, and others). Irreversible thermodynamics refers to these potentials in a general way as driving forces and shows that any of the fluxes can be caused by any of the forces as long as both are of the same mathematical nature (either vectors or tensors). This is summarized in Table I, in which the forces are listed along the horizontal row and the fluxes along the vertical column [3]. The relations connecting fluxes with forces are also indicated. It is found that fluxes described by the relations listed along the diagonal—stress–strain, Fourier conduction, and Ficks diffusion—are

TABLE I

Fluxes	Forces		
	Deformation velocities (tensor)	Temperature gradient (vector)	Mass fraction gradient, pressure gradient, body force (vector)
Momentum p_{ij} (tensor)	Stress–strain relation		
Energy ε_i (vector)		Fourier conduction	Diffusion-thermo (Dufour effect, 1873)
Mass $j_{m,i}$ (vector)		Thermal diffusion (Sorret effect)	Fick's diffusion

generally of a larger order of magnitude than the others. There are, however, situations in which these other fluxes cannot be neglected. This was recently demonstrated by analytical and experimental work carried out concurrently at the Massachusetts Institute of Technology ([4,5]) and at the University of Minnesota ([6,7]). In the present discussion, these effects will not be included, and the interested reader is referred to the literature.

The stress–strain relation, which connects momentum fluxes with deformation velocities, depends on the nature of the fluid involved. For a Newtonian fluid, they are described by the following equation:

$$p_{ij} = \mu\left(\frac{\partial v_i}{\partial x_j} + \frac{\partial v_j}{\partial x_i}\right) - \mu' \frac{\partial v_i}{\partial x_i}\, \delta_{ij} \tag{13}$$

The term δ_{ij} is called Kronecker delta and is defined as

$$\delta_{ij} = 1 \quad \text{for} \quad i = j$$
$$= 0 \quad \text{for} \quad i \neq j$$

The bulk viscosity $\varkappa = \mu' - \frac{2}{3}\mu$ is sometimes used instead of the viscosity μ'. Different relations hold for non-Newtonian fluids. Gases are considered Newtonian fluids, so that Eq. (13) will be used in the following discussion.

Fick's diffusion equation describes the connection between the mass fluxes and mass fraction gradients:

$$j_{m,i} = \varrho D_{m,e}\, \partial w_m/\partial x_i \tag{14}$$

where $D_{m,e}$ denotes effective mass diffusion coefficients describing the diffusion of component m into the mixture. These are connected in a complicated way with the binary diffusion coefficients describing diffusion of the various components into each other. Mass fluxes are, according to Table I, also caused by pressure gradients and by body forces as long as body forces of different magnitude act on the components of the mixture. Such body forces are created, e.g., by an electric field when some of the components in the mixture are electrically charged. Corresponding situations will not be included in the following discussion, nor will the effect of pressure gradients, which is generally negligible in boundary layers. For the complete equations describing the mass fluxes caused by all of these forces, reference should be made to the literature [3].

The energy flux ε_i finally is described by the following equation:

$$\varepsilon_i = -k_f \frac{\partial T}{\partial x_i} + \sum i_m j_{m,i} \tag{15}$$

where k_f denotes the thermal conductivity; the index f is added to indicate that the first term in Eq. (15) does not include diffusional energy transport, which is given separately as the second term. This conductivity is often referred to as frozen conductivity.

3. LAMINAR BOUNDARY LAYER EQUATIONS

At sufficiently large Reynolds numbers [and large values of (RePr) for liquid metals], the viscous regions in fluids are of a boundary layer type in the sense that the extent of these boundary layers is small in the direction normal to the flow as compared to the flow direction. This holds for boundary layers on a surface as well as for channel flow or for free shear layers, types which are shown in Fig. 3. In writing down the boundary layer equations, we will make a change in nomenclature in the form

$$x_i \to x, \qquad x_2 \to y, \qquad x_3 \to z, \qquad v_i \to u, \qquad v_2 \to v, \qquad v_3 \to w$$

to be in agreement with the customary notation. Here y may indicate the coordinate normal to the surface and v the corresponding velocity component. In boundary layers, certain terms of the conservation equations can be neglected because changes normal to the surface are large compared to changes in the other directions. Correspondingly, velocity components normal to the surface are small compared to the other velocity components,

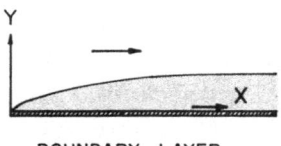

BOUNDARY LAYER

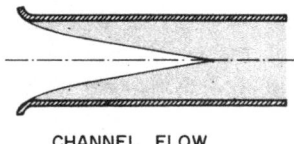

CHANNEL FLOW

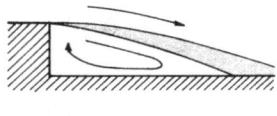

SEPARATED SHEAR LAYER

Fig. 3. Boundary-layer-type flow.

and the same statement holds for the variation of the pressure p:

$$\frac{\partial}{\partial y} \gg \frac{\partial}{\partial x}, \qquad \frac{\partial}{\partial y} \gg \frac{\partial}{\partial z}, \qquad u \gg v, \qquad w \gg v, \qquad p \approx p(x, z)$$

These boundary layer simplifications lead from the general conservation Eqs. (8)–(12) with the constitutive Eqs. (13)–(15) to the following relations for a two-dimensional, steady, laminar boundary layer in forced flow ($g_i = 0$), to which the following discussion will be restricted:

$$\frac{\partial(\varrho u)}{\partial x} + \frac{\partial(\varrho v)}{\partial y} = 0 \tag{16}$$

$$\varrho u \frac{\partial u}{\partial x} + \varrho v \frac{\partial u}{\partial y} = -\frac{dp}{dx} + \frac{\partial}{\partial y}\left(\mu \frac{\partial u}{\partial y}\right) \tag{17}$$

$$\varrho u \frac{\partial i^{\circ}}{\partial x} + \varrho v \frac{\partial i^{\circ}}{\partial y} = -\frac{\partial \varepsilon_y}{\partial y} + \frac{\partial}{\partial y}\left(u\mu \frac{\partial u}{\partial y}\right) \tag{18}$$

$$\varrho u \frac{\partial w_m}{\partial x} + \varrho v \frac{\partial w_m}{\partial y} = -\frac{\partial j_{m,y}}{\partial y} + \psi_m \tag{19}$$

Solutions of this system of equations will be obtained following a procedure which was introduced by Shvab [8] and Zeldovich [9] and refined by Lees [1]. For this purpose, the energy equation (18) is transformed. In boundary layers, the pressure p is considered constant on normals to the surface (in direction y). Correspondingly, the enthalpy i varies with temperature and composition. For gases, the relation $i = \Sigma\, w_m i_m$ describes the dependence of mixture enthalpy on composition. The differential of the enthalpy can now be written

$$di = \frac{\partial i}{\partial T}\, dT + \Sigma\, \frac{\partial i}{\partial w_m}\, dw_m = c_{p,f}\, dT + \Sigma\, i_m\, dw_m$$

where $c_{p,f}$ is the "frozen" specific heat, referring solely to a temperature change of the gas mixture. Combining Eqs. (14) and (15) results in

$$\varepsilon_y = -k_f \frac{\partial T}{\partial y} - \varrho\, \Sigma\, D_{m,e} i_m \frac{\partial w_m}{\partial y} \tag{20}$$

Combining the last two equations gives

$$\varepsilon_y = \frac{k_f}{c_{p,f}} \frac{\partial i}{\partial y} + \frac{k_f}{c_{p,f}} \Sigma \left(1 - \frac{\varrho D_{m,e} c_{p,f}}{k_f}\right) i_m \frac{\partial w_m}{\partial y}$$

Defining a frozen Prandtl number by

$$\mathrm{Pr}_f = \mu c_{p,f}/k_f$$

and a frozen Lewis number of component m by

$$\mathrm{Le}_{m,f} = \varrho D_{m,e} c_{p,f}/k_f$$

leads to the following equation:

$$\varepsilon_y = -\frac{\mu}{\mathrm{Pr}_f} \frac{\partial i}{\partial y} + \frac{\mu}{\mathrm{Pr}_f} \Sigma\, (1 - \mathrm{Le}_{m,f}) i_m \frac{\partial w_m}{\partial y} \tag{21}$$

Introduction of Eq. (21) into Eq. (18) leads to

$$\varrho u \frac{\partial i^{\circ}}{\partial x} + \varrho v \frac{\partial i^{\circ}}{\partial y} = \frac{\partial}{\partial y}\left(\frac{\mu}{\mathrm{Pr}_f} \frac{\partial i^{\circ}}{\partial y}\right) + \frac{\partial}{\partial y}\left[\left(1 - \frac{1}{\mathrm{Pr}_f}\right)\mu u \frac{\partial u}{\partial y}\right]$$
$$+ \frac{\partial}{\partial y}\left[\frac{\mu}{\mathrm{Pr}_f} \Sigma\, (\mathrm{Le}_{m,f} - 1) i_m \frac{\partial w_m}{\partial y}\right] \tag{22}$$

4. SOLUTIONS OF THE LAMINAR BOUNDARY LAYER EQUATIONS

Solutions to the Eqs. (16), (17), (19), and (22) will now be discussed for the special situation in which $Pr_f = 1$ and $Le_{m,f} = 1$. In this case, the energy equation (22) simplifies very radically because the last two terms drop out The Prandtl number and Lewis number of composite gases have values which are not too far from 1 when lightweight gases like hydrogen or helium are excluded as components. An ideal gas mixture with $Pr_f = 1$ and $Le_{m,f} = 1$ therefore appears as a reasonable approximation to real gas mixtures. The effects of deviations from the ideal mixture will be discussed later.

From an engineering standpoint, we are mainly interested in the magnitude of the heat flux and the mass flux in the gas in the immediate vicinity of the solid surface on which the boundary layer develops.

4.1. Heat Transfer

The heat flux per unit time and area is given by the following equation:

$$q_w = \varepsilon_w + \varrho_w v_w i_w \tag{23}$$

when a mass flux exists through the surface which, per unit area and time, has the value $\varrho_w v_w$. It is convenient to define a heat transfer coefficient with the flux term ε_w only and to refer it to an enthalpy difference as driving force, according to the equation

$$\varepsilon_w = \frac{\mu_w}{Pr_{f,w}} \left(\frac{\partial i}{\partial y} \right)_w = h_i(i_w - i_s^\circ) \tag{24}$$

A Nusselt number, Nu, can then be expressed through the equation

$$Nu = \frac{h_i L c_{p,f,w}}{k_{f,w}} = \left[\frac{\partial(i/(i_w - i_s^\circ))}{\partial(y/L)} \right]_w \tag{25}$$

In high-temperature gases we usually have the situation that the thermodynamic transport properties involved vary throughout the boundary layer much more with temperature than with concentration. If we consider as an approximation that the properties vary with temperature only, $\varrho = \varrho T$, $\mu = \mu(T)$, and $Pr_f = Pr_f(T)$, then Eqs. (16), (17), and (22) are independent of the mass fraction w_m and can be solved by themselves. They are not different from the equations of a one-component fluid. It is well known that the solution becomes especially simple if the product $\varrho\mu$ and the Prandtl number can be considered as constant. In this case, the constant-property solutions for the Nusselt number also hold for the fluid with the above-prescribed property variation. This also applies to rotationally-symmetrical boundary layers, for which only Eq. (16) changes its form. For a rotationally-symmetrical stagnation-point region and a wall surface with constant temperature, e.g., the following relations hold:

$$\text{Nu}_0 = 0.76\text{Re}^{1/2}\text{Pr}^{0.4} \qquad \text{For} \quad \varrho_w v_w = 0 \qquad (26)$$

$$\text{Nu} = 0.76\varphi\text{Re}^{1/2}\text{Pr}^{0.4} \qquad \text{For} \quad \varrho_w v_w \neq 0 \qquad (27)$$

The factor φ in Eq. (27) describes the influence of mass ejection or suction through the surface on the Nusselt number. It is represented in Fig. 4

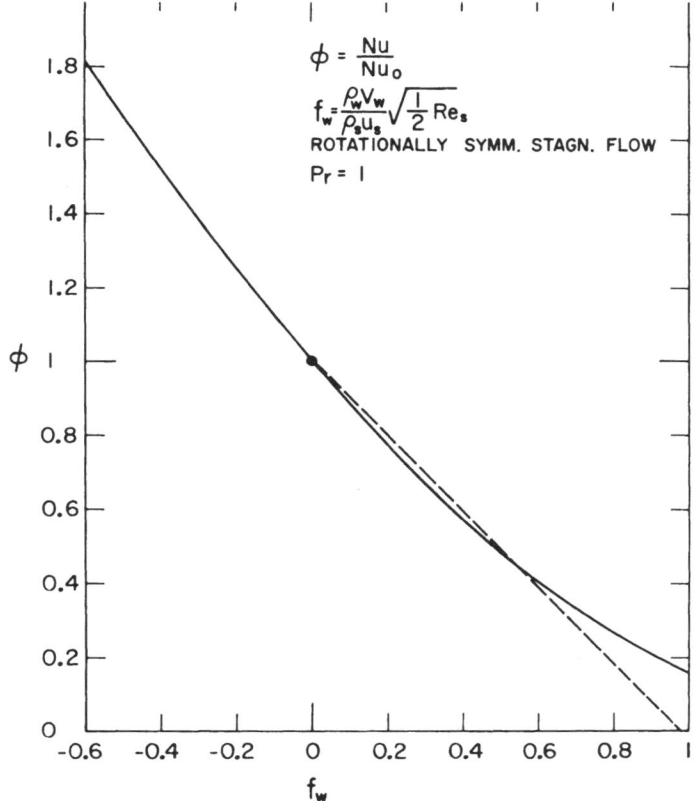

Fig. 4. Ratio of Nusselt number Nu with mass transfer to Nusselt number Nu_0 without mass transfer.

and for mass ejection can be approximated fairly well by the linear relation

$$\varphi = 1 - 0.72(\varrho_w v_w)/(\varrho_s u_s)(\text{Re}_s)^{1/2} \tag{28}$$

provided the blowing parameter f_w has values not exceeding 0.7.

4.2. Mass Transfer

The mass transfer is described by

$$\varrho u \frac{\partial w_m}{\partial x} + \varrho v \frac{\partial w_m}{\partial y} = \frac{\partial}{\partial y} \left(\frac{\mu \text{Le}_{m,f}}{\text{Pr}_f} \frac{\partial w_m}{\partial y} \right) + \psi_m \tag{29}$$

which follows readily from Eq. (19). The total mass release at the solid surface is described by the equation

$$\dot{m}_w = \varrho_w v_w$$

and the mass flux of component m in the gas in the immediate vicinity of the surface by equation

$$\dot{m}_{m,w} = j_{m,w} + \varrho_w v_w w_{m,w} \tag{30}$$

Again, it is convenient to define a mass transfer coefficient solely with the mass release term $j_{m,w}$, given according to the equation

$$j_{m,w} = -\varrho_w D_{m,e,w}(\partial w_m/\partial y)_w = h_M(w_{m,w} - w_{m,s}) \tag{31}$$

and to define a Sherwood number Sh as

$$\text{Sh} = \frac{h_M L}{\varrho_w D_{m,e,w}} = \left[\frac{\partial [w_m/(w_{m,w} - w_{m,s})]}{\partial(y/L)} \right]_w \tag{32}$$

In the absence of chemical reactions ($\psi_m = 0$) and for a gas with $\text{Pr}_f = \text{Le}_f = 1$, it immediately follows from the similarity between the equations describing the temperature and mass fraction fields that the Sherwood number as defined by Eq. (32) is equal to the Nusselt number as defined by Eq. (25) provided similarity also exists in the boundary conditions:

$$\text{Sh} = \text{Nu} \tag{33}$$

From the definitions of both parameters, one sees immediately that the equality also holds for the coefficients

$$h_i = h_M \qquad (34)$$

an equation usually referred to as the Lewis relation.

In the presence of chemical reactions, the similarity between Eqs. (22) and (29) for $Pr_f = Le_f = 1$ is disturbed by the mass source terms ψ_m. These terms, however, can be eliminated when the mass conservation equations (29) are written in terms of a species mass fraction $\tilde{w}_m$. Such a species mass fraction considers the mass of an atomic species (oxygen, nitrogen, carbon, etc.) regardless of whether it is contained in the mixture in atomic or molecular form or whether it is contained in any compound. Chemical species are not created or destroyed by chemical processes. This means that no mass source term appears in the mass conservation equations if written in species mass fractions:

$$\varrho u \frac{\partial \tilde{w}_m}{\partial x} + \varrho v \frac{\partial \tilde{w}_m}{\partial y} = \frac{\partial}{\partial y} \left(\frac{\mu Le_{m,f}}{Pr_f} \frac{\partial \tilde{w}_m}{\partial y} \right) \qquad (35)$$

For a gas with $Pr_f = 1$ and $Le_{m,f} = 1$, the similarity between the energy equation and the species mass fraction equations is again established, and as a consequence a properly-defined Sherwood number is again equal to the Nusselt number.

4.3. Example

The method of solving a combined heat and mass transfer process with chemical reactions may now be demonstrated by an example which considers an air stream over the stagnation-point region of an object consisting of carbon under conditions that a steady, rotationally-symmetrical, laminar boundary layer exists and that the temperature is high enough to cause chemical reactions in the boundary layer as well as on the surface which may be kept at a uniform temperature. The following chemical reactions will be considered:

$$2O \rightleftarrows O_2, \quad C + O \rightleftarrows CO, \quad C + \tfrac{1}{2} O_2 \rightleftarrows CO$$

$$2N \rightleftarrows N_2, \quad C + N \rightleftarrows CN, \quad C + \tfrac{1}{2} N_2 \rightleftarrows CN$$

It will also be assumed that the free stream of air has a temperature sufficiently high so that oxygen and nitrogen exist there in atomic form as well as in molecular form. This is indicated in Fig. 5. All of the oxygen may be consumed by surface reactions, so that only the components CO, CN,

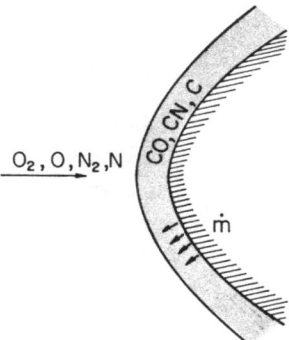

Fig. 5. Heat and mass transfer of high-temperature air on a carbon surface in rotationally-symmetrical stagnation flow.

and C exist in the gas immediately at the surface. The Prandtl and Lewis numbers may have the value 1. The species mass fractions of oxygen, nitrogen, and carbon are described by the following equations:

$$
\begin{aligned}
\tilde{w}_O &= w_O + w_{O_2} + \tfrac{4}{7} w_{CO} \\
\tilde{w}_N &= w_N + w_{N_2} + \tfrac{7}{13} w_{CN} \\
\tilde{w}_C &= w_C + \tfrac{3}{7} w_{CO} + \tfrac{6}{13} w_{CN}
\end{aligned}
\tag{36}
$$

A relation for the species mass fraction of oxygen at the surface is obtained from the consideration that the surface is impermeable to oxygen:

$$
m_{\tilde{O},w} = -\varrho D (\partial \tilde{w}_O / \partial y)_w + \dot{m}_w \tilde{w}_{O,w} = 0
$$

Introducing a mass transfer coefficient and considering the fact that this coefficient is equal to the heat transfer coefficient h_i through the Lewis relation [Eq. (34)] transforms this equation to

$$
h_i (\tilde{w}_{O,w} + \tilde{w}_{O,s}) + \dot{m}_w \tilde{w}_{O,w} = 0
\tag{37}
$$

Equation (28) can be transformed to the expression

$$
h_i = h_{i,0} - 0.547 \dot{m}_w
$$

where $h_{i,0}$ denotes the heat transfer coefficient based on enthalpies at a mass flow rate $\dot{m}_w = 0$. Dividing the above equation by $\dot{m}_w$ leads to the relation

$$
h_i / \dot{m}_w = (h_{i,0} / \dot{m}_w) - 0.547 = B
\tag{38}
$$

Introducing this parameter B into Eq. (37) results in

$$(B + 1)\tilde{w}_{0,w} - B\tilde{w}_{0,s} = 0$$

The species mass fractions are now replaced by the component mass fractions, with the result

$$\tfrac{4}{7}(B + 1)w_{CO,w} = B(w_{0,s} + w_{O_2,s})$$

The mass fraction of CO at the wall surface is thus obtained as

$$w_{CO,w} = \frac{7B}{4(B + 1)}(w_{0,s} + w_{O_2,s}) = \frac{7B}{4(B + 1)} 0.23 \qquad (39)$$

In the same way, the mass fraction of CN at the wall surface is obtained from the consideration that the wall is also impermeable to nitrogen:

$$w_{CN,w} = [13B/7(B + 1)]0.77 \qquad (40)$$

Ablation of carbon from the surface provides the mass flux $\dot{m}_w$ according to the relation

$$\dot{m}_{\tilde{C}} = \dot{m}_w = -\varrho D\left(\frac{\partial \tilde{w}_C}{\partial y}\right)_w + \dot{m}_w \tilde{w}_{C,w}$$

Introduction of the heat transfer coefficients leads to

$$\dot{m}_w = h_i(\tilde{w}_{C,w} - \tilde{w}_{C,s}) + \dot{m}_w \tilde{w}_{C,w} \qquad (41)$$

With the parameter B, one obtains

$$\tilde{w}_{C,w} = 1/(B + 1)$$

keeping in mind that no carbon exists in the air stream outside of the boundary layer. The mass fraction of the gaseous component carbon is then described by the relation

$$w_{C,w} = \tilde{w}_{C,w} - \tfrac{3}{7}w_{CO,w} - \tfrac{6}{13}w_{CN,w} \qquad (42)$$

In this way, the composition of the gas at the wall surface has been determined. Equation (42) can be used to calculate the parameter B and from it the mass release rate $\dot{m}_w$, since $w_{C,w} \approx 0$ when the combustion occurs at the surface (assuming the chemical equilibrium to be close to complete combustion), or $w_{C,w}$ is determined by an equilibrium relation of carbon in solid and gaseous state.

Calculations of this form have to be made in the analysis of the ablation

ccoling process. In this case, it is especially important to know the heat
flux to the surface which is reduced by the ablation process. The heat flux
w_thin the boundary layer in the immediate vicinity of the surface is

$$q_w = \varepsilon_w + \dot{m}_w i_w$$

$$= (h_i + \dot{m})(w_{CO,w} i_{CO,w} + w_{CN,w} i_{CN,w} + w_{C,w} i_{C,w}) - h_i i_s^\circ \qquad (43)$$

All terms in this equation are known when the total enthalpy i_s° in the free
stream and the component enthalpies $i_{m,w}$ at the wall surface with the surface
temperature t_w are prescribed. The first term will be known from the conditions in the free stream; the second is determined by heat conduction into
the interior of the solid material and by the effects of thermal radiation,
and has to be obtained from an energy balance including the ablating
material as well as the boundary layer. It may have been observed that no
assumption was made in the preceding analysis about the speed of the homogeneous chemical reactions within the boundary layer. The heat and mass
transfer is therefore the same whether this speed is large—leading to chemical
equilibrium within the boundary layer—or small—leading to the chemically-
frozen condition. Only the reaction rates of the heterogeneous reactions at
the surface influence the heat and mass transfer. In the preceding example,
the reaction rate of C with O and N was assumed large. Otherwise, finite
mass fractions of O and N would exist at the surface.

5. TURBULENT BOUNDARY LAYERS

The Eqs. (13), (14), and (15) describing the momentum, energy, and
mass fluxes have to be enlarged by terms describing the fluxes caused by
turbulent exchange. Conventionally, these fluxes are connected with the
corresponding driving forces by turbulent diffusivities ε_v, ε_i, and ε_w for
momentum, energy, and mass, respectively. The boundary layer equations
again take on a convenient form when a turbulent Prandtl number

$$\mathrm{Pr}_t = \varepsilon_v / \varepsilon_i \qquad (44)$$

and turbulent Lewis numbers

$$\mathrm{Le}_t = \varepsilon_w / \varepsilon_i \qquad (45)$$

are introduced so that ε_v, Pr_t, and Le_t represent the turbulent exchange of
momentum, energy, and mass. For the situation where $\mathrm{Pr}_t = 1$ and $\mathrm{Le}_t = 1$
in addition to $\mathrm{Pr} = \mathrm{Le} = 1$, one finds again that the boundary-layer energy
equation and the boundary layer equations expressing the component mass

fractions are similar. As a consequence, the Sherwood number defined by Eq. (32) is equal to the Nusselt number defined by Eq. (25).

In the flow of a real gas, it was found experimentally that the turbulent Prandtl number varies with position in the flow field and probably also with the Reynolds number and the Prandtl number of the fluid. The values for the turbulent Prandtl number, however, are closer to the value 1 than the value of the laminar Prandtl number for the fluid involved. All indications which we have at present point to the fact that the turbulent Lewis number has a value 1. It is therefore expected that relations describing the Nusselt and Sherwood numbers for a situation $Pr = 1$ and $Le = 1$ are better approximations for the actual flow of gases than corresponding relations for laminar flow.

6. CORRECTION EQUATIONS FOR $Pr \neq 1$ AND $Le \neq 1$

It has been mentioned that the Prandtl and Lewis numbers of gas mixtures do not deviate strongly from the value 1 as long as very light-weight gases like hydrogen or helium are excluded. This suggests that we use relations for an ideal gas with $Pr = 1$ and $Le = 1$ as a first approximation and apply corrections for the deviations of the parameters from the value 1. Couette flow of a two-component gas mixture has been analyzed by Monaghan [10]. Zero mass release from either of the two walls bounding the Couette flow was specified. The temperature of the moving wall was assumed sufficiently high that dissociation of the gas occurred. Local chemical equilibrium was postulated. The heat flux from the stationary wall into the fluid could be expressed by the following equation:

$$q_w = \frac{k}{c_p b}\left[(i_w - i_s^\circ) + (1 - Pr)\frac{v_s^2}{2} + (1 - Le)i_D(w_{a,s} - w_{a,w})\right] \quad (46)$$

where b denotes the distance of the two walls bounding the Couette flow, the index w refers to the stationary wall, the index s refers to the wall moving in its own plane with the velocity v_s; i_D is the heat of dissociation per unit mass of the atomic species; and $w_{a,s}$ and $w_{a,w}$ indicate the mass fractions of the atomic species at the two wall surfaces. The form of this equation suggests a generalization which will describe conditions in boundary layer flows:

$$\varepsilon_w = h_i[(i_w - i_s^\circ) + (1 - r_v)\tfrac{1}{2}v_s^2 + (1 - r_{ch})i_D(w_{a,s} - w_{a,w})] \quad (47)$$

The equation defines two recovery factors: r_v, describing the recovery of kinetic energy, and r_{ch}, describing the recovery of chemical energy. The index s now refers to conditions at the outer border of the boundary layer; here h_i is the heat transfer coefficient for a gas with $Pr = Le = 1$. Without chemical reactions and with zero mass release from the surface, the relations describing the recovery factor for kinetic energy in boundary layers are well known. They are approximated by the equation

$$r_v = (Pr)^{1/2} \tag{48}$$

for laminar flow and

$$r_v = (Pr)^{1/3} \tag{49}$$

for turbulent flow. The effect of dissociation in a laminar air boundary layer at a three-dimensional stagnation point has been studied by Fay [11]. The equation

$$r_{ch} = Le^{0.52} \tag{50}$$

was found to approximate the results of this analysis well for chemical equilibrium, and the equation

$$r_{ch} = Le^{2/3} \tag{51}$$

for chemically-frozen conditions within the boundary layer and complete recombination on a cold catalytic wall.

Conditions for boundary layer flow over a surface from which mass is released have been studied to a lesser extent. Lees [1] provides the following estimate for the chemical recovery factor: $r_{ch} = Le^n$, with n having values between $\frac{2}{3}$ and 1 for moderate blowing at the surface and frozen catalytic wall, for laminar or turbulent boundary layers. The expression $i_D(w_{a,s} - w_{a,w})$ can be replaced by $\Sigma i_{ch,m}(w_{m,s} - w_{m,w})$, in which $i_{ch,m}$ indicates the enthalpy change of component m created in the chemical process.

NOTATION

A	area
c	specific heat
D	mass diffusion coefficient
f	blowing parameter
g	body force
h_i	heat transfer coefficient based on enthalpies
h_M	mass transfer coefficient
i	enthalpy

j	mass flux per unit area and time
k	thermal conductivity
L	characteristic dimension
Le	Lewis number
$\dot{m}$	mass flux per unit area and time
Nu	Nusselt number
Pr	Prandtl number
p	stress
q	heat flux per unit area and time
Re	Reynolds number
r	recovery factor
Sh	Sherwood number
T	absolute temperature
t	temperature
u	velocity component
V	volume
v	velocity component
w	velocity component
w	mass fraction
$\tilde{w}$	species mass fraction
x	coordinate
y	coordinate
z	coordinate

Greek Letters

ε	energy flux
μ	viscosity
ϱ	density
τ	time
φ	correction factor
ψ	mass source per unit volume and time

Superscript

$\circ$	total

Subscripts

0	zero mass flow
ch	chemical
D	diffusion

e effective
f frozen state
i coordinate direction
j coordinate direction
m component in a fluid mixture
p constant pressure
s outside boundary layer
t turbulent
w at wall surface
Properties without index are mixture properties

REFERENCES

1. L. Lees, "Convective Heat Transfer with Mass Addition and Chemical Reactions," *Combustion and Propulsion*, Third AGARD Colloquium, Pergamon Press, New York, pp. 451–498; reprinted in: *Recent Advances in Heat and Mass Transfer*, J. P. Hartnett, ed., McGraw-Hill Book Co., New York (1961), pp. 161–207.
2. S. Chapman and T. G. Cowling, *The Mathematical Theory of Non-Uniform Gases*, Cambridge University Press, Cambridge, Massachusetts (1953).
3. J. O. Hirschfelder, C. F. Curtiss, and R. B. Bird, *Molecular Theory of Gases and Liquids*, John Wiley and Sons, New York (1954); R. B. Bird, W. E. Stewart, and E. N. Lightfoot, *Transport Phenomena*, John Wiley and Sons, New York (1960).
4. J. R. Baron, "Thermodynamic Coupling in Boundary Layers," *Am. Rocket Soc. J.* **32**, 1053–1059 (1962).
5. H. Thomann and J. R. Baron, "Experimental Investigation of Thermal Diffusion Effects in Laminar and Turbulent Shear Flow," *Int. J. Heat Mass Transfer* **8**, 455–466 (1965).
6. E. M. Sparrow, "Recent Studies Relating to Mass Transfer Cooling," in: *Proceedings of the 1964 Heat Transfer and Fluid Mechanics Institute*, Stanford University Press, Stanford, California (1964), pp. 1–18.
7. E. R. G. Eckert, "Diffusion Thermo Effects in Mass Transfer Cooling," in: *Proceedings of the Fifth U.S. National Congress of Applied Mechanics*, The American Society of Mechanical Engineers (1966), pp. 639–650.
8. V. A. Shvab, "Relation between the Temperature and Velocity Fields of the Flame of a Gas Burner," *Sb. Issled. Gorenia Naturalnogo Topliva Gas. Energ.*, Moscow–Leningrad (1948).
9. Y. B. Zeldovich, "On the Theory of Combustion of Initially Unmixed Systems," *Zh. Tekh. Fiz.* **19** (10), 1199–1210 (1944); translated as NACA Technical Memorandum No. 1296 (June 1950).
10. R. J. Monaghan, L. F. Crabtree, and B. A. Woods, "Features of Hypersonic Heat Transfer," in: *Advances in Aeronautical Sciences*, Vol. 1, Proceedings of the First International Congress in the Aeronautical Science, Madrid 8–13 September, 1958, Pergamon Press, New York (1959), pp. 287–313.
11. J. A. Fay, F. R. Riddell, and N. H. Kemp, "Stagnation Point Heat Transfer in Dissociated Air Flow," *Jet Propulsion*, **27**, 672–674 (1957).

Chapter 3

HYPERSONIC VISCOUS FLOWS

Arthur Henderson, Jr.

Head, Flow Analysis Section
NASA—Langley Research Center

1. INDUCED-PRESSURE EFFECTS*

1.1. Introduction

At hypersonic Mach numbers, significant pressures are induced on aerodynamic surfaces by both viscosity and bluntness effects. Viscosity-induced pressures are caused by the growth of the boundary layer. They are accounted for in the manner proposed by Prandtl ([1]) many years ago when he suggested taking the computed boundary-layer displacement thickness as the boundary of a new body for which the external flow is then calculated. Bluntness-induced pressures result from complicated flow field interactions integral with the highly-curved nose shock and attendant strong entropy and vorticity gradients associated with blunt noses in hypersonic flow; this interaction is essentially inviscid.

The induced-pressure problem treated here is the one of most practical interest: that for combined bluntness–viscosity interaction in continuum flow, where the viscous contribution is treated by the now-classical "strong" or "weak" interaction theory [a historical review and bibliography of the bluntness- and "classical" viscosity-induced pressure problems is given in ([2])]. In the "weak-interaction" regime, the pressure increase caused by the growth of the boundary layer is small enough that the boundary layer remains essentially parabolic in shape; whereas in the "strong-interaction" regime, the pressures induced by the boundary layer are so large that the shape of the boundary layer itself is affected.

Only the two-dimensional case is treated here; the approach used for the viscosity- induced-pressure problem is common to most of the literature:

* Symbols used in this chapter are defined on pp. 124–125.

use Prandtl's concept of a displacement thickness and account for its effect by application of the inviscid tangent-wedge approximation.

More elaborate methods of treating the induced-pressure problem are available in the literature; for instance, Lewis and Whitfield [3] obtained solutions to the combined bluntness–viscosity induced-pressure problem on hemispherically-blunted cones by numerically iterating the boundary-layer displacement thickness and inviscid characteristics flow field solutions; whether the added complexity required to obtain a more accurate first-order boundary layer solution is justified for most cases remains to be established, and will not be considered here.

1.2. Basic Considerations

Before looking at the induced-pressure problem in more detail, a brief consideration of the correlation parameters involved is in order. Underlying most of the boundary-layer displacement-effect work is the inviscid hypersonic similarity law, which states that the pressures at corresponding points on two similarly-shaped slender bodies in hypersonic flow are identical if, at the two points, the product of Mach number and local body slope, $M(dy/dx)$, is constant.

A physical analog to this law is illustrated in Fig. 1. Two marbles are shown, each rolling toward its own wedge. The upper marble will rise a height h in the length l_1, with the velocity V_1, while the lower marble will rise the same height h in the longer length $l_2 = Al_1$ but with the greater velocity $V_2 = AV_1$. The ratio of lengths and velocities is such that both marbles rise the same height h in the same length of time; i.e., they both experience the same change in velocity, and, consequently, each marble will impart the same amount of momentum to its particular wedge. If the marbles are thought of as air molecules and the wedges as corresponding

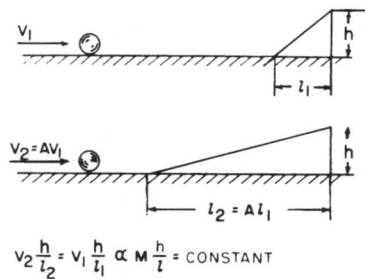

$$v_2 \frac{h}{l_2} = V_1 \frac{h}{l_1} \propto M \frac{h}{l} = \text{CONSTANT}$$

Fig. 1. Physical concept behind the hypersonic similarity law.

slopes on two similar bodies, a direct analogy with the hypersonic similarity law is apparent.

When, following Prandtl, the effective body is the boundary-layer displacement thickness, the hypersonic similarity parameter becomes $M \, d\partial^*/dx$. It can be shown that

$$p/p_\infty \propto M_\infty \, d\delta^*/dx \qquad \text{For weak interaction} \qquad (1a)$$

$$\propto (M_\infty \, d\delta^*/dx)^2 \qquad \text{For strong interaction} \qquad (1b)$$

(the former is the first-order term obtained from expanding the oblique-shock equations for weak disturbances under hypersonic conditions, whereas the second may be obtained from the exact oblique-shock equations, assuming both hypersonic flow and that the flow deflection and shock angles are small enough that sines and tangents of these angles can be interchanged with little loss in accuracy).

By relating δ^* to x through the reference-temperature relationship, Lees [4] [see also Hayes and Probstein [5]] shows that (1a) and (1b) both reduce to

$$p/p_\infty \propto (\gamma - 1)M_\infty^3 (C_\infty/R_{\infty,x})^{1/2} = (\gamma - 1)\bar{\chi}_\infty \qquad (2)$$

That is, starting with

$$\delta^* \propto (\mu_r x/\varrho_r u_\infty)^{1/2} \qquad (3)$$

referring the displacement thickness to free-stream conditions and using the proportionality which exists between the reference temperature within the boundary layer and the total temperature, it is found that

$$d\delta^*/dx \propto (\gamma - 1)M_\infty^2 (C_\infty/R_{\infty,x})^{1/2}(p_w/p_\infty)^{-1/2} \qquad (4)$$

For the weak-interaction case, we assume the slope of the boundary layer is so small that p_w/p_∞ is near unity, for which

$$\frac{p}{p_\infty} \propto M_\infty \frac{d\delta^*}{dx} \propto (\gamma - 1)M_\infty^3 \left(\frac{C_\infty}{R_{\infty,x}}\right)^{1/2} \propto (\gamma - 1)\bar{\chi}_\infty \qquad (5)$$

For strong interaction, substituting (1b) into (4) gives

$$d\delta^*/dx \propto [(\gamma - 1)M_\infty(C_\infty/R_{\infty,x})^{1/2}]^{1/2} \qquad (6)$$

and

$$\frac{p}{p_\infty} \propto M_\infty^2 \left(\frac{d\delta^*}{dx}\right)^2 \propto (\gamma - 1)M_\infty^3 \left(\frac{C_\infty}{R_{\infty,x}}\right)^{1/2} \propto (\gamma - 1)\bar{\chi}_\infty \qquad (7)$$

The controlling parameter for the viscosity-induced-pressure problem is thus seen to be $\bar{\chi}_\infty$ for both the strong and weak interactions, and this would be expected to apply in the intermediate regime as well.

The inviscid bluntness-induced pressures can be predicted with good accuracy by exact characteristics calculations, but these are generally not suitable for engineering calculations. On the other hand, the much simpler and less accurate "blast-wave" theory ([6,7]) is of fundamental importance. The value of the blast-wave theory lies not in its analytical solution, which, because of highly-restrictive conditions, has limited utility, but rather in the parameter developed in the analysis. This so-called blast-wave parameter has proven very useful in correlating both experimental and analytical results.

Application of the correlation parameter to exact characteristic calculations using either the sonic-wedge or sonic-cone leading edge followed by a flat slab or circular cylinder, respectively, has yielded equations which are quite reliable for predicting inviscid bluntness-induced pressures.

The two-dimensional blast wave parameter for $\alpha = 0°$ is

$$\frac{p}{p_\infty} \propto \left(\frac{M_\infty^3 C_{D,N}}{x/t} \right)^{2/3} \tag{8}$$

1.3. Flat Plate $\alpha = 0$

That the magnitude of both primarily viscosity- and primarily bluntness-induced pressures can be significant is illustrated in Fig. 2. The parameter Ω is associated with the relative degrees to which combined bluntness and viscosity effects influence induced pressure, and will be discussed subsequently. The schlieren photographs in Fig. 3 illustrate the main features of primarily bluntness- and primarily viscosity-dominated flow fields. Figures 4 and 5 show the efficiency with which the viscosity and bluntness correlating parameters just discussed collapse data, including that shown in Fig. 2, into a single narrow band.

The theoretical predictions are both from ([8]). For the viscous case (Fig. 4),

$$p/p_\infty = 0.83 + \tfrac{3}{4}[\tfrac{1}{2}\gamma(\gamma + 1)]^{1/2} G\bar{\chi}_\infty \tag{9}$$

where

$$G = 1.648\left(\frac{\gamma - 1}{2}\right)\left(\frac{T_w}{T_R} + 0.352\right) \tag{10}$$

[A more accurate expression for G is given in Appendix A of ([73]); it differs from Eq. (10) by about 4%.]

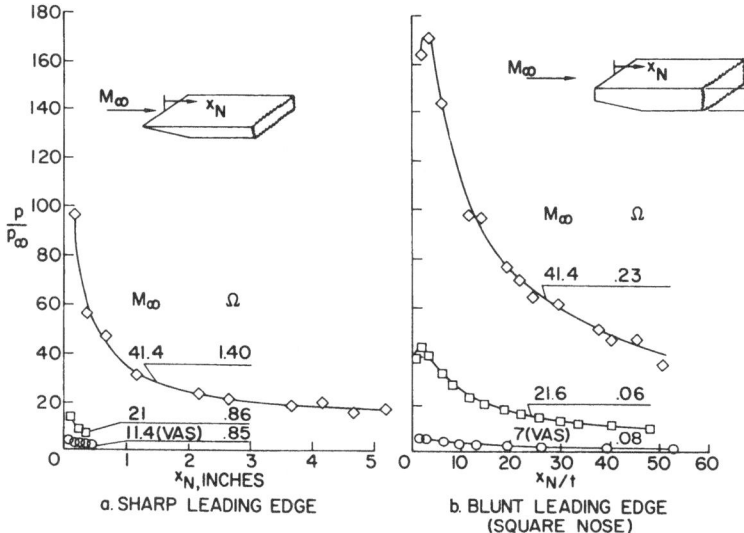

Fig. 2. Effect of Mach number on induced pressure.

For the primarily bluntness case (Fig. 5),

$$\frac{p}{p_\infty} = \frac{1}{2}\left\{\left[\left(\frac{p_b}{p_\infty}\right)^2 + \frac{\gamma(\gamma+1)(1-n)^2\Omega^2}{2}\left(\frac{M_\infty^3 C_{D,N}}{x/t}\right)\right]^{1/2} + \frac{p_b}{p_\infty}\right\} \quad (11)$$

where

$$\frac{p_b}{p_\infty} = 0.187\left[\gamma^{1/2}(\gamma - 1)\frac{M_\infty^3 C_{D,N}}{x/t}\right]^{2/3} + 0.74 \quad (12)$$

In Eq. (11), $n = \frac{2}{3}$ for primarily bluntness-induced pressures and $n = \frac{1}{2}$ for pure viscosity-induced pressures. Equation (9) is the analytical relationship between the viscous interaction correlation parameter and the induced pressure for the strong-interaction case.

Equation (12) relates the blast-wave correlation parameter [Eq. (8)] to the inviscid bluntness-induced pressure. It is emphasized that this is not a blast-wave theory solution, however. Rather, it is an analytical expression obtained from using the blast-wave parameter to correlate many inviscid characteristics solutions. Thus by coupling the parameter uncovered by a theory of questionable validity with a highly reliable numerical calculation method, very useful analytical results are obtained.

That viscosity- and bluntness-induced effects always occur simultaneously has long been recognized; initial attempts to account for the com-

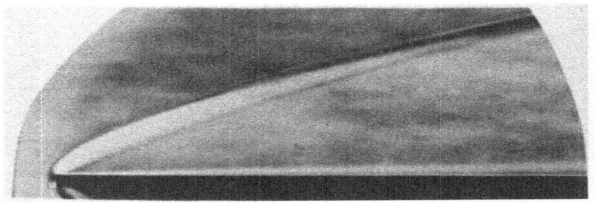

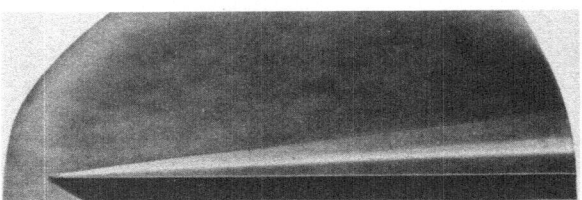

Fig. 3. Schlieren photographs for flow over flat plates; $M_\infty = 21$, $\alpha = 0°$. (a) Blunt leading edge. (b) Sharp leading edge.

bined effect consisted of simply linearly adding equations such as (9) and (12). Later, Cheng et al. ([9]) and Bertram and Blackstock ([8]) solved the combined bluntness–viscosity-induced-pressure problem; Eq. (11) is the latter's solution; the solution of Cheng et al. is numerical. Although the two approaches differ considerably, the final results are in substantial agreement.

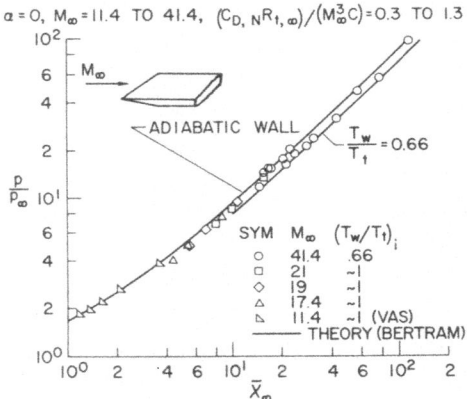

Fig. 4. Correlation of two-dimensional viscosity-induced pressure.

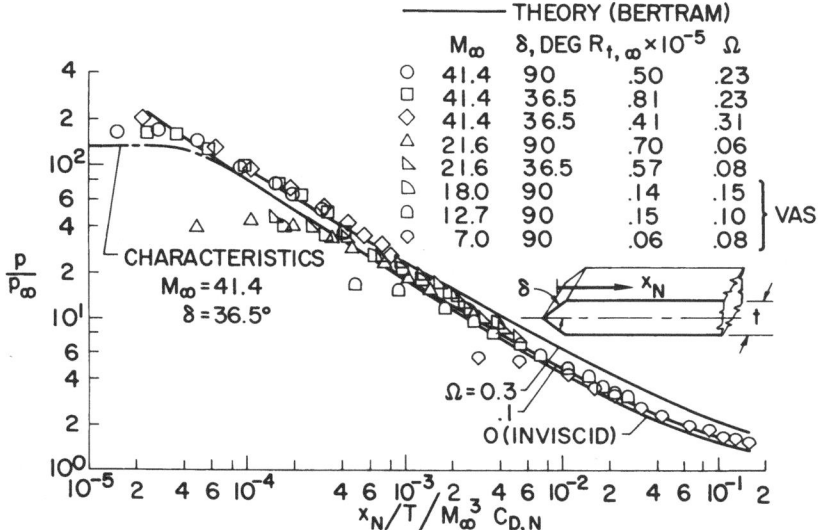

Fig. 5. Two-dimensional correlation of bluntness-induced pressures; $M_\infty = 7\text{-}41$, $\alpha = 0°$.

The parameter Ω is defined by the following identity:

$$\frac{M_\infty^3 C_{D,N}}{x/t} \equiv \frac{(G\bar{\chi})^2}{\Omega^2} \tag{13}$$

In order to present the entire spectrum of induced-pressure effects (pure viscous, mixed, pure inviscid) in a unified manner on a single plot requires that both viscous and inviscid contributions be expressed in the same functional form. The parameter Ω allows the inviscid blast-wave parameter to be artificially expressed as a function of the viscous interaction parameter. In addition, its value apportions the proper weight to the two parameters (viscous and inviscid) in determining their relative significance in the combined bluntness–viscosity problem. It appears in both developments ([8],[9]).

The combined bluntness–viscosity correlation parameters developed in the analysis of Cheng et al. afford a unification of the viscosity and bluntness problems in a manner which is not explicitly apparent in Eq. (11). However, the analytical result of Bertram and Blackstock, Eq. (11), can be recast into functions of Cheng's parameters,

$$\frac{\varepsilon^2}{\Omega^4}\frac{\Delta p}{\gamma p_\infty} \quad \text{and} \quad \left[\frac{\varepsilon^2}{\Omega^4}(G\bar{\chi})\right]^{-1/3}$$

and Eq. (11) becomes

$$\frac{\varepsilon^2}{\Omega^4}\frac{\Delta p}{\gamma p_\infty} = 0.0935\left(\frac{\gamma+1}{\gamma}\right)^{2/3}\left[\frac{\varepsilon^2}{\Omega^4}(G\bar{\chi})\right]^{4/3}$$

$$\times\left\{\left[1 + 32.2\left(\frac{\gamma}{\gamma+1}\right)^{1/3}\left[\frac{\varepsilon^2}{\Omega^4}(G\bar{\chi})\right]^{-2/3}\right]^{1/2} + 1\right\} \qquad (14)$$

Here $\Delta p/p_\infty = (p - p_\infty)/p_\infty$ has been used rather than p/p_∞ as in [9], and $G\bar{\chi}$ is slightly different than the corresponding parameter of Cheng et al.

Equation (14) accounts for variation in γ, whereas the numerical analysis of Cheng et al. is for $\gamma \to 1$. Recently, Kemp [10] has modified the solution of Cheng et al.; he showed that removing the $\gamma \to 1$ restriction gave significantly better agreement with his recent experimental induced-pressure studies at $M_\infty = 42$ in helium.

The degree to which Cheng's modified correlation successfully unifies the bluntness and viscosity problems into a single "flat-plate problem" is shown in Fig. 6; the Bertram–Cheng theoretical prediction is seen to be fairly good. Equation (14) gives the pure viscosity solution when $\Omega \to \infty$,

$$\frac{\varepsilon^2}{\Omega^4}\frac{\Delta p}{\gamma p_\infty} = 0.530\left(\frac{\gamma+1}{\gamma}\right)^{1/2}\frac{\varepsilon^2}{\Omega^4}(G\bar{\chi}) \qquad (15)$$

and the pure bluntness solution for $\Omega \to 0$,

$$\frac{\varepsilon^2}{\Omega^4}\frac{\Delta p}{\gamma p_\infty} = 0.187\left(\frac{\gamma+1}{\gamma}\right)^{2/3}\left[\frac{\varepsilon^2}{\Omega^4}(G\bar{\chi})\right]^{4/3} \qquad (16)$$

1.4. Blunt Wedge

Although the wedge is closely related to the flat plate at zero incidence (the latter being a special case of the former), theoretical correlation procedures that have been developed treat the two problems differently [see [9,11]]. As an example, the correlation of pressures on blunt-leading-edge wedges is shown in Fig. 7 at Mach numbers of 21.6 and 41.4 for a flow deflection angle range of 2–6° with nearly-constant leading-edge Reynolds number. The plates all had sonic-wedge leading edges, and the characteristics-theory is for the same shapes. Although Mach-41 data were obtained at both 0° and 6° flow deflection angles, only the 6° case can be shown because of the nature of the correlation parameters. Theories shown are the hypersonic small-disturbance theory from [11] and the characteristics theory for $M_\infty = 21.4$. The inability of the correlation parameters to absorb the zero-flow-deflection case is a deficiency which can be corrected by replacing

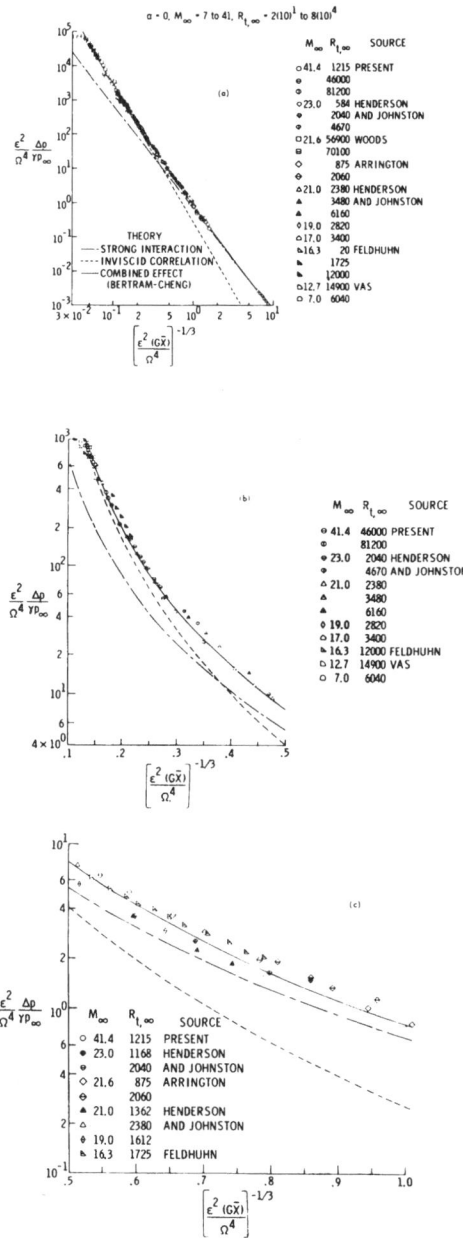

Fig. 6. (a) Unified two-dimensional hypersonic induced-pressure correlation. (b) Enlargement of mutual bluntness–viscous interaction region of (a). (c) Enlargement of mutual bluntness–viscous interaction region of (a).

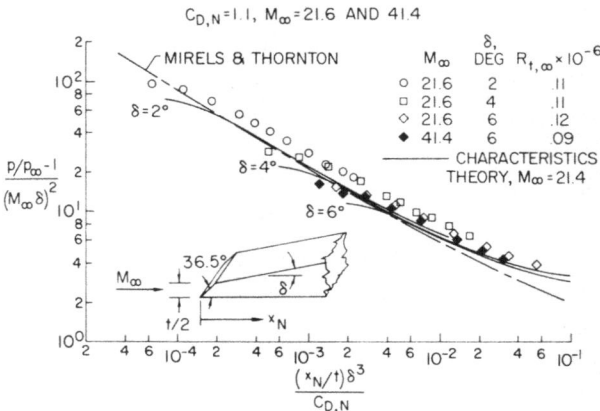

Fig. 7. Correlation of two-dimensional blunt-wedge pressures.

the flow-deflection angle by another parameter which is related to δ, but does not approach zero as δ does. One approach, suggested by hypersonic slender-body theory and used by Traugott ([12]), replaces δ by the ratio of the asymptotic wedge pressure to the maximum pressure on the blunt nose.

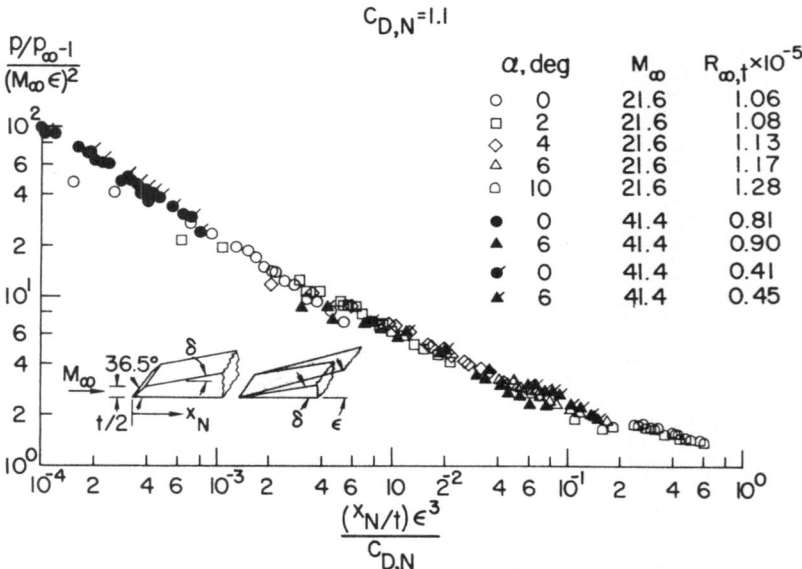

Fig. 8. Correlation of two-dimensional blunt-wedge pressures using asymptotic shock angle ε.

Another approach, suggested by oblique-shock theory (assuming hypersonic flow) replaces the flow deflection angle δ by the asymptotic shock angle ε. The success of the latter approach is shown in Fig. 8.*

2. CLOSED-FORM LOCAL SIMILARITY

2.1. Local Similarity Defined

The local-similarity approximation is frequently used to represent the solution to laminar boundary layer problems which, in their more exact form, are described by partial differential equations. The similarity assumption, besides reducing the governing partial differential equations to ordinary differential equations, reduces the number of controlling space coordinates to a single parameter; velocity profiles at all streamwise stations are then made congruent when plotted against this parameter, which thus becomes a natural correlating parameter.

Before considering closed-form local-similarity solutions, a brief review of the implications of "local," as distinguished from "exact," similarity is in order.

Historically, the first exact similarity solution to the laminar boundary layer equations was that of Blasius for incompressible flow over a flat plate with zero pressure gradient [see ([13]), for example]. By selecting natural scaling parameters for his particular problem, he was able to represent the dimensionless velocity profile u/U_∞ as a universal function of the similarity parameter η, where

$$\eta = y(U_\infty/\nu x)^{1/2} \qquad (17)$$

[the natural scaling parameter for y was δ, which, from previous solutions, was known to be proportional to $(\nu x/U_\infty)^{1/2}$; thus $y/\delta \propto y(U_\infty/\nu x)^{1/2} = \eta$]. An integral part of his solution was the definition of a stream function which indentically satisfied the equation of continuity

$$\frac{\partial u}{\partial x} + \frac{\partial v}{\partial y} = 0 \qquad (18)$$

and reduced the momentum equation from the partial differential equation

$$u\frac{\partial u}{\partial x} + v\frac{\partial u}{\partial y} = \nu\frac{\partial^2 u}{\partial y^2} \qquad (19)$$

* This correlation is due to R. D. Wagner of Langley Research Center.

to the ordinary differential equation

$$ff'' + 2f''' = 0 \tag{20}$$

(primes denote differentiation with respect to η). He put the stream function

$$\psi(x, y) = (\nu x U_\infty)^{1/2} f(\eta) \tag{21}$$

where $\partial\psi/\partial y = u$, $-\partial\psi/\partial x = v$; and the coefficient of $f(\eta)$ is chosen so that

$$\frac{\partial\psi}{\partial y} = \frac{\partial\psi}{\partial\eta}\frac{\partial\eta}{\partial y} = u = U_\infty f'(\eta) \tag{22}$$

The universal velocity profiles u/U_∞ are one of several boundary layer functions which are tabulated against the similarity parameter η [Eq. (20) is solved numerically].

Later, Falkner and Skan [74] examined the requirements for exact similarity when a pressure gradient is impressed upon the incompressible boundary layer. The boundary layer equations are now

$$u\frac{\partial u}{\partial x} + v\frac{\partial u}{\partial y} = U\frac{dU}{dx} + \nu\frac{\partial^2 u}{\partial y^2}$$

$$\frac{\partial u}{\partial x} + \frac{\partial v}{\partial y} = 0 \tag{23}$$

with $u = v = 0$ at $y = 0$ and $u = U(x)$, $v = 0$ at $y = \infty$.

With U now a function of x, rather than a constant as for Blasius' case, the scaling factors must be appropriately modified and the solution to (23) is not, in general, independent of x.

Again continuity is satisfied by a stream function; in this case,

$$\psi(x, y) = (\nu L U_\infty)^{1/2}[U(x)/U_\infty]g(x)f(\xi, \eta) \tag{24}$$

where

$$\xi = x/L, \qquad \eta = [y/g(x)](U_\infty/\nu L)^{1/2} \tag{25}$$

The coefficient of $f(\xi, \eta)$ is chosen so that now

$$\partial\psi/\partial y = u = U(x)f'(\xi, \eta) \tag{26}$$

The dimensionless scale factor $g(x)$ is as yet undetermined.

With the stream function, Eq. (24), and the transformed coordinates, Eq. (25), Eq. (23) becomes

$$f''' + \alpha(x)ff'' + \beta(x)(1 - f'^2) = \frac{U(x)}{U_\infty} [g(x)]^2 \left(f' \frac{\partial f'}{\partial \xi} - f'' \frac{\partial f}{\partial \xi} \right) \quad (27)$$

where

$$\alpha(x) = \frac{Lg(x)}{U_\infty} \frac{d}{dx} [U(x)g(x)], \qquad \beta(x) = \frac{L}{U_\infty} [g(x)]^2 \frac{dU(x)}{dx} \quad (28)$$

For similar solutions to exist, f must be independent of ξ and depend only on η; the right-hand side of Eq. (27) must therefore vanish. Furthermore, both $\alpha(x)$ and $\beta(x)$ must be constants. It can be shown that, since both $\alpha(x)$ and $\beta(x)$ must be constant, $\alpha(x)$ may be arbitrarily set equal to unity with no loss in generality, and that then

$$\beta = \frac{2m}{m+1}, \qquad \frac{U}{U_\infty} \propto \left(\frac{2}{m+1} \frac{x}{L} \right)^m, \qquad g(x) = \left(\frac{2}{m+1} \frac{x}{L} \frac{U_\infty}{U} \right)^{1/2} \quad (29)$$

Having determined the form of $U(x)$ and of the scale factor $g(x)$ for similarity to exist, the stream function and the transformed similarity coordinate are seen to be

$$\psi(x, y) = \left(\frac{2}{m+1} \nu x U \right)^{1/2} f(\eta)$$

$$\eta = y \left(\frac{m+1}{2} \frac{U}{\nu x} \right)^{1/2} \quad (30)$$

and the differential equation (27) becomes

$$f''' + f'' + \beta(1 - f'^2) = 0 \quad (31)$$

Solutions to Eq. (31) are tabulated as functions of η for various values of β, the Falkner–Skan velocity gradient parameter (Blasius' solution is obtained when $\beta = 0$).

Thus similarity is exactly satisfied only if $U \propto x^m$.* If the streamwise velocity distribution is of any other functional form, then both α and β will be functions of x and the significant component of the stream function, f, will no longer be a function of η only; i.e., the full equation (27) should

---

* For an apparently inconsequential exception, see p. 120 of ([13]) and footnote on p. 3 of ([14]).

be used. Use of equation (31) with a nonpower-law velocity distribution is inconsistent.

Nonetheless, similarity solutions to Eq. (31) have been applied to other than power-law velocity distributions with good results [15]. One approach [15] is to break the velocity distribution up into segments, each of which is represented by a different power-law curve. In this way, similar solutions are pieced together to obtain an approximate solution to the nonsimilar boundary layer problem. Another approach is to assume local similarity [16], i.e., at each point on the velocity distribution curve, find the local effective power law by letting

$$m_{\text{local}} = d(\ln U)/d(\ln x) \tag{32}$$

and approximate the boundary layer solution with the tabulated solutions to the similarity equation, Eq. (31). A method for evaluating the error involved in using similar solutions for nonsimilar problems is given in [17].

2.2. Compressible Similar Solutions

Because compressibility effects generally couple the momentum and energy equations, compressible laminar boundary-layer equations are more difficult to solve than incompressible equations, where the momentum equation alone determines the velocity distribution.

One approach to the compressible problem involved a search for a transformation which, when applied to the compressible equations, would reduce them to exactly the incompressible form just discussed. In 1949, Stewartson [18] found such a transformation for Prandtl number 1.0 and zero surface heat transfer (the latter condition decouples the momentum and energy equations). The value of this approach is that now the already tabulated solutions to the incompressible problem can be applied to the compressible flow case.

Later, Cohen and Reshotko [14] applied Stewartson's transformation to the laminar compressible boundary-layer equations with heat transfer and pressure gradient present, and introduced the similarity requirement. The similarity requirement is more restrictive in the compressible case than for incompressible flow. For transformed flows of the Falkner–Skan type, velocity distributions of the form $U \propto x^m$ satisfy exact similarity only when $M_\infty \to \infty$. However, the local-similarity approximation discussed in the previous section makes these solutions more widely applicable than a strict adherence to exact similarity requirements would allow.

Since Cohen and Reshotko's paper ([14]) and their tabulated solutions form the backbone of the closed-form local-similarity solutions, a discussion of the use of their paper is in order.

The uninitiated frequently find the use of compressible similarity-theory solutions baffling because the solutions are tabulated as functions of quantities in the transformed plane where similarity is achieved rather than in the physical plane. The basic approach to use is to formulate the desired quantity in the physical plane and then to transform it into quantities which have been tabulated in the transformed plane. Following ([14]), the transformations and other functions of interest here are:

1. The modified Stewartson transformation

$$X = \int_0^x \lambda \frac{\rho_e a_e}{\rho_0 a_0}\, dx, \qquad Y = \frac{a_e}{a_0} \int_0^y \frac{\varrho}{\varrho_0}\, dy \tag{33}$$

where

$$\lambda = \frac{\mu_w}{\mu_0} \frac{T_0}{T_w} \tag{34}$$

2. The enthalpy function for an ideal gas,

$$S = \frac{C_p T + \frac{1}{2} u^2}{C_p T_0} - 1 \tag{35}$$

3. The relation between the real and transformed velocities,

$$U = (a_0/a_e) u \tag{36}$$

where u is the real velocity.

4. The transformed similarity parameter,

$$\eta = Y \left(\frac{m+1}{2} \frac{U_e}{\nu_0 X} \right)^{1/2} \tag{37}$$

where m is the exponent in the transformed power-law velocity,

$$U_e \propto X^m \tag{38}$$

The velocity ratio is given by

$$U/U_e = u/u_e = f'(\eta) \tag{39}$$

The momentum thickness will be obtained in some detail:

$$\theta = \int_0^\infty \frac{\varrho u}{\varrho_e u_e} \left(1 - \frac{u}{u_e}\right) dy \tag{40}$$

$$dy = \left(\frac{dY}{dy} \frac{d\eta}{dY}\right)^{-1} d\eta$$

$$= \frac{a_0}{a_e} \frac{p_0}{p_e} \frac{T}{T_0} \left(\frac{2}{m+1} \frac{v_0 X}{U_e}\right)^{1/2} d\eta \tag{41}$$

here the subscript 0 refers to local stagnation conditions; thus $\varrho_0/\varrho = (p_0/p_e)(T/T_0)$; p_e is used rather than p, since constant static pressure is assumed through the boundary layer. Then θ becomes

$$\theta = \frac{T_e}{T_0} \frac{a_0}{a_e} \frac{p_0}{p_e} \left(\frac{2}{m+1} \frac{v_0 X}{U_e}\right)^{1/2} I_{\theta, tr} \tag{42}$$

where

$$I_{\theta, tr} = \int_0^\infty f'(1 - f') d\eta \tag{43}$$

The integral is exactly one of the tabulated functions [see Eq. (38) and Table II of ([14])]; its solution is therefore known.

It is more usual to express the above in the dimensionless form $(\theta/x) \times (R_{\infty, x})^{1/2}$. After some algebraic manipulation, including isolation of reference parameters on the left-hand side, and use of the relation

$$\lambda(p_e a_e / p_0 a_0) = dX/dx \tag{44}$$

Eq. (42) can be written

$$\frac{(\theta/x)(R_{\infty, x})^{1/2}}{(T_e/T_\infty)[C_w(u_\infty/u_e)(p_\infty/p_e)]^{1/2}} = \left(\frac{m+1}{2} \frac{x}{X} \frac{dX}{dx}\right)^{-1/2} I_{\theta, tr} \tag{45}$$

Following a similar procedure, it can be shown that

$$\frac{(\delta^*/x)(R_{\infty, x})^{1/2}}{(T_0/T_\infty)[C_w(u_\infty/u_e)(p_\infty/p_e)]^{1/2}}$$

$$= \left(\frac{m+1}{2} \frac{x}{X} \frac{dX}{dx}\right)^{-1/2} \left[I_{\delta^*, tr} + \left(1 - \frac{T_e}{T_0}\right)I_{\theta, tr}\right] \tag{46}$$

where

$$I_{\delta^*, tr} = \int_0^\infty (1 - f' + S) d\eta \tag{47}$$

$$\frac{C_{f,\infty}(R_{\infty,x})^{1/2}}{[C_w(p_e/p_\infty)]^{1/2}(u_e/u_\infty)^{3/2}} = 2\left(\frac{m+1}{2} \frac{x}{X} \frac{dX}{dx}\right)^{1/2} f_w'' \tag{48}$$

$$\frac{S_{t,\infty}(R_{\infty,x})^{1/2}}{[C_w(p_e/p_\infty)(u_e/u_\infty)]^{1/2}} = \left(\frac{m+1}{2} \frac{x}{X} \frac{dX}{dx}\right)^{1/2}\left(-\frac{S_w'}{S_w}\right)(P_r)^{-0.6} \tag{49}$$

$I_{\delta^*,tr}$, f_w'', and S_w'/S_w are additional tabulated functions in [14]. The factor $(P_r)^{-0.6}$ compensates partially for the $P_r = 1.0$ assumption in the solutions.

The crucial, and still undefined, portion of each of these expressions is the quantity $\frac{1}{2}(m+1)(x/X)\, dX/dx$. If the assumption that isentropic flow exists along the edge of the boundary layer is permissible, then the stage is set for Section 2.3.

2.3. Closed-Form Local-Similarity Solutions

Equation (30) of [14] can be written

$$m = \frac{1}{u_e} \frac{du_e}{dx} \frac{T_0}{T_e} X\left(\frac{dX}{dx}\right)^{-1} \tag{50}$$

The temperature and velocity are related through

$$C_p T_e + \tfrac{1}{2}u_e^2 = C_p T_0 \tag{51}$$

giving

$$\frac{1}{u_e} \frac{du_e}{dx} = -\frac{(d/dx)(T_e/T_0)}{2[1 - (T_e/T_0)]}$$

Using the perfect gas equation of state, $p = \varrho RT$, along with the isentropic relation $p\varrho^{-\gamma} = \text{const}$, dX/dx and X are, respectively,

$$\frac{dX}{dx} = \lambda\left(\frac{p_e}{p_0}\right)^{(3\gamma-1)/2\gamma}$$

$$X = \lambda \int_0^x \left(\frac{p_e}{p_0}\right)^{(3\gamma-1)/2\gamma} dx \tag{52}$$

(λ is assumed independent of x). When $(p_e/p_0)^{(3\gamma-1)/2\gamma}$ is integrable in closed form, then $\frac{1}{2}(m+1)(x/X)\, dX/dx$ is solved and a closed-form solution exists for the various boundary layer functions [Eqs. (45), (46), (48), (49)].

The above is essentially the approach used in [19] for $p \propto x^n$; Crawford [20], used it for a more general class of modified power-law pressure distributions of the form $p \propto (x + A)^n$. An interesting example of one of the

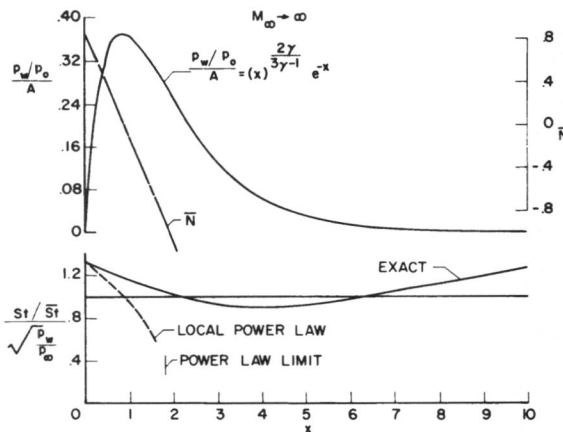

Fig. 9. Typical application of closed-form similarity solution.

many types of pressure distributions for which M. H. Bertram (unpublished) has obtained results by this approach is shown in Fig. 9.

A note of caution: Because local similarity is assumed, the validity of the method is not assured for all integrable pressure functions; the range of validity has yet to be established. Also, certain integrable functions yield solutions in the complex plane; these are of no value.

3. BOUNDARY LAYER TRANSITION

An introduction to the boundary layer transition problem is illustrated in Fig. 10, which shows a portion of the available data on the variation of the Reynolds number for the beginning of transition based on local conditions versus the local Mach number. The data are for both sharp and slightly-blunted ($r_N < \frac{1}{2}$ in.) cones and were obtained in wind tunnels, ballistic ranges, and free flight. Local inviscid sharp nose conditions have been used for both parameters. It is clear that the many additional parameters known to affect boundary layer transition (and perhaps others as yet unknown) make an uncorrelated plot such as this of no value to the designer. Some of the parameters known to affect transition are: Mach number, unit Reynolds number, heat transfer, wing blunting, wing sweep, angle of attack, stagnation temperature, two- or three-dimensional flow, pressure gradients, surface roughness, free-stream turbulence, noise, and vibration. One or more of these parameters would be affecting any data point shown.

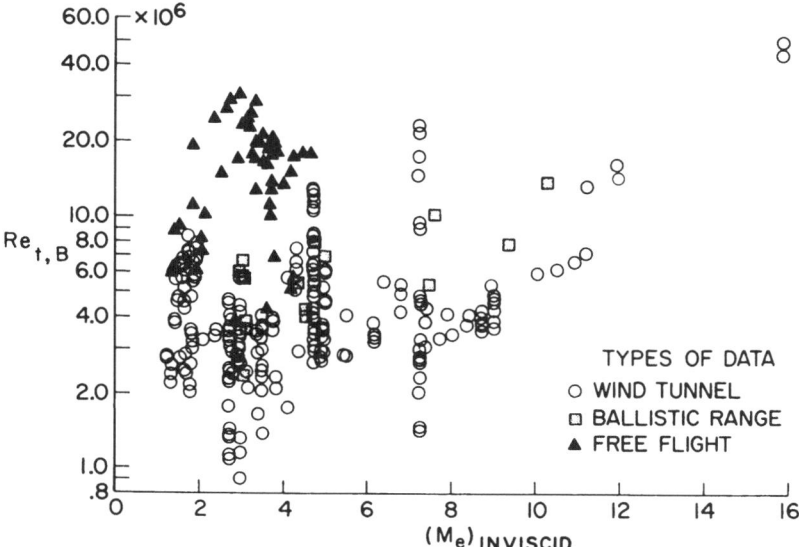

Fig. 10. Typical transition data. Cones.

Before attempting to assess our current understanding of the boundary layer transition phenomenon, we will consider some of these parameters and what we know (or think we know) about their influence on transition, with particular emphasis on the hypersonic regime.

1. Hypersonic boundary layers are very stable; moderate surface roughness, contrary to subsonic and supersonic experience, has negligible effect on promoting transition. This is illustrated in Fig. 11, where the effect on tripping transition of a single row of spheres of various diameters is shown; the spheres must be greater than one boundary-layer thickness to be effective. Note that roughness may even delay transition, perhaps through a mechanism similar to that associated with slight leading-edge blunting ([21]).

2. The location of the so-called critical layer (or region of maximum amplification of disturbances) moves from deep within the boundary layer at low speeds to near the edge of the boundary layer at hypersonic speeds, as illustrated in Fig. 12. Results of various investigations of the location of the critical layer in laminar boundary layers are shown for air ([22]) and helium ([23]) (the difference between the air and helium subsonic–supersonic relative flow regions is that theory was used for air δ, whereas experiment was used in the helium case). The variation of critical layer position with

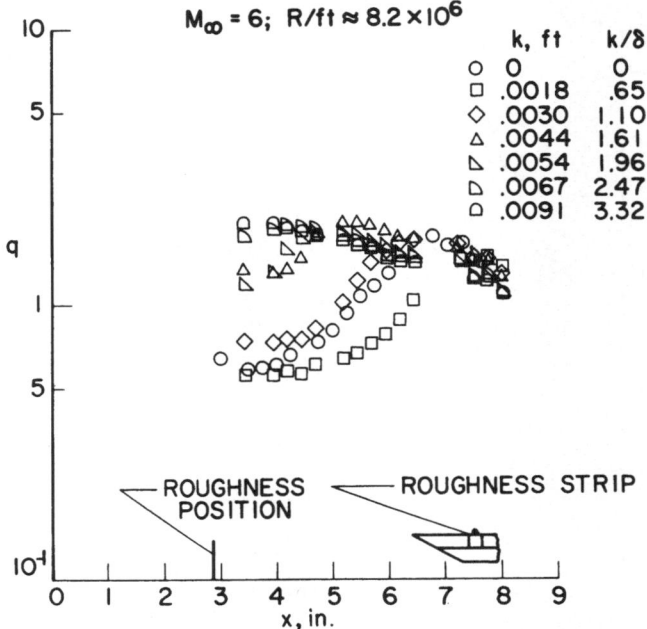

Fig. 11. Effect of roughness on transition. From ([75]).

Mach number may be partially responsible for the insensitivity of the boundary layer to small roughness at hypersonic Mach numbers.

3. The Reynolds number for transition increases rapidly with M_e, as shown in Figs. 13 and 14. Figure 13 is from a study in the Langley 22-in. helium tunnel ([23]). Here the transition Reynolds number increases about exponentially with M_e. The data of Fig. 14, from several facilities, have been

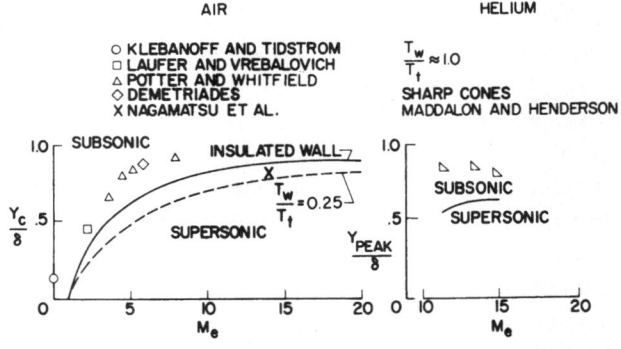

Fig. 12. Locations of maximum disturbance in boundary layer.

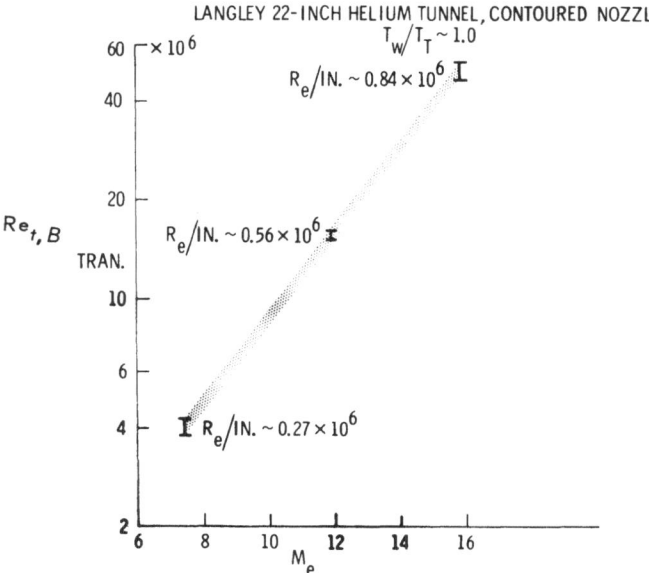

Fig. 13. Variation of transition Reynolds number with local Mach number.

referred to the same unit Reynolds number, and show essentially the same trend of $Re_{t,B}$ with M_e [24]. Although $Re_{t,B}$ is depicted as increasing as M_e^4, an exponential curve would fair through the data just as well.

4. Transition takes place over increasingly greater distances as M_e increases; several researchers, after examining large bodies of data, have noted that the increment $(Re_t - Re_{t,B})$ falls within 50–150% of $Re_{t,B}$, with a mean of about 100%; i.e., the length of transitional flow roughly equals the length of laminar flow. At $M_e \approx 16$, $(Re_t - Re_{t,B})$ would be on the order of $50(10)^6$. The importance of the large increment in transition Reynolds number is related to the problem of obtaining the virtual origin for the ensuing turbulent boundary layer; this will be treated later.

5. Figure 15 illustrates the effect of nose blunting on boundary layer transition on cones [25,26]. Bluntness effects are weak at low $R_{D,SC}$ (the nose-bluntness Reynolds number); above a certain value of $R_{D,SC}$, as $R_{D,SC}$ increases, the transition Reynolds number increases to a maximum and then decreases as the transition moves close to the nose. Some of the variables which contribute to this behavior are local variation in unit Reynolds number, Mach number, and pressure gradient; the intensity of these variables is associated primarily with geometry and free-stream Mach

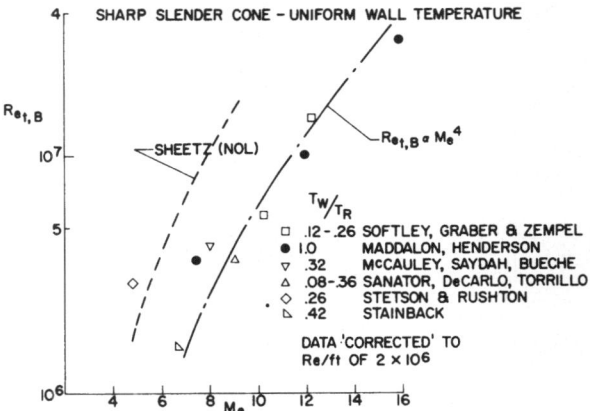

Fig. 14. Effect of local Mach number on transition Reynolds number.

number. As M_∞ increases, the level of induced pressure, its streamwise
gradient, and the downstream extent over which it is propagated increase.
The combination of local-static-pressure variation with local-stagnation
and pitot-pressure variation at the edge of the boundary layer (the latter
two caused by the varying entropy gradient through the strongly-curved
bow shock) cause the variation in local unit Reynolds number and Mach
number.

6. Under actual flight conditions, superimposed on these bluntness
effects will be the effects of angle of attack ([26]) illustrated in Fig. 16. The
ratio of transition distance to transition distance on the sharp cone at
$\alpha = 0$ is plotted against α for three values of nose bluntness; results for
both windward and leeward surfaces are shown. The favorable effect of

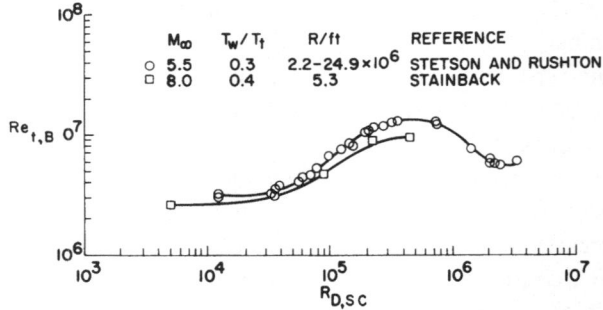

Fig. 15. Effect of nose bluntness on boundary layer transition for cones.

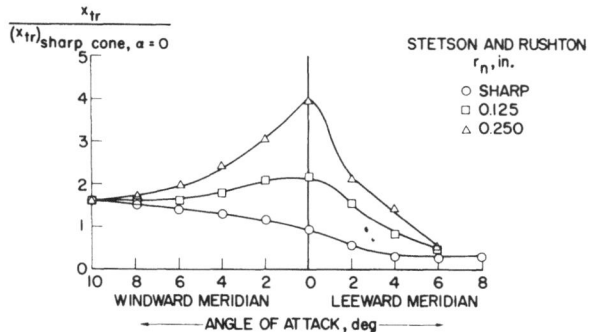

Fig. 16. Effect of angle of attack on transition for cones;
$M_\infty = 5.5$.

nose blunting on transition at $\alpha = 0$ tends to wash out at relatively low angles of attack (6–8°).

The effect of α on the sharp cone is typical of most of the available data; transition is delayed on the windward and promoted on the leeward ray. The favorable effect on the windward ray may be caused by the cross-flow's thinning the boundary layer and thereby reducing the momentum thickness loss Reynolds number ([27]), a parameter which has been fairly successful in correlating low-to-moderate Mach-number transition data.

7. Recent studies have indicated that at high local Mach numbers, the strong hypersonic Mach-number effect on transition shown in Figs. 13 and 14 may overpower some of the other, more subtle effects found at lower speeds. This is illustrated in Fig. 17, where transition location on a

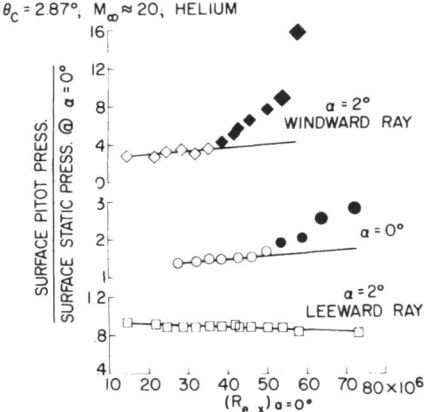

Fig. 17. Effect of angle of attack on cone transition.

2 87° half-angle cone ([28]) at $M_\infty = 20$ is shown at $\alpha = 0°$ and 2°. Detection is with a surface pitot probe. The effect is opposite to that of Fig. 16; now transition is promoted on the windward and delayed on the leeward ray. This reversal in transition trend is tentatively ascribed to the strong effect of M_e; at $\alpha = 0°$, $M_e \approx 16$, while on the windward and leward rays, it is about 13 and 19, respectively. This change is believed enough to reverse the trend normally found in transition at lower Mach numbers, where the changes in Mach number due to α are relatively small.

8. One of the more controversial parameters associated with boundary layer transition is the unit-Reynolds-number effect—is it a fundamental phenomenon, a ground-facility-generated nuisance, or a combination of the two? The problem posed by the apparent unit-Reynolds-number effect is illustrated in Fig. 18, which is a selective version of the data of Fig. 10. Here, the effect of nose bluntness on local properties, discussed previously, has been accounted for by the method of ([25]). Unit Reynolds number appears to have a strong influence on $Re_{t,B}$ from wind tunnels in the lower hypersonic range; its effect appears to diminish as M_e increases. However, these trends are based on meager data, and the picture may change as further studies are made. That more than unit Reynolds number is significant is shown by the large difference between wind-tunnel and free-flight $Re_{t,B}$ for the same unit-Reynolds-number range. Two possible additional factors are the differing stagnation temperature levels in the wind tunnel and free flight, and the aerodynamic noise created by the turbulent tunnel-wall boundary layer, which is generally absent in free flight (although on a flight vehicle, aerodynamic noise from the turbulent boundary layer on the fuselage could affect transition on the wing, engine nacelle, or control surfaces, for instance). The possibility that stagnation temperature could have

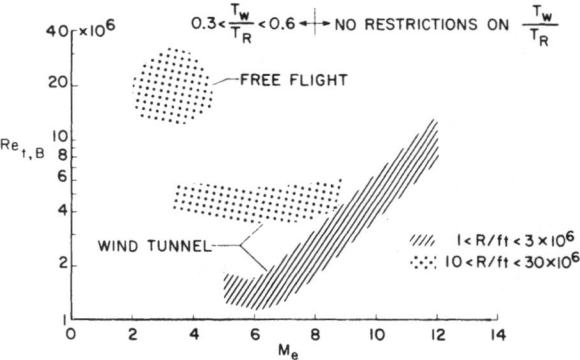

Fig. 18. Effect of M_e and R/ft on transition for cones.

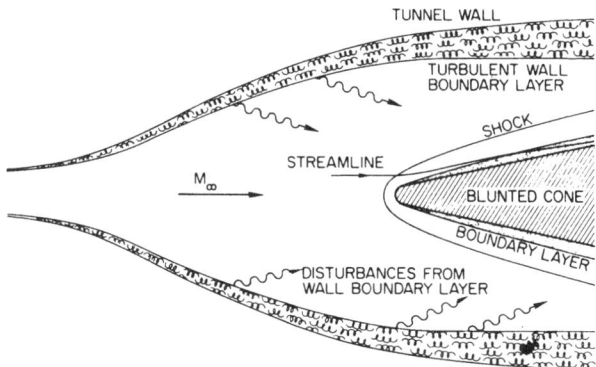

Fig. 19. Some factors affecting transition.

an influence is suggested by Mack's ([29]) recent work on boundary layer stability theory. Although this relationship between the onset of boundary layer instability and of boundary layer transition is nebulous at best, it is intuitively acceptable that parameters which have a major influence on stability would also affect transition.

Figure 19 depicts the manner in which a model is subjected to several of the potential sources of unit-Reynolds-number effects. The bluntness and boundary layer displacement effects discussed previously would be present in free flight as well, while the aerodynamic noise radiated to the model would depend on the condition of the tunnel-wall boundary layer, whether laminar, transitional, or turbulent; however, the turbulent state is the most common. Turbulent boundary layers are noisiest at low Reynolds numbers, where the skin friction coefficient is highest ([30]).

Pate and Schueler ([31]) examined the existing literature [see bibliography in ([31])] for clues as to the probable source of disturbances which might influence model boundary-layer transition in high supersonic and hypersonic wind tunnels. Of the three types of disturbances Kovasznay ([76]) indicates may occur in wind tunnels (vorticity fluctuations, entropy fluctuations, and sound waves), they conclude that sound waves generated by turbulent tunnel-wall-boundary layers are the most likely source. Vorticity fluctuations are eliminated from consideration by Laufer's work ([77]); he showed vorticity to have insignificant effect on model boundary-layer transition Reynolds number above $M \approx 2.5$. Entropy fluctuations (temperature spottiness) were felt to be unimportant because stagnation chamber mixing sections and turbulence damping screens are generally used to reduce these fluctuations to small levels in supersonic tunnels. (Many hypersonic tunnels have no screens or other mixing devices in their settling chambers; thus entropy fluctuations cannot yet be ruled out as a potential source of hypersonic tunnel disturbances).

Pate and Schueler noted that the more important factors affecting the intensity of sound radiated from turbulent tunnel-wall boundary layers include: Mach number, boundary layer thickness, mean shear, and tunnel size. Using these factors as a guide, previously published transition Reynolds numbers on two-dimensional models in eight different facilities at Mach numbers from 3 to 8 were correlated as a function of (1) test-section turbulent tunnel-wall boundary-layer displacement thickness, (2) average tunnel-wall turbulent skin-friction coefficient, and (3) test section circumference. The empirical correlation is

$$\mathrm{Re}_t = 0.0141(C_F)^{-2.55}\left[0.56 + 0.44\,\frac{C_1}{C}\right]\!\left(\frac{\delta^*}{C}\right)^{-1/2} \tag{53}$$

In their most recent experimental work ([31]), they explicitly checked the tunnel circumference effect by examining the variation of Re_t with unit Reynolds number, using hollow cylinders at $M_\infty = 3$ in three different wind tunnels with square test sections measuring 12×12 in., 40×40 in., 16×16 ft. (Fig. 20). For any unit Reynolds number, Re_t increases with tunnel size, as anticipated. The degree to which the parameters in Eq. (53) correlate the previously published and new data just discussed is illustrated in Fig. 21. Equation (53) represents all the data fairly well. Note that the correlation equation is independent of Mach number and unit Reynolds number and depends only on parameters known to influence the radiation of aerodynamic noise.

That noise contributes to the usual variation of Re_t with unit Reynolds number is illustrated in Fig. 22. The top sketch shows the relative location

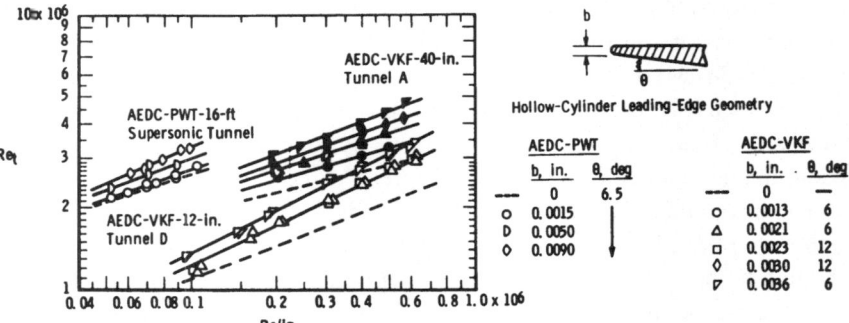

Fig. 20. Transition on hollow cylinder in three wind tunnels; $M_\infty = 3$.

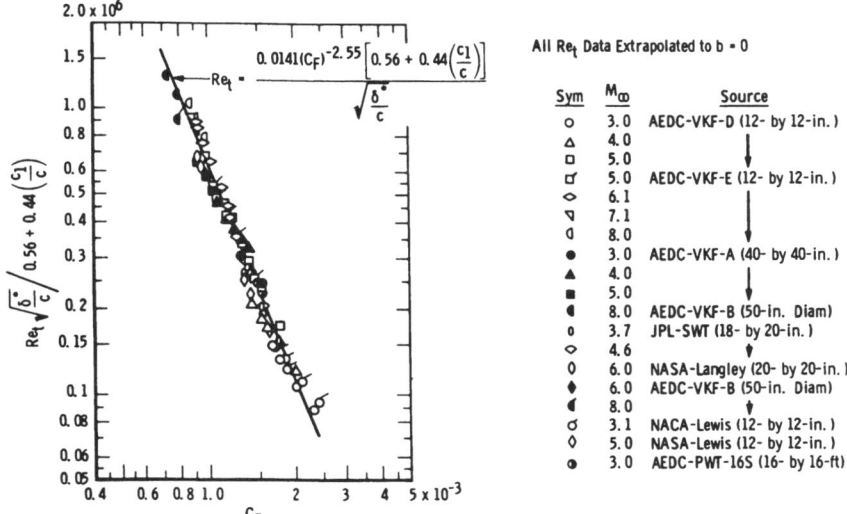

Fig. 21. Correlation of transition Reynolds number.

of the shroud used to generate laminar and turbulent boundary layers in close proximity to the two models used in this study—the previously mentioned hollow cylinder, and a sharp, flat plate with microphones on the surface and designated "noise model." Results of both the sound pressure and transition study show clearly that the manner in which Re_t varies with unit Reynolds number depends upon the intensity of the aerodynamic noise.

Pate and Schueler make a good case for blaming the aerodynamic noise which radiates from turbulent tunnel-wall boundary layers for the unit-Reynolds-number effect upon transition Reynolds number which is almost universally found in wind tunnels [for an exception, see ([32]), where injecting helium into the tunnel wall boundary layer appears to eliminate the unit-Reynolds number effect, at least when the test models are cones].

It is therefore somewhat surprising to find that a similar effect of unit Reynolds number upon Re_t has been found in an aeroballistic range by Potter ([33]). A schematic of his experimental arrangement is given in Fig. 23. He avoided ablation by using the low hypersonic Mach number $M_\infty \approx 5$. Cone half-angle was $10°$ and $T_w/T_{a,w}$ was about 0.18; noses were sharp; unavoidably, $\alpha \neq 0$, but on the average was $2°$. Results of the study are shown in Fig. 24 compared to typical wind-tunnel data; the familiar variation of Re_t with unit Reynolds number is clearly evident. Potter used a microphone at the model-viewing station to assure that firing and sabot

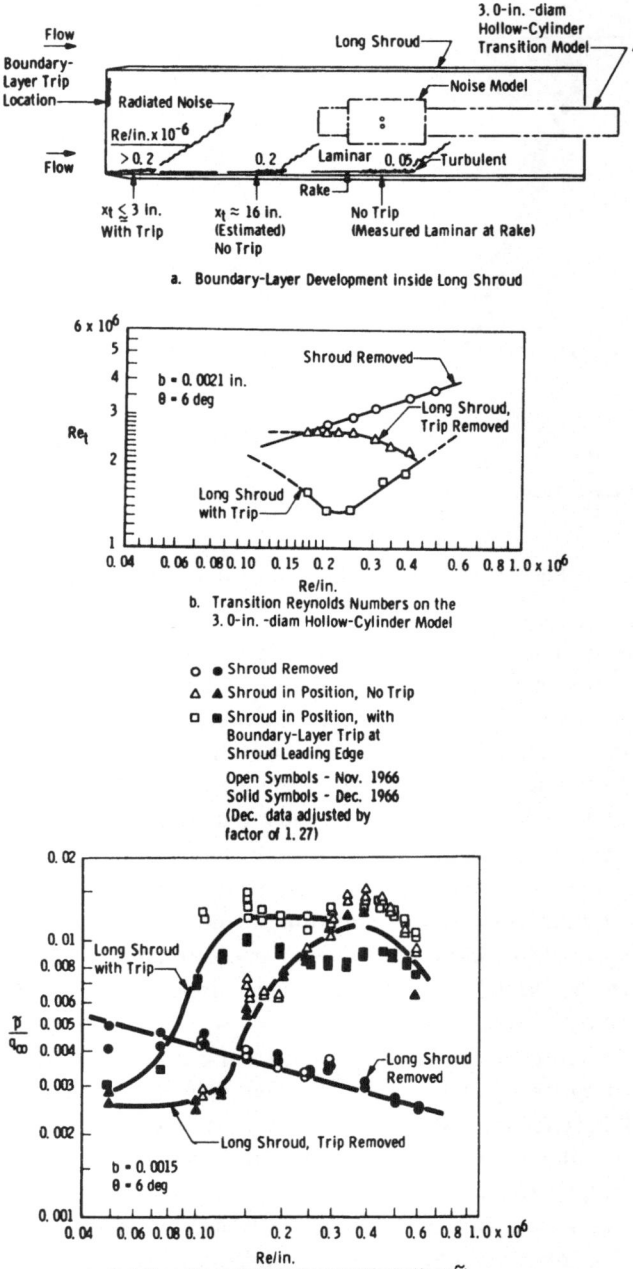

a. Boundary-Layer Development inside Long Shroud

b. Transition Reynolds Numbers on the
3. 0-in. -diam Hollow-Cylinder Model

c. Root-Mean-Square Radiated Pressure Fluctuations (p̃)

Fig. 22. Comparisons of transition Reynolds numbers and root-mean-square radiated
pressure fluctuations at $M_\infty = 3.0$.

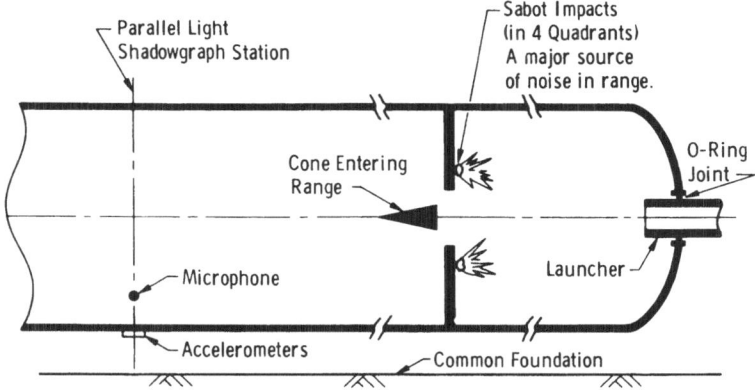

Fig. 23. Schematic of aeroballistics arrangement used by Potter ([33]).

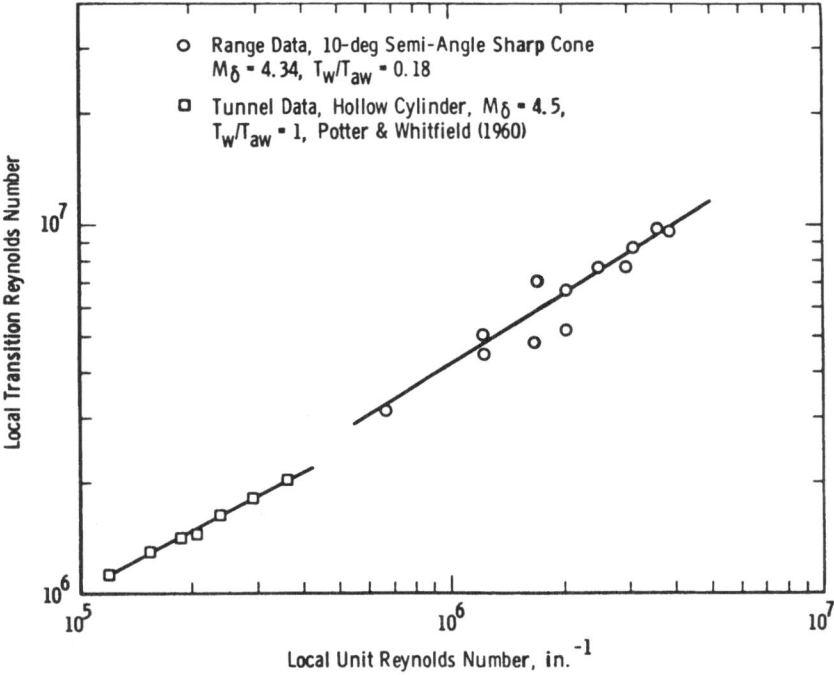

Fig. 24. Unit-Reynolds-number effect in aeroballistics range.

impact noise, which was transmitted through the range wall and radiated into the range centerline, did not affect the model.

Potter's work suggests that the unit-Reynolds-number effect may be a fundamental phenomenon, thus lending support to the view held by some that boundary layer stability theory has something to tell us about transition [29,34,35]. Reshotko [34] points out that under ideal conditions (i.e., sharp leading edge, no roughness, etc.), the parameters of significance for determining the stability of a laminar boundary layer on a flat plate are

$$Re_c = Re_c(M, T_w/T_R, \beta\mu/\varrho U^2) \tag{54}$$

It is the latter parameter which may be of fundamental relevance to the unit-Reynolds-number problem.

Reshotko indicates that this parameter, $\beta\mu/\varrho U^2$, may also influence the "sometimes" observance of transition reversal as the wall-to-recovery temperature ratio is decreased; results indicating the "sometimes" nature of the phenomena are shown in Fig. 25 [36-39]. As in too many boundary

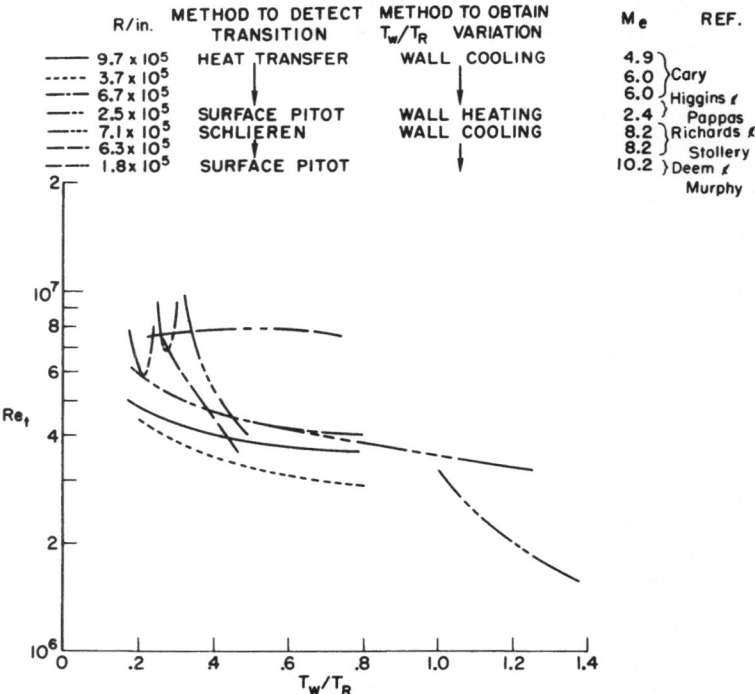

Fig. 25. Effect of wall cooling on boundary layer transition.

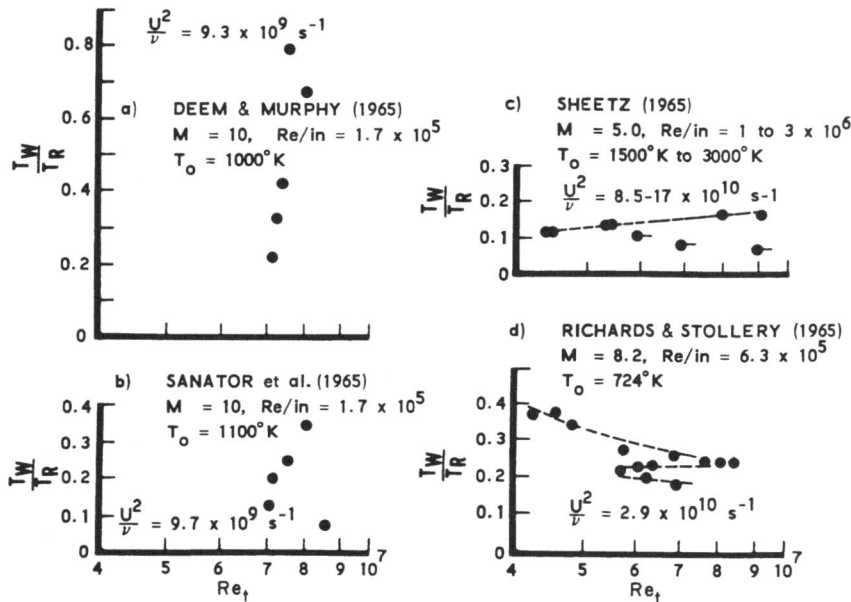

Fig. 26. Variation of transition Reynolds number with wall-to-recovery temperature ratio. From Reshotko ([34]).

layer transition investigations, it is clear that the behavior shown must depend on more parameters than were intentionally varied during the studies. Figure 26 ([34]) shows essentially the figures found in ([38]) except that Reshotko has calculated the corresponding values of $U^2\varrho/\mu$, that portion of the dimensionless parameter $\beta\mu/\varrho U^2$ which could contribute to the unit-Reynolds-number effect. Referring to Mack's ([29]) recent stability-theory work and his (Mack's) discovery of the importance of the higher modes, Reshotko points out that the higher is the value of $U^2\varrho/\mu$, the less sensitive the boundary layer is to higher-mode instability. Since only the first-mode disturbances are stabilized by wall cooling (the second and higher modes are actually destabilized by wall cooling), studies conducted under conditions of high $U^2\varrho/\mu$ favor the transition-reversal phenomenon as opposed to those with lower values of $U^2\varrho/\mu$, where wall cooling would be expected to be ineffective or actually promote transition (according to stability theory). Although there is not a one-to-one relationship between the experimentally-observed transition behavior shown in Fig. 26 and the reasoning suggested by stability theory, there is sufficient resemblance that continued study of stability theory is warranted.

With conflicting information from careful studies indicating that, the sc-called unit-Reynolds-number effect is caused by noise radiated from turbulent tunnel-wall boundary layers on the one hand, and that it is a fundamental phenomenon which has all along been predicted by stability theory on the other, the feeling grows that it may not be a question of either/or; instead, it may be more proper to ask how much of one and how much of the other are simultaneously affecting the results. Considerations such as these, along with the possible importance of the higher modes, make it clear that future hypersonic boundary-layer transition experiments must be more definitive than has been standard practice until now. A new generation of transition studies must be made. Hypersonic hot-wire anemometry (or other equivalent instrumentation) must be an integral feature of all such studies—knowledge of free-stream turbulence level, noise level, spectral distributions, etc., must accompany future investigations. Perhaps then the knowledge we now have, coupled with that gained from these new studies, will be transformed into a deeper understanding of the transition phenomenon.

4. TURBULENT BOUNDARY LAYER

4.1. Effect of Wall-Temperature Ratio

Turbulent-boundary-layer prediction methods still rely heavily on empiricism; the final form of any of the various methods is determined by experimental data; the usefulness of the resulting expression is then determined by examining how well it predicts other data. The importance of accurate and reliable experiment in both formulating and assessing turbulent-boundary-layer prediction methods is obvious. Unfortunately, not all data are uniformly valid, and generally it is difficult to evaluate the reliability of an experiment from the publication in which it appears.

Probably the most comprehensive and systematic evaluation of turbulent-boundary-layer theories is that undertaken by Spalding and Chi ([40,41]). They make no attempt to estimate the accuracy of the experimental data against which the theories were compared; they chose rather a statistical approach. They culled 491 experimentally-measured local and average skin-friction data points on sharp-leading-edge flat plates and hollow cylinders from all the available literature (up to about 1961 or 1962). Ranges of Mach number, Reynolds number, and wall-to-recovery temperature ratio were, respectively, 0–10, $3(10)^5$–$(10)^8$, and 0.2–1.7. They wrote computer programs for 20 different turbulent-boundary-layer theories and then systematically compared the prediction of each theory with each of the 491 data points. The root mean square of the quantity

$$(C_{f,\text{exp}} - C_{f,\text{theory}})/C_{f,\text{theory}}$$

for all data points was found for each theory; this determined the order of merit of each of the 20 theories. The three "best" ([42-44]) had rms errors of 11–12% and were all based on a mixing length theory. Kutafeladze and Leontev ([44]) used the Prandtl mixing length, while Van Driest ([42]) and Wilson ([43]) used von Kármán's.

A second objective of Spalding and Chi ([40]) was to develop a new calculation procedure based upon the accumulated theoretical and experimental knowledge which would permit making skin-friction calculations for a wide variety of conditions with but a few minutes of work—this they have done [also ([45])], and with an rms error of about 10%.

As higher Mach numbers and their associated low wall-to-recovery temperature ratios began to be important [the X-15 flight experiments reported in 1961 ([46]), focused attention on this range], it was found that

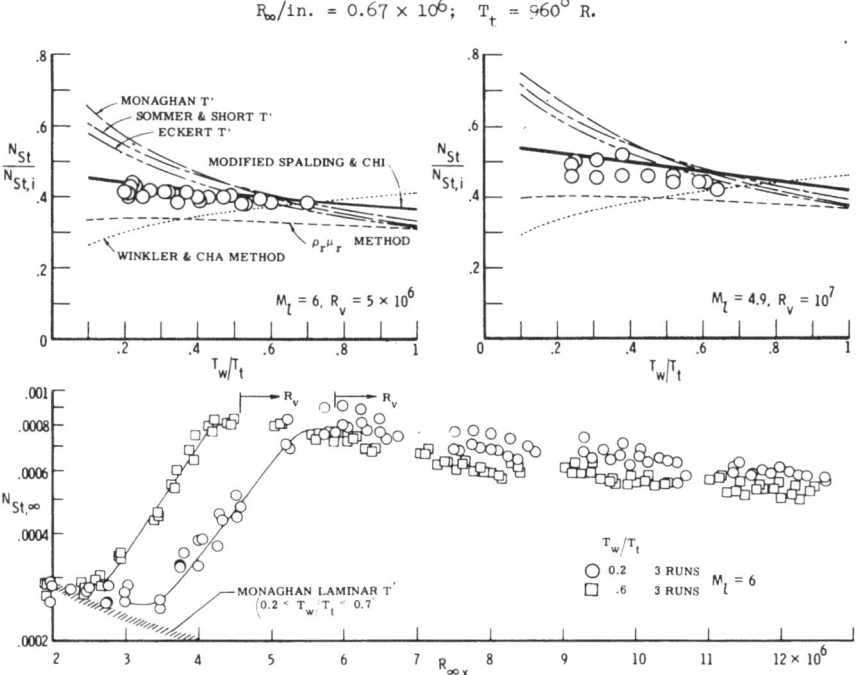

Fig. 27. Effect of wall temperature on turbulent heat transfer to a sharp, flat plate at $M_\infty = 6$.

nore of the turbulent-boundary-layer theories then existing was adequate. In 1965, Bertram and Neal ([47]) presented new wind-tunnel data which confirmed the flight data and also showed that the prediction method developed by Spalding and Chi was superior to any other available at the time. That this situation has not changed is illustrated by the more recent wind-tunnel data shown in Fig. 27. The data were obtained by Bertram et al. ([48]) on a sharp, flat plate cooled internally by liquid nitrogen. Little effect on the Stanton number is found from changing the level of wall temperature. The experimental trend agrees with that predicted by the Spalding–Chi method ([40]) as modified to heat transfer in Appendix A of ([47]) and Hank's $\varrho_r \mu_r$ method [Appendix B of ([49])]. However, the level of the data favors the prediction by the Spalding–Chi method. Clearly, the T' or reference-temperature methods ([50-52]) significantly overestimate the heat transfer at the low wall temperatures, and the Winkler–Cha method ([53]) underestimates the heating. Examples of the data from which ratios were obtained are given in the lower part of Fig. 27, where the small scatter indicates the uniformity of the data for a given run and the repeatability of runs.

4.2. Virtual Origin

In all their work, both that shown in Fig. 27 [which is typical of all cold-wall data given in ([48])] and that in ([47]), where superiority of the Spalding–Chi method over other methods for a wider range of conditions than had previously been examined was initially established, Bertram and his colleagues used the peak heating location as the virtual origin of the turbulent boundary layer, as did Coles ([54]) in his earlier work. Bertram selected this location only after checking a large body of data for various assumed virtual origins, including: (1) leading edge, (2) beginning of transition, and (3) several locations in the transition region. The assumed effective origin of the turbulent boundary layer which was most reasonably consistent for the bulk of the data lay ahead of the peak heating location by about 20% of the transition distance. Since differences between this location and the peak heating were generally minor, the peak heating location was chosen as the virtual origin for convenience. This is not a universally-accepted location. As the Mach-number range of interest increases into the hypersonic, the problem of compiling and examining data from various sources is compounded by this lack of a universally-accepted method of locating the virtual origin.

Lack of uniformity was frequently of little consequence at low speeds, where usually the laminar and transitional flows occupied such a small portion of the model under study that assuming the virtual origin was at the leading edge gave little error except in the immediate neighborhood of the start of turbulent flow—this region was generally small compared to the entire model and the error was acceptable. At high hypersonic speeds, on the other hand, the regions of laminar, transitional, and turbulent flow are liable to be more equitably distributed; good prediction in the vicinity of the beginning of turbulent flow is then important, as the region may cover a good portion of the model. At a local Mach number of 16, the laminar and transitional regions might each be expected to occupy increments of about $50(10)^6$ in Reynolds number ([23]), e.g., whereas typical low-speed values may be more like $0.5(10)^6$.

One of the more attractive methods of locating virtual origin at low speeds is the so-called momentum-matching procedure; it is a well-accepted refinement over using the leading edge at low speeds. Here, it is assumed that transition from laminar to turbulent flow occurs instantly at a point (a transition region is assumed to be nonexistant). The momentum thickness of the laminar boundary layer at the transition point is determined; the momentum thickness distribution of the turbulent flow is calculated; and the origin of the turbulent flow is located at such a position that the point where $\theta_{turb} = \theta_{lam}$ coincides with the θ_{lam} point; this origin is the so-called virtual origin of the turbulent boundary layer.

The momentum-matching procedure has worked well at low speeds, and some find it attractive in the hypersonic range as well. It has the merit of being less arbitrary than other assumed virtual-origin locations; the assumption of a transition point rather than the extensive transition region encountered at hypersonic speeds does appear incongruous, however. Richards ([55]) used a variation of the momentum-matching procedure in his recent studies of transitional and turbulent flow on a flat plate at $M_\infty = 8.2$. He experimentally measured the variation of momentum thickness with x in the turbulent region and extrapolated this forward into the transitional region to $\theta = 0$; this virtual origin nearly coincides with the beginning of transition. Using this virtual origin, the Spalding and Chi method is found to underpredict the turbulent heating by 40% (the Sommer and Short reference-temperature and Diessler–Loeffler mixing-length theories also underpredict experiment, but by lesser amounts).

It is thus quite apparent that until virtual-origin selection is standardized, one man's good theory will be another man's poor one. We do not yet know enough about transition, turbulence, and their interrelationships to establish by any means other than empiricism (trial and error)

the most rational choice for virtual-origin location. This writer favors
peak heating; using this location, it has been demonstrated that a fairly
reliable and easy-to-use method for predicting turbulent heating is available
—that developed by Spalding and Chi ([40,41]).

4.3. Transformation of Compressible Boundary Layer Profiles*

One approach to the compressible turbulent boundary-layer problem
which has received considerable attention in recent years has been to seek
a transformation which when applied to the compressible turbulent boundary-layer equations will yield identically the better-known incompressible
turbulent boundary-layer equations. In this manner, the more extensive
knowledge for the incompressible turbulent boundary layer can, in theory,
be extended to the compressible-flow case of interest. Typical investigations
([56-62]) have achieved some measure of success in defining transformations
for the turbulent boundary layer. Coles ([61]) proposed an approach to the
transformation of the compressible turbulent boundary-layer equations
in which the compressible and the constant-density flows are assumed to
be related by three scaling parameters. The first relates the stream functions
of the two flows, the second is a multiplicative factor of the Dorodnitsyn–
Howarth scaling of the normal coordinate, and the third relates the streamwise coordinates of the two flows. An additional assumption pertaining to
the invariance of a Reynolds number characterizing the law of the wall
region of the boundary layer is necessary to complete the transformation.
This assumption, called the "substructure hypothesis" by Coles, provides
a substitute for the reference state utilized with many theoretical approaches.
Coles' transformation has been extended by Crocco ([60]) and modified as
well as applied to practical cases by Baronti and Libby ([62]). It is with the
analysis of Baronti and Libby that the remainder of this section is concerned.

Baronti and Libby modified Coles' substructure hypothesis (they
introduced a sublayer hypothesis) and applied the transformation technique
by point-by-point mapping of supersonic velocity profiles into the incompressible plane. It should be noted that the transformation theory is applicable only for two-dimensional or axisymmetrical flow with and without
heat transfer or streamwise pressure gradient. This analysis does not define

---

* The author is indebted to M. H. Bertram, A. M. Cary, Jr., and A. H. Whitehead, Jr.,
  for their contribution to this section.

completely the constant-density flow corresponding to the compressible case, since the velocity profiles, once transformed, correspond to some unknown $\bar{x}$-station in the constant-density flow.

Baronti and Libby employed the conventional incompressible equations for the universal velocity profile such that the boundary layer profile is composed of two distinct regions, a law-of-the wall region near the wall and a wake or velocity-defect region consisting of the major portion of the boundary layer. The equations governing each of these regions are, respectively,

$$\bar{u}/\bar{u}_\tau = f(\bar{\xi}) \qquad \text{and} \qquad (\bar{u} - \bar{u}_l)/\bar{u}_\tau = F(\bar{y}/\bar{\delta}, \bar{x})$$

where $\bar{u}_\tau = (\bar{\tau}_w/\bar{\varrho}_w)^{1/2}$ and $\bar{\xi} = \bar{y}\bar{u}_\tau/\bar{\nu}$. The law of the wall is conventionally expressed as:

$$\bar{u}/\bar{u}_\tau = \bar{\xi} \qquad\qquad 0 \leq \bar{\xi} \leq \bar{\xi}_f \text{ (sublayer)}$$

$$= A \ln b\bar{\xi} \qquad \bar{\xi}_f \leq \bar{\xi} \leq \bar{\xi}_l$$

where $\bar{\xi}_f$ and $\bar{\xi}_l$ are the values of $\bar{\xi}$ at the edge of the laminar sublayer and the outer limit of the region of application of the law of the wall, respectively. The coefficients A and b are 2.43 and 7.5, respectively, as taken from Clauser [63], so that $\bar{\xi}_f = 10.6$. The outer limit for the application of the law of the wall is taken as the end of the logarithmic portion of the boundary layer profile on a scale of $\bar{u}/\bar{u}_\tau$ plotted against $\bar{\xi}$. Simplified equations for the direct application of the analysis of Baronti and Libby to velocity profiles for compressible flow may be found in [64].

The process of applying the transformation theory through the law of the wall is actually an iterative one, since the value of the skin friction in the incompressible plane is necessary in order to transform the corresponding compressible velocity profile to the incompressible plane. In actual practice, the procedure is to assume values of the wall skin friction in an incompressible plane until acceptable agreement of the velocity profiles with the constant-density result is achieved. The success of the transformation may then be judged by observing how well the transformed velocity profile correlates with the incompressible results, and comparing the resulting compressible skin friction estimate with that measured or predicted by a reliable theory. Once the incompressible skin friction has been determined from the law-of-the-wall analysis, a comparison with the velocity-defect law is directly obtainable.

Baronti and Libby [62] applied the transformation to velocity profiles for compressible flows up to Mach 9 for adiabatic wall and moderate heat-transfer conditions. In general, their results indicated good correlation of the compressible velocity profiles for the law of the wall in the incompressible plane, and the values of skin friction resulting from the transformation compared well with those measured in most of the investigations cited. However, when a correlation was attempted with the velocity-defect law, the results indicated that a compressible velocity profile under a uniform flow would transform into the incompressible plane and show the characteristics of an incompressible velocity profile under the influence of a pressure gradient. Tennekes [65] has suggested that this discrepancy may be a result of a distortion of the velocity-defect region of the boundary layer by the Dorodnitsyn–Howarth density scaling of the normal coordinate.

Here, to reduce the compressible velocity profiles to the incompressible form, the same procedures were used as were used by Baronti and Libby, including the use of the Crocco relation to calculate the density integral through the boundary layer, but the range of Mach number and heat transfer is extended. Experimental velocity profiles were calculated using measured temperature profiles where available; otherwise, the Crocco relation was used. Illustrations of the correlation of the transformed compressible boundary-layer profiles according to the law of the wall with the classical incompressible results are shown in Fig. 28. Since C_f was not measured directly for most of the profiles presented, the skin-friction results obtained from the transformation for all the cases were normalized by the skin-friction coefficient predicted by the method of Spalding and Chi [40]. In each case, the Spalding–Chi prediction was based on the measured R_θ and T_w/T_t.

The transformation of profiles obtained on tunnel walls in nominal zero-pressure-gradient flow as shown in Fig. 28a provides a good correlation for Mach numbers from 2.5 to 8.* The skin-friction results from these profiles compare favorably with the Spalding–Chi predictions. For still higher Mach numbers, in the range 15–20, the profiles shown in Fig. 28b appear to correlate well with the incompressible results, but the extent of the logarithmic part of the law-of-the-wall region of the profile is small in comparison with the lower-Mach-number profiles.[†] An inspection of the

---

* Mach 2.49 and 4.44 profiles from unpublished measurements by Stallings and Couch in the Langley Unitary wind tunnel; Mach 6.0 and 6.8 from [47]; Mach 7.95 from unpublished measurements by Feller in Langley 18-in. variable-density wind tunnel; all with dp/dx essentially zero.

† Mach 15.6 profile from [66]; Mach 20.2 profile from [64]; Mach 18.4 profile from unpublished measurements by Clark, but a similar profile is in [67].

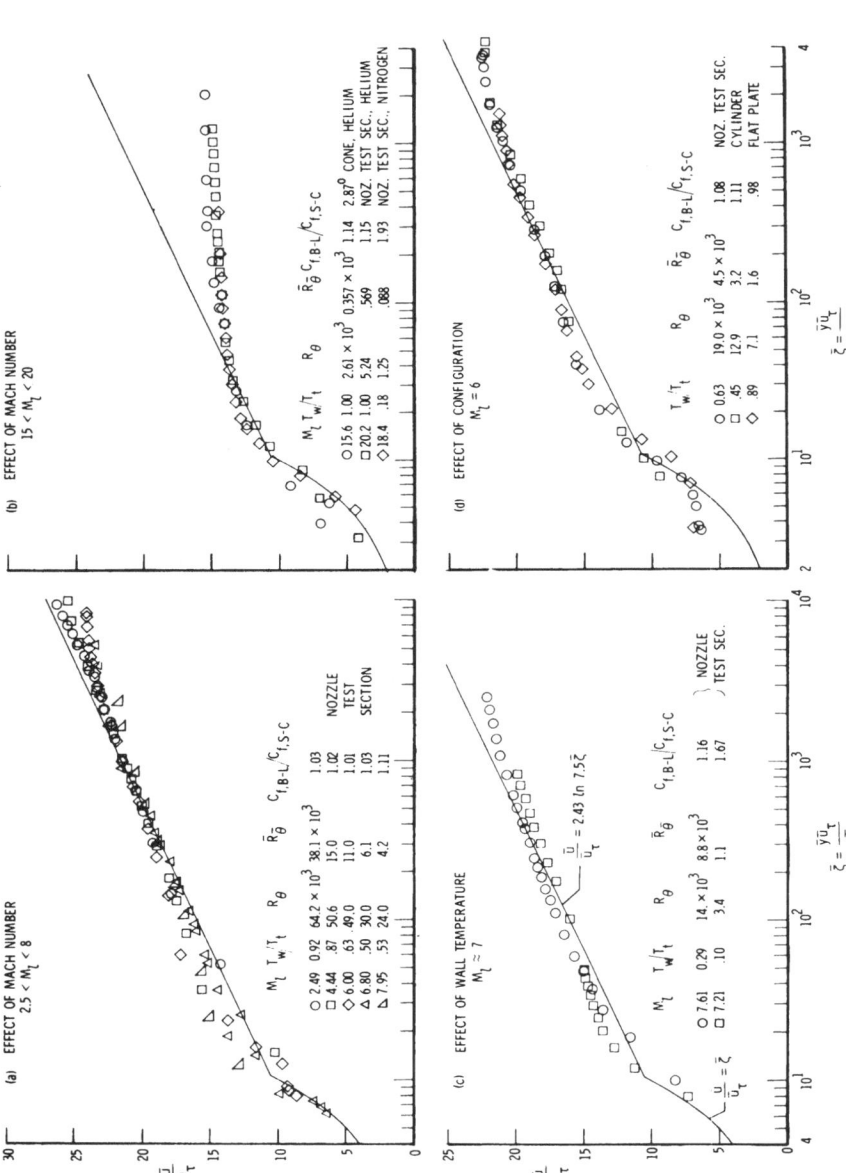

Fig. 28. Compressible turbulent velocity profiles in the incompressible plane according to the transformation of Coles [61] and of Baronti and Libby [62].

compressible velocity profiles indicates that, in general, as Mach number increases, the laminar sublayer thickness as well as the extent of the wake or velocity-defect becomes larger. As a result, there appears to be a corresponding decrease in the extent of the logarithmic law-of-the-wall region. Since the wall shear obtained from the transformation is dependent upon a curve fit in the logarithmic law-of-the-wall region, a physical limit of the application of the transformation in the present form may thus exist. However, for profiles with thick laminar sublayers, for which sublayer velocity measurements are accurate, the transformation can be applied directly in conjunction with the sublayer part of the law of the wall. The skin-friction results from the transformation of the high-Mach-number profiles show more deviation than those for the lower-Mach-number profiles. It should be remarked that the nitrogen profile presented in Fig. 28b is believed to be transitional by the experimenters ([67]). However, note that there is no particular difference between this transformed profile and the other presented at considerably higher values of R_θ. Note the low value of T_w/T_t, however.

Most of the profiles presented thus far are for moderate values of wall-temperature ratio. Transformed profiles for Mach numbers near 7 ([68]) and low values of wall-temperature ratio are given in Fig. 28c. Correlation is as good as was found for the previous profiles in Fig. 28a, but the resulting values of skin friction are significantly greater than Spalding–Chi predictions.

An illustration of the effect of previous history of the boundary layer on the results of the transformation is shown in Fig. 28d. The transformed profiles correlate nicely, and the skin-friction results compare favorably with the prediction of Spalding–Chi even though each of the boundary layers developed under different conditions.

A compilation of the skin-friction results obtained from the transformation technique is presented in Fig. 29a. The skin-friction results from the transformation are referenced to the skin friction predicted by the Spalding–Chi method and presented as a function of the ratio of wall temperature to total temperature for each particular case. The data include all the experiments analyzed by Baronti and Libby ([62]), results cited in Fig. 28, and additional results ([69–72]). It has been shown in several investigations [e.g., ([47]) and in the first part of this section] that the method of Spalding and Chi can be expected to give accurate skin-friction predictions on flat plates and cones at least up to Mach 9 and over the entire range of T_w/T_t for the data in Fig. 29.

In general, the skin-friction results from the transformation appear to be consistently higher than those predicted by the Spalding–Chi method.

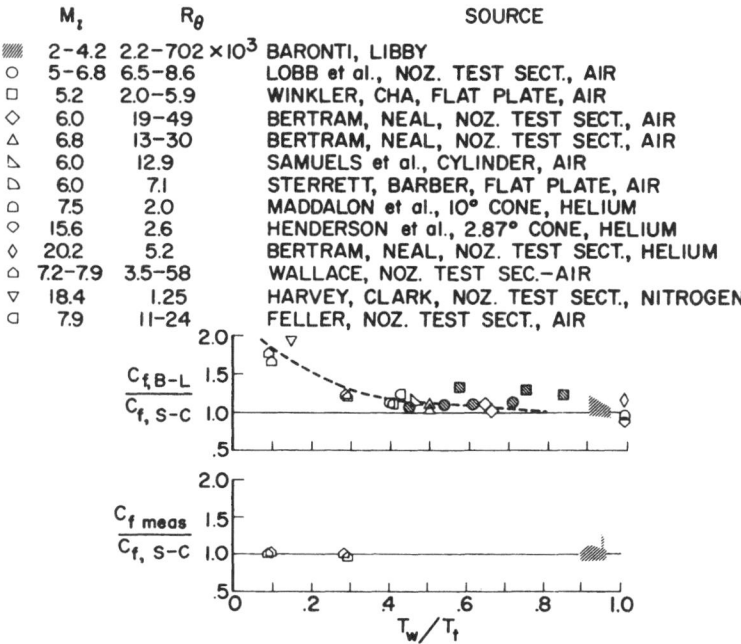

	M_t	R_θ	SOURCE
▨	2–4.2	2.2–702 ×10³	BARONTI, LIBBY
○	5–6.8	6.5–8.6	LOBB et al., NOZ. TEST SECT., AIR
□	5.2	2.0–5.9	WINKLER, CHA, FLAT PLATE, AIR
◇	6.0	19–49	BERTRAM, NEAL, NOZ. TEST SECT., AIR
△	6.8	13–30	BERTRAM, NEAL, NOZ. TEST SECT., AIR
◁	6.0	12.9	SAMUELS et al., CYLINDER, AIR
▷	6.0	7.1	STERRETT, BARBER, FLAT PLATE, AIR
◠	7.5	2.0	MADDALON et al., 10° CONE, HELIUM
♢	15.6	2.6	HENDERSON et al., 2.87° CONE, HELIUM
◇	20.2	5.2	BERTRAM, NEAL, NOZ. TEST SECT., HELIUM
◠	7.2–7.9	3.5–58	WALLACE, NOZ. TEST SEC.–AIR
▽	18.4	1.25	HARVEY, CLARK, NOZ. TEST SECT., NITROGEN
◻	7.9	11–24	FELLER, NOZ. TEST SECT., AIR

Fig. 29. Skin-friction results; Baronti–Libby transformation.

Although the overprediction is in the 10% range for adiabatic and moderately-cooled walls, the error is large for extreme cooling conditions. Wallace ([68]) obtained direct skin-friction measurements on the nozzle wall at the same locations and for the same flow conditions as the profile data. The measured skin-friction results shown in Fig. 29b are in good agreement with the Spalding–Chi predictions, as are the results from other investigations in which direct measurements of skin friction were made. It thus appears that the transformation as applied is not generally applicable even for the logarithmic portion of the law of the wall. The discrepancy in wall shear at low wall-to-total temperature ratios does not appear to be a function of Mach number or Reynolds number, and is assumed to result from a deficiency in the transformation theory itself.

Nonetheless, the partial success of the transformation theory for the moderate-to-high wall-temperature ratios is encouraging, and investigators are continuing their work. The potential rewards are high, for if successful, the considerable body of well-known and dependable incompressible flow results can be brought to bear on the often intractable compressible-flow turbulent boundary-layer problem.

NOTATION

A	constant
C_∞	$\mu_r T_\infty / \mu_\infty T_r$
C_w	$\mu_w T_\infty / \mu_\infty T_w$
c	circumference
c_1	reference circumference $= 48$ in.
C_D	drag coefficient
C_F	mean turbulent skin-friction coefficient
D	nose diameter
G	see Eq. (10)
M	Mach number
$\bar{N}$	exponent in local power-law fit to pressure
n	exponent in power-law variation of surface pressure with x
p	pressure
Re	Reynolds number
Re_c	critical Reynolds number
Re_t	Reynolds number for end of transition
$\text{Re}_{t,B}$	Reynolds number for beginning of transition
r	radius
St	Stanton number
T	temperature
U	streamwise velocity
x	distance along axis
y	distance normal to axis

Greek Letters

α	angle of attack
β	Falkner–Skan velocity gradient parameter, or characteristic frequency in dimensionless frequency parameter of boundary layer stability theory
γ	ratio of specific heats
δ	plane flow deflection angle or boundary layer thickness
δ^*	boundary layer displacement thickness
ε	$(\gamma - 1)/(\gamma + 1)$
θ	boundary layer momentum thickness
μ	dynamic viscosity
ϱ	density

τ thickness ratio

$\bar{\chi}$ viscous interaction parameter $= M^3(C)^{1/2}(R)^{-1/2}$

Ω see Eq. (13)

Subscripts

0	stagnation conditions
∞	free stream
b	nose bluntness induced
c	critical layer
e	edge of boundary layer
i	incompressible
N	nose
r	reference condition
R	recovery
SC	sharp cone conditions
t	leading-edge thickness
w	wall

REFERENCES

1. L. Prandtl, "The Mechanics of Viscous Fluids," in: *Aerodynamic Theory*, Vol. III, W. F. Durand, ed. (1934), pp. 89–90; reprinted by Dover Publications, New York (1963).
2. A. Henderson, Jr., R. D. Watson, and R. D. Wagner, Jr., "Fluid Dynamic Studies to $M = 41$ in Helium," *AIAA J.* **4** (12) (December 1966).
3. C. H. Lewis and J. D. Whitfield, "Theoretical and Experimental Studies of Hypersonic Viscous Effects," AGARDograph 97 (May 1965).
4. L. Lees, "Hypersonic Flow," in: *Fifth International Aeronautical Conference*, Institute of Aeronautical Sciences, New York (1955), pp. 241–276.
5. W. D. Hayes and R. F. Probstein, *Hypersonic Flow Theory*, Academic Press, New York and London (1959), Chapter 9.
6. H. K. Cheng and A. J. Pallone, "Inviscid Leading-Edge Effect in Hypersonic Flow," *J. Aeron. Sci.* **23**, 700–702 (1956).
7. J. Lukasiewicz, "Hypersonic Flow-Blast Analogy," Arnold Engineering Development Center, AEDC-TR-61-4 (June 1961).
8. M. H. Bertram and T. A. Blackstock, "Some Simple Solutions to the Problem of Predicting Boundary-Layer Self-Induced Pressures," NASA TN D-798 (1961).
9. H. K. Cheng, J. G. Hall, T. C. Golian, and A. Hertzberg, "Boundary-Layer Displacement and Leading-Edge Bluntness Effects in High-Temperature Hypersonic Flow," *J. Aerospace Sci.* **28**, 353–381 (May 1961).
10. J. H. Kemp, Jr., "Hypersonic Viscous Interaction on Sharp and Blunt Inclined Plates," AIAA paper No. 68-720 presented at the AIAA Fluid and Plasma Dynamics Conference, Los Angeles, California, June 24–26, 1968.

11. H. Mirels and P. R. Thornton, "Effect of Body Perturbations on Hypersonic Flow over Slender Power-Law Bodies," NASA TR R-45 (1959).

12. S. C. Traugott, "A Transonic Experiment at Hypersonic Speed," *AIAA J.* **2** (9), 1521–1527 (September 1964).

13. H. Schlicting, *Boundary Layer Theory*, Pergamon Press, New York (1955).

14. C. B. Cohen and E. Reshotko, "Similar Solutions for the Compressible Laminar Boundary Layer with Heat Transfer and Pressure Gradient," NACA TR 1293 (1956).

15. A. M. O. Smith, "Rapid Laminar Boundary-Layer Calculations by Piecewise Application of Similar Solutions," *J. Aeron. Sci.* **23** (10), 901–912 (October 1956).

16. H. A. Stine and K. Wanlass, "Theoretical and Experimental Investigation of Aerodynamic-Heating and Isothermal Heat-Transfer Parameters on a Hemispherical Nose with Laminar Boundary Layer at Supersonic Mach Numbers," NACA TN 3344 (1954).

17. C. F. Dewey, Jr. and J. F. Gross, "Exact Similar Solutions of the Laminar Boundary Layer Equations," in: *Advances in Heat Transfer*, Vol. 4, J. P. Hartnett and T. F. Irvine, Jr., eds., Academic Press, New York (1967).

18. K. Stewartson, "Correlated Incompressible and Compressible Boundary Layers," *Proc. Roy. Soc. (London)*, Ser. *A*, **200** (A1060), 84–100 (December 22, 1949).

19. M. H. Bertram and W. V. Feller, "A Simple Method for Determining Heat Transfer, Skin Friction, and Boundary Layer Thickness for Hypersonic Laminar Boundary Layer Flows in a Pressure Gradient, NASA Memo 5-24-59L (1959).

20. Davis H. Crawford, "Applications of Similar Solutions for Calculation of Laminar Boundary-Layer Characteristics in the Presence of a Pressure Gradient," *AIAA J. (Tech. Notes)* **5** (4), 799–801 (April 1967); NASA TN D-4367.

21. P. F. Brinich, "Effect of Leading Edge Geometry on Boundary Layer Transition at Mach 3.1," NACA TN-3659 (March 1956).

22. H. T. Nagamatsu, B. C. Graber, R. E. Sheer, Jr., "Critical Layer Concept Relative to Hypersonic Boundary Layer Stability," General Electric Rept. No. 66-C-192 (June 1966).

23. D. V. Maddalon, and A. Henderson, Jr., "Boundary Layer Transition at Hypersonic Mach Numbers," *AIAA J.* **6** (3), 424–431 (March 1968).

24. E. J. Softley, B. C. Graber, and R. E. Zempel, "Transition of the Hypersonic Boundary Layer on a Cone, Part I—Experiments at $M_\infty = 12$ and 15," General Electric Report R67SD39 (November 1967).

25. C. P. Stainback, "Effects of Unit Reynolds Number, Nose Bluntness, Angle of Attack, and Roughness on Transition on a 5° Half-Angle Cone at Mach 8," NASA TN D-4961, January 1969.

26. Kenneth F. Stetson, and H. George Rushton, "A Shock Tunnel Investigation of the Effects of Nose Bluntness, Angle of Attack and Boundary Layer Cooling on Boundary Layer Transition at a Mach Number of 5.5," AIAA paper No. 66-495 presented at Aerospace Meeting, Los Angeles, June 27–29, 1966.

27. H. Hidalgo, "The Influence of Aerodynamics and Reentry Heating on Reusable Space Launch Vehicles," IDA Research Paper P-230 (June 1966).

28. D. V. Maddalon and A. Henderson, Jr., "Hypersonic Transition Studies on a Slender Cone at Small Angles of Attack," *AIAA (Tech. Notes)* **6** (1), 176 (January 1968).

29. L. M. Mack, "The Stability of the Compressible Laminar Boundary Layer According to a Direct Numerical Solution," AGARDograph 97 (1965), pp. 483–501.

30. J. Laufer, J. E. Ffowcs-Williams, and S. Childress, "Mechanism of Noise Generation in the Turbulent Boundary Layer," AGARDograph 90 (November 1964).

31. S. R. Pate and C. J. Schueler, "Effects of Radiated Aerodynamic Noise on Model Boundary Layer Transition in Supersonic and Hypersonic Wind Tunnels," AEDC-TR-67-236 (March, 1968).

32. G. G. Mateer and H. K. Larson, "Unusual Boundary Layer Transition Results on Cones in Hypersonic Flow," *AIAA Preprint* No. 68-40 (January 1968).

33. J. L. Potter, *AIAA J.* 6 (10), 1907–1911, (October, 1968).

34. Eli Reshotko, "Stability Theory as a Guide to the Evaluation of Transition Data," General Electric Technical Information Series, No. 67-SD-330, (October 1967).

35. A. L. Nagel, "Compressible Boundary Layer Stability by Time Integration of the Navier–Stokes Equations and an Extension of Emmon's Transition Theory to Hypersonic Flow," Boeing Scientific Research Lab., Rept. 119 (D1-82-D655) (September 1967).

36. A. M. Cary, Jr., "Turbulent Boundary Layer Heat Transfer and Transition Measurements for Cold Wall Conditions at $M = 6$," *AIAA J.* (*Tech. Notes*) 6 (5), 958, 959 (May 1968).

37. R. W. Higgins and C. C. Pappas, "An Experimental Investigation of the Effect of Surface Heating on Boundary Layer Transition on a Flat Plate in Supersonic Flow," NACA TN-2351 (April 1951).

38. B. E. Richards and J. J. Stollery, "Further Experiments on Transition Reversal at Hypersonic Speeds," *AIAA J.* 4 (12), 2224–2226 (December 1966).

39. R. E. Deem, C. R. Erickson, and J. S. Murphy, "Flat Plate Boundary-Layer Transition at Hypersonic Speeds," FDL-TDR-64-129, AF Flight Dynamics Laboratory, Wright-Patterson Air Force Base, Ohio (October 1964).

40. D. B. Spalding and S. W. Chi, "The Drag of a Compressible Turbulent Boundary Layer on a Smooth Flat Plate with and without Heat Transfer," *J. Fluid Mech.* 18 (I), 117–145 (January 1964).

41. D. B. Spalding and S. W. Chi, "Skin Friction Exerted by a Compressible Fluid Stream on a Flat Plate," *AIAA J.* (*Tech. Notes*) 1 (9), 2160 (September 1963).

42. E. R. Van Driest, "50 Years of Boundary Layer Theory," H. Borfler and W. Tollmien, eds., Braunschweig, F. Vieweg u. Sohn (1955), p. 257.

43. R. E. Wilson, *J. Aeron. Sci.* 17, 585 (1950).

44. S. S. Kutafeladze and A. I. Leontev, "Discussion of Heat and Mass Transfer," *Akad. Nauk BSSR, Minsk*, 1 (1961).

45. L. Neal, Jr. and M. H. Bertram, "Turbulent-Skin-Friction and Heat Transfer Charts Adopted from the Spalding and Chi Method," NASA TN D-3969 (May 1967).

46. Richard D. Banner, A. E. Kuhl, and R. D. Quinn, "Preliminary Results of Aerodynamic Heating Studies on the X-15 Airplane," NASA TM X-638 (1962).

47. M. H. Bertram and L. Neal, Jr., "Recent Experiments in Hypersonic Turbulent Boundary Layers," NASA TM X-56335, presented at the Agard Specialists' Meeting on Recent Developments in Boundary Layer Research Sponsored by the Fluid Dynamics Panel of AGARD, Naples, Italy, May 10–14, 1965.

48. M. H. Bertram, A. M. Cary, Jr., and A. H. Whitehead, Jr., "Experiments with Hypersonic Turbulent Boundary Layers on Flat Plates and Delta Wings," presented at AGARD Specialists' Meeting on Hypersonic Boundary Layers and Flow Fields, London, England, May 1–3, 1968.

49. A. L. Nagel, H. D. Fitzsimmons, and L. B. Doyle, "Analysis of Hypersonic Pressure and Heat Transfer Tests on Delta Wings with Laminar and Turbulent Boundary Layer," NASA CR-535 (1966).

50. R. J. Monaghan, "On the Behavior of Boundary Layers at Supersonic Speeds," in: *Fifth International Aeronautical Conference*, Rita J. Turino and Caroline Taylor, eds., Institute of Aeronautical Sciences, Los Angeles, California (1955), pp. 277–315.

51. E. R. G. Eckert, "Engineering Relations for Friction and Heat Transfer to Surfaces in High Velocity Flow," *J. Aeron. Sci.* **22** (8), 585–587 (August 1955).

52. S. C. Sommer and P. J. Short, "Free-Flight Measurements of Turbulent Boundary Layer Skin Friction in the Presence of Severe Aerodynamic Heating at Mach Numbers from 2.8 to 7.0," NACA TM-3391 (1955).

53. E. M. Winkler and M. H. Cha, "Investigation of Flat Plate Hypersonic Turbulent Boundary Layers with Heat Transfer at a Mach Number of 5.2," NAVORD Rept. 5631 (September 1959).

54. D. Coles, "Measurements of Turbulent Friction on a Smooth Flat Plate in Supersonic Flow," *J. Aeron. Sci.* **21** (7), 433–448 (July 1954); also see Jet Propulsion Lab. Rept. No. 20-71 (1953).

55. B. E. Richards, "Transitional and Turbulent Boundary Layers on a Cold Flat Plate in Hypersonic Flow," *Aeron. Quart.* **18** (3), 237–258 (August 1967).

56. O. R. Burggraf, "The Compressibility Transformation and Turbulent-Boundary-Layer Equations," *J. Aerospace Sci.* **29** (4) 434–439 (April 1962).

57. Arthur Mager, "Transformation of the Compressible Turbulent Boundary Layer," *J. Aeron. Sci.* **25** (5), 305–311 (May 1968).

58. D. A. Spence, "Velocity and Enthalpy Distributions in the Compressible Turbulent Boundary Layer on a Flat Plate," *J. Fluid Mech.* **8** (3), 368–387 (July 1960).

59. D. E. Coles, "Measurements in the Boundary Layer on a Smooth Flat Plate in Supersonic Flow; III. Measurements in a Flat-Plate Boundary Layer at the Jet Propulsion Laboratory," Jet Propulsion Laboratory Rept. No. 20-71 (Contract No. DA-04-495-Ord 18), Calif. Inst. of Tech. (June 1, 1953).

60. L. Crocco, "Transformations of the Compressible Turbulent Boundary Layer with Heat Exchange," *AIAA J.* **1** (12), 2723–2731 (December 1963).

61. D. E. Coles, "The Turbulent Boundary Layer in a Compressible Fluid," U. S. Air Force Project Rand Rept. R-403-PR, The Rand Corp. (September 1962).

62. P. O. Baronti and P. A. Libby, "Velocity Profiles in Turbulent compressible Boundary Layers," *AIAA J.* **4** (2), 193–202 (February 1966).

63. F. H. Clauser, "Turbulent Boundary Layers in Adverse Pressure Gradients," *J. Aeron. Sci.* **21**, 91–103 (1954).

64. R. D. Watson and A. M. Cary, Jr., "The Transformation of Hypersonic Turbulent Boundary Layers to Incompressible Form," *AIAA J.* (*Tech. Notes*) **6** (6), 1202–1203 (June 1967).

65. H. Tennekes, "Law of the Wall for Turbulent Boundary Layers in Compressible Flow," *AIAA J.* **5** (3), 489–492 (March 1967).

66. A. Henderson, Jr., R. S. Rogallo, W. C. Woods, and C. R. Spitzer, "Exploratory Hypersonic Boundary-Layer Transition Studies," *AIAA J.* (*Tech. Notes*) **3** (7) 1363–1364 (July 1965).

67. F. L. Clark, J. C. Ellison, and C. B. Johnson, "Recent Work in Flow Evaluation and Techniques of Operation for the Langley Hypersonic Nitrogen Facility in Hyper-velocity Techniques," in: *Advanced Experimental Techniques for Study of Hyper-*

velocity Flight, Vol. 1, 5th Symposium, University of Denver, Denver, Colorado (1967), pp. 347–373.

68. J. E. Wallace, "Hypersonic Turbulent Boundary Layer Studies at Cold Wall Conditions," in: *Proceedings of 1967 Heat Transfer and Fluid Mechanics Institute*, Stanford University Press, Stanford, California (1967).

69. K. R. Lobb, E. M. Winkler, and J. Persh, "NOL Hypersonic Tunnel No. 4 Results; VII: Experimental Investigation of Turbulent Boundary Layers in Hypersonic Flow," NAVORD Rept. 262 (March 1955).

70. R. D. Samuels, J. B. Peterson, Jr. and J. B. Adcock, "Experimental Investigation of the Turbulent Boundary Layer at a Mach Number of Six with Heat Transfer at High Reynolds Numbers," NASA TN D-3857 (1967).

71. J. R. Sterrett and J. B. Barber, "A Theoretical and Experimental Investigation of Secondary Jets in a Mach 6 Freestream with Emphasis on the Structure of the Jet and Separation ahead of the Jet," presented at the Separated Flows Specialists' Meeting Fluid Dynamics Panel–AGARD, Brussels, Belgium, 1966.

72. D. V. Maddalon, R. S. Rogallo, and A. Henderson, Jr., "Transition Measurements at Hypersonic Mach Numbers," *AIAA J.* (*Tech. Notes*) **5** (3), 590–591 (March 1967).

73. M. H. Bertram, "Hypersonic Laminar Viscous Interaction Effects on the Aerodynamics of Two-Dimensional Wedge and Triangular Planform Wings," NASA TN D-3523 (1967).

74. V. M. Falkner and S. W. Skan, "Some Approximate Solutions of the Boundary Layer Equations," *Phil. Mag.*, **12**, 865 (1931).

75. P. F. Holloway and J. R. Sterrett, "Effect of Controlled Surface Roughness on Boundary Layer Transition and Heat Transfer at Mach Numbers of 4.8 and 6.0," NASA TN D-2054.

76. L. S. G. Kovasznay, "Turbulence in Supersonic Flow," *J. Aeron. Sci.*, **20** (10) 657–674 (October, 1953).

77. J. Laufer, "Factors Affecting Transition Reynolds Numbers on Models in Supersonic Wind Tunnels," *J. Aeron. Sci.* **21** (7) 497–498 (July, 1954).

Chapter 4

HYPERSONIC GAS DYNAMICS OF SLENDER BODIES

H. K. Cheng

Professor, Department of Aerospace Engineering
University of Southern California

INTRODUCTION

This article is devoted to a study of current work on theoretical gas dynamics associated with hypersonic flow past slender bodies. From the viewpoint of nonlinear fluid mechanics, the problems are attractive in that they yield to asymptotic analysis. The many flow regimes and distinct layers, on the other hand, make challenging studies of singular perturbations. The relevance of the present study to the engineering problems of hypersonic flight may be found in the discussions in ([1-5]).

The material is arranged in three sections. Section 1 deals with inviscid hypersonic flows and related problems. It presents a theory on entropy wake, and offers a critique of recent work on blast-wave analogy; some current work on strong blowing and on minimizing drag will also be noted. In Section 2, the problem of the outer-edge singularity of a hypersonic boundary layer and its relation to external vorticity and shock-heating effects is delineated. A number of recent works on needle-like bodies in viscous hypersonic flow, and a formulation of the three-dimensional interaction problem are reviewed in Section 3.

The scope of the article is necessarily restricted, and work on flow stability, turbulent shear layer, nonequilibrium flow chemistry, and other topics in related fields can not be included. Regretably, two highly interesting developments have also been omitted to conserve space, namely, the problem of local strong interaction and separation in hypersonic flow, and the slender-body problem in the merged-layer and other rarefied gas dynamic regimes.*

---

* A fuller version of this chapter is given in Report USCAE No. 108 ([6]), in which some of the discussion on new work is substantially augmented.

1. INVISCID FLOWS AND RELATED PROBLEMS

The first two subsections are concerned primarily with the leading-edge bluntness effects and the related problem of an intense explosion. In Section 1.1, we discuss the entropy layer associated with the bluntness and the asymptotic theory underlying the blast-wave analogy. In Section 1.2, a theory of gas dynamic explosion under the influence of a driving piston is presented. This problem corresponds to that of a slightly blunted slender body. The theory deals with the structure of the entropy wake and its interaction with the shock layer.

A clarification in terminology is desirable at this juncture. Following Hayes and Probstein's more recently adopted definitions ([3]), the term "entropy layer" will be used to denote a thin region next to the body where the blast-wave analogy gives incorrect descriptions of the entropy, whereas the term "entropy wake" will be used to denote the high-entropy field which is generated by the portion of the stronger shock upstream, or at an earlier time, irrespective of the validity of blast-wave analogy. The term "entropy layer" used in the earlier works of Chernyi ([1]) and Cheng et al. ([8]) will therefore be identified with the entropy wake in this article.

1.1. Entropy and Speed Defects

The blast-wave analogy for the steady hypersonic flow downstream of a blunt nose is not a (uniformly) valid one near the aftbody surface, because the shock strength is initially infinite in the blast-wave solution, which does not correctly describe the finite entropy field in a hypersonic flow. With this in mind, a few words on the background of the problem may be helpful. Sychev ([9]) and Yakura ([10]) analyzed the flow fields behind a power-law shock $y_s \propto x^{2/(3+\nu)}$ (where $\nu = 0$ and 1 for plane and axisymmetrical flows, respectively) in a uniform stream for $M_\infty \to \infty$ (see Fig. 1). Their solutions to this "inverse problem," after accounting properly for the entropy in the entropy layer, show that a blunt-nosed body supporting such a shock is one growing downstream (at large x) like

$$y_b \propto x^{2/(3+\nu)\gamma} \tag{1}$$

where γ is the specific-heat ratio. Since this body appears to grow at a rate almost as fast as the blast-wave radius $y_s \propto x^{2/(3+\nu)}$, it became questionable whether the asymptotic flow field over a flat plate or straight rod (in a direct problem) can be correctly represented by the blast-wave analogy.

The defects resulting from the geometry of the nose shock in the direct

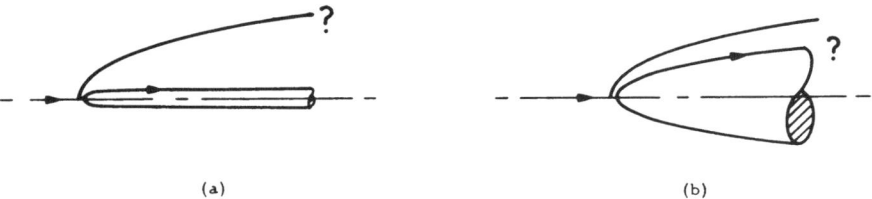

(a)   (b)

Fig. 1. (a) Direct and (b) inverse problems in the study of asymptotic solutions at large
distances on a straight aftbody in the hypersonic limit.

problem was studied by Freeman [11] for a sphere-cylinder (Fig. 1) within
the framework of the (Newtonian) thin shock-layer theory, namely,
$M_\infty\beta \gg 1$ and $\varepsilon \equiv (\gamma - 1)/2\gamma \ll 1$, where β is the shock angle. In Freeman's theory, the blast-wave solution in the Newtonian limit [8,12] is indeed
recovered far downstream. The shock layer in this instance may be regarded
as a continuation of the Newtonian "free layer." We will come back to
Freeman's theory in a later stage of our discussion. Vaglio-Laurin [13]
and Guiraud and co-workers [14,15] independently attack the direct problem
for large $M_\infty\beta$ and a large x without assuming $\varepsilon \ll 1$, stipulating an algebraic form for the perturbed shock

$$y_s \sim Ax^{2/(3+\nu)}[1 + \alpha x^{-k\sigma} + \cdots] \qquad (2)$$

where $\sigma \equiv 2(1 + \nu)(\gamma - 1)/(3 + \nu)\gamma$, and k, A, and α are constants not
a priori known. Following Yakura [10] and Van Dyke [16], these authors
assumed an outer region where the HSDT holds as a leading approximation,
and an inner region in which the streamlines come from the blunter part,
i.e., $dy_s/dx = O(1) \neq 0$, of the shock. In terms of a dimensionless shock
radius y_s based on the nose radius a, and a stream function ψ based on
$\varrho_\infty u_\infty^2 a^{1+\nu}$, the two regions (at large x) are $\psi = O(1)$ and $\psi = O(y_s^{1+\nu})$,
corresponding to the two forms of independent variable used—namely,

$$\psi \quad \text{and} \quad \Omega \equiv \psi/\psi_s \propto \psi/y_s^{1+\nu} \qquad (3)$$

As far as procedure is concerned, the analyses of Vaglio-Laurin [13] and
Guiraud et al. [14] could both be viewed as an inverse problem like Yakura's
or Van Dyke's [see [16]], being carried out to the next order for large x,
inasmuch as the shock form has been prescribed by Eq. (2).

Intuition would suggest the constant k of Eq. (2) to be unity (and α
to be negative), since this is what is required to induce a streamline deflection
near the axis so as to cancel off the Sychev–Yakura body of Eq. (1).

With $k = 1$, the outer solution gives a streamline behavior near the axis

$$y \sim Ax^{2/(3+\nu)}\{\Omega^{(\nu-1)/\gamma} + \text{const } x^{-\sigma} \ln \Omega + \cdots\}$$

$$\sim Ax^{2/(3+\nu)\gamma}\left\{\psi^{(\nu-1)/\gamma} + \text{const}\left[\ln \psi - \frac{2(1+\nu)}{3+\nu} \ln x\right] + \cdots\right\} \quad (4)$$

However, matching of Eq. (4) with the inner solution is impossible because the term associated with $\ln x$ is absent in the corresponding outer limit of the inner solution. This fact was overlooked in the earlier study by Vaglio-Laurin [13], was noticed by Guiraud, and subsequently confirmed by Messiter [17]. Thus Guiraud and Messiter assume $k = 2$ and obtain the perturbation solution, including the value of α in Eq. (2), but invoke an identity

$$\text{Finite Part} \int_0^\infty (\sin \beta)^{2/\gamma} y_s^\nu \, dy_s = 0 \quad (5)$$

which, without a proof, would impose physically a rather stringent requirement on the shock shape, and hence the upstream aftbody shape.

Interestingly, in a domain corresponding to (1) $\gamma \to 1$ and (2) $x \to \infty$, Guiraud recovers from his result the Newtonian result of Freeman. The agreement may not be, however, a meaningful check on the validity of Guiraud's result, because it is seen readily from the exposition of Guiraud et al. [14] that the integral of Eq. (5) always occurs as a product with $(\gamma - 1)/(\gamma + 1)$. Hence in the limit $\gamma \to 1$, the assumption of Eq. (5) may not be necessary.

The prospect of $k = 1$ for the flat plate and the rod, on the other hand, has not been strictly closed. The difficulty with $k = 1$ may have been overcome through the introduction of an intermediate (third) layer, as suggested by Vaglio-Laurin [see the comment at the end of [14]]. It must be pointed out that for $k = 1$, regardless of whether or not it is the solution to the direct problem under study, the inner and outer regions never overlap each other. This observation follows from Eq. (4), which gives a region of nonuniformity

$$\frac{x^\sigma}{\ln x} \Omega^{(\gamma-1)/\gamma} \propto \frac{\psi^{(\gamma-1)/\gamma}}{\ln x} = O(1) \quad (6a)$$

For this region, use of

$$\zeta = \text{const} \mid x^\sigma/\ln \Omega \mid^{\nu/(\gamma-1)} \Omega \quad (6b)$$

as one of the independent variables is appropriate. In addition, the study should not overlook the possibility of a shock with the form

$$y_s \approx A x^{2/(3+\nu)} [1 + \alpha(x^{-\sigma}/\ln x) + \cdots] \tag{7}$$

The foregoing discussion is concerned with straight aftbody. As is well known from HSDT, under infinite shock strength, a power-law body $y_b \propto x^{m_b}$ will support a similar shock $y_s \propto x^{m_s}$ with $m_s = m_b$, provided $m_b > 2/(3 + \nu)$ [7,18]. The HSDT does not furnish meaningful solutions to $m_b \lesssim 2/(3 + \nu)$. Freeman [19] proposes that the effect of the power-law body with $m_b \lesssim 2/(3 + \nu)$ at large x may be treated as a perturbation of a blast-wave solution $[m_s = 2/(3 + \nu)]$. Freeman's idea is illustrated in Fig. 2.

Inasmuch as perturbation analyses [13,19], including the interesting recent study by Stewartson and Thompson [20], have not proven conclusive, it may be more fruitful to study the problem first from the Newtonian viewpoint, as was initiated by Freeman [11]. Although the procedure employed by Freeman [11] may be regarded as *ad hoc*, a formal expansion procedure, similar to that shown in Section 1.2, is applicable. From the field structure and higher Newtonian approximation, the proper form for Eq. (2) may then be inferred. Freeman's model and principal results for a sphere-cylinder will be discussed below. Certain forms of the results are due, in part, to Guiraud *et al.* [14].

Beginning with the shock layer on a sphere, Freeman's picture en-

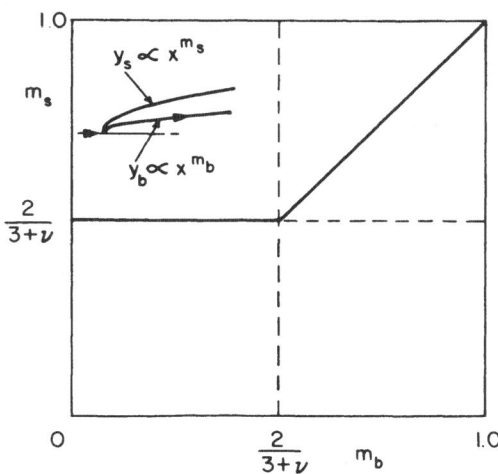

Fig. 2. Shock exponent plotted against body exponent. After Freeman [11].

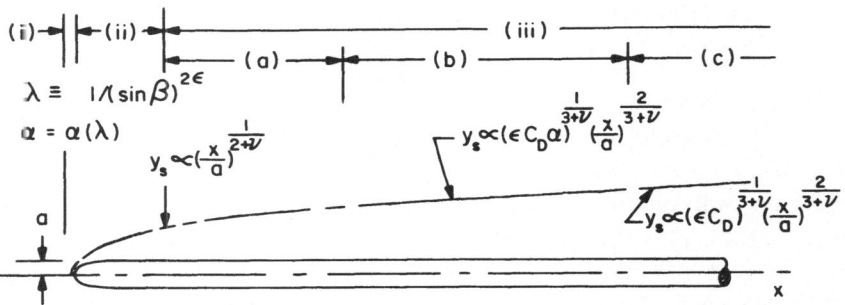

Fig. 3. The transition pattern from the attached shock layer to the Newtonian blast wave for a sphere-cylinder. After Freeman [11].

compasses the free layer, two distinct transitional regions, and the HSDT region in the Newtonian limit ($M_\infty\beta \to \infty$, $\gamma \to 1$). Figure 3 illustrates such construction. Here (i) is the classical shock layer, (ii) the free layer, and (iii) the transition to Newtonian blast wave. The last region has three subdivisions: (a) incipient transition, (b) blast-wave-like transition, (c) Newtonian blast wave.

Freeman [11] was concerned primarily with the x-dependences of the shock and the pressure. The nonuniformity of the Newtonian approximation for the free layer is first found in part (a) of region (iii) (Fig. 3), where $p/p_\infty u_\infty^2$ under the free layer is found to change from an order $1/\exp|\,O(1/\varepsilon)\,|$ to order ε^2. Characterizing the transition pattern is the *slow* variable

$$\lambda \equiv 1/(\sin\beta)^{2\varepsilon} \tag{8}$$

where β is the local shock angle. Region (ii) and parts (a), (b), and (c) of region (iii) are delimited by the range of λ as $\lambda - 1 = 0(\varepsilon)$, $\varepsilon \ll \lambda - 1 \ll 1$, $1 < \lambda = 0(1)$, and $1 \ll \lambda$, respectively.

It suffices to study the equation governing the shock for region (iii), where $\beta \ll 1$. The simplest way of obtaining the results is to begin with the integral identity under the Newtonian approximation:

$$D(x) = 2\pi^\nu \int_{y_b}^{y_s} \left(\frac{p}{\gamma-1} + \varrho\,\frac{v^2}{2}\right) y^\nu\,dy - \frac{\varrho_\infty u_\infty}{1+\nu} \int_0^{y_s} \frac{(u-u_\infty)^2}{u}\,d\psi$$
$$D_N \sim \frac{2\pi^\nu}{1+\nu}\,\frac{p_b}{\gamma-1}\,(y_s^{1+\nu} - y_b^{1+\nu}) - \pi^\nu \varrho_\infty u_\infty \int_0^{y_s} \frac{(u-u_\infty)^2}{u}\,y_*^\nu\,dy_* \tag{9a}$$

where $d\psi = (2\pi y_*)^\nu\,dy_*$ and the asymptotically correct substitution $D(x)$

$\approx D_N$ has been used. Consistent with this is an approximate Bernoulli equation

$$u^2 = u_\infty^2 \left[1 - \left(\frac{p}{\varrho_\infty u_\infty^2} \right)^{2\varepsilon} (\sin \beta_*)^{2-4\varepsilon} + O(\beta^2) \right] \tag{9b}$$

where β_* is the value of β at $y_s = y_*$. So long as p/p_s is not exponentially large or small in ε, one has

$$\left(\frac{p}{\varrho_\infty u_\infty^2} \right)^\varepsilon = \frac{1}{\lambda} \left[1 + O\left(\varepsilon \ln \frac{p}{p_s} \right) \right] \sim \frac{1}{\lambda} \tag{10}$$

Therefore, for region (iii), one arrives at a pressure–area relation formally similar to that of ([8,12]):

$$\frac{p_b}{\varrho_\infty u_\infty^2} [y_s^{1+\nu} - y_b^{1+\nu}] = \varepsilon C_D a^{1+\nu} \alpha(\lambda) \tag{11}$$

where p_b is the surface pressure and

$$C_D \equiv D_N / \tfrac{1}{2} \varrho_\infty u_\infty^2 \pi^\nu a^{1+\nu}$$

$$\alpha(\lambda) \equiv 1 + \frac{1+\nu}{C_D a^{1+\nu}} \int_0^{y_s} \frac{\{ 1 - [1 - (\sin \beta_*)^{2-4\varepsilon}/\lambda^2]^{1/2} \}^2}{[1 - (\sin \beta_*)^{2-4\varepsilon}/\lambda^2]^{1/2}} y_*^\nu \, dy_*$$

It can be shown that, subject to an error of the order ε, the integral of $\alpha(\lambda)$ is contributed only by the range of y_* corresponding to $1/\beta_* = O(1)$; in other words, $\alpha(\lambda)$ depends only on the shock geometry of the upstream part of the free layer and the attached shock layer; therefore it is independent of the aftbody shape, and is a function of λ alone. The function $\alpha(\lambda)$ for a spherical nose cap has been analytically determined in ([11]). With the Busemann formula for p_b, Eq. (11) is a second-order ordinary differential equation for the shock radius y.

As one approaches upstream (assuming $y_b/x^{2/(3+\nu)} \to 0$ as $x \to 0$), Eq. (11) yields an integral

$$y_s' y_s^{2(1+\nu)} = \Gamma + 2\varepsilon C_D a^{1+\nu} y_s^{1+\nu} \alpha(0) \tag{12}$$

where $\alpha(0) \neq 1$ generally, and Γ is a constant of integration. This behavior allows matching with the free layer and thereby provides the initial data for the integration of Eq. (11). At far downstream, $x \to \infty$ and $\lambda \to \infty$, i.e., in part (c) of region (iii),

$$\alpha(\lambda) \sim 1 + (k_N/\lambda^4) \tag{13}$$

where k_N is a constant determined by the nose shape alone,

$$k_N \equiv \frac{1}{C_D a^{1+\nu}} \int_0^\infty (\sin \beta_*)^4 y_*^\nu \, dy^* \atop (\text{SL--FL})$$

which confirms Cheng's pressure–area relation [8] and indicates that the departure from the HSDT belongs to order $\beta^{8\varepsilon}$. The "entropy layer" in this domain is a *sublayer* of the "entropy wake."

It is essential to observe that the correction given by Eqs. (11) and (13) appears to associate primarily with the velocity defect, but not directly with the entropy defect [see Eqs. (9a,b)]. On this ground, Hayes and Probstein[1] appear to have raised doubts concerning these results. On the other hand, the entropy-layer effect, which is proportional to $\beta^{4\varepsilon}$, could appear in the higher-order Newtonian approximations, e.g., $\varepsilon\beta^{4\varepsilon}$.

The foregoing development is applicable to an aftbody of arbitrary shape. In the case of a straight aftbody, it is easy to show from Eq. (11) that, consistent with the approximation, there is a similitude for the entire transition region (iii)—namely,

$$\varepsilon C_D y_s^{1+\nu} = f(\tilde{x})g(\tilde{x}^\varepsilon)$$
$$p_b/\varepsilon^2 C_D^2 \varrho_\infty u_\infty^2 = h(\tilde{x})g^2(\tilde{x}^\varepsilon)$$

(14)

where $\tilde{x} \equiv (\varepsilon C_D)^{(2+\nu)/(1+\nu)}(x/a)$ and the function g can be determined after f and h are solved. In fact, throughout parts (b) and (c) of region (iii) (Fig. 3), in this case, the entropy layer is blast-wave-like; i.e.,

$$y_s = [\alpha(\lambda)]^{1/(3+\nu)}(y_s)_{\text{NBW}}$$
$$p_b = [\alpha(\lambda)]^{2/(3+\nu)}(p_b)_{\text{NBW}}$$

(15)

where the subscript NBW refers to the Newtonian blast wave and the slow variable λ of $\alpha(\lambda)$ may be evaluated from $\lambda \sim (dy_s/dx)_{\text{NBW}}^{-2\varepsilon}$.

1.2. The Shock-Layer/Entropy-Wake Interaction

The problem of inviscid hypersonic flow around a blunt-nosed slender body of two-dimensional or axisymmetrical *arbitrary* shape has been treated by Chernyi [[1]; also see [7]] and by Cheng and co-workers [8,12]. The basic framework of these studies are defined by the HSDT and the blast-wave analogy; introduction of the Newtonian limit $M_\infty\beta \to \infty$, $\varepsilon \to 0$ [which is explicit in [8,12], but implicit in [1]] simplifies the problem to a tractable form. The Newtonian limit, which is also known as the "snowplow model" in the unsteady problem, reduces the study of the interaction between shock layer and entropy wake to the solution of an ordinary differential equation.

As a theory, however, both the work of Chernyi and that of Cheng and co-workers are incomplete in that their methods fail to describe the field structure, and become arbitrary beyond the stage of leading approximation. Below, we discuss a recent study of Cheng and Kirsch [21], where structures in the entropy wake and shock layer are obtained which, in turn, lead to unambiguous determination of the shock and pressure to the order ε.

In passing, we may recall that both Chernyi and Cheng and co-workers applied their methods to the study of slightly-blunted wedge and cone. Strong pressure undershoot was reported in their work, from which a slight reduction of cone drag by blunting appear possible, at least for a $\gamma = 1.40$ [1,7]. The results of Cheng and co-workers [8,12] also show an oscillatory decay in the bluntness effect far downstream. It is important to point out that the oscillatory decay and drag reduction were actually found in the region where the shock layer begins to reattach to the body. The attachment invalidates the assumption implicit in their models, and conclusions on the drag reduction and oscillatory decay should be taken with caution if γ is not very close to unity.

The problem is analyzed as an unsteady one in the transverse plane with $t = x/u_\infty$. Initially, the field is assumed to be dominated by a Taylor–Sedov blast wave; the analysis is concerned with the subsequent strong interaction of the shock layer and the (low-density) entropy wake under the influence of a symmetrically-expanding contact surface (the aftbody). Figure 4 illustrates the problem to be analyzed, where the particle trajectories shown may, of course, be interpreted as streamlines.

For the outer region (shock layer), asymptotic solutions in ascending

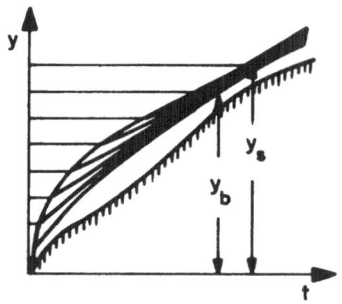

Fig. 4. Illustration of the shock-layer/entropy-wake interaction problem.

powers of ε can be obtained in a manner similar to the slender-body Newtonian of Cole (^{22}) except that the shock shape $y = y_s(t)$ is not *a priori*
known. Owing to the predominance of the blast solution at small time, one
has the singular behavior

$$dy_s/dt \propto y_s^{-(1+\nu)/2}, \qquad y_s \to 0$$

As a result, a region of nonuniformity associated with a logarithmic singularity arises in the outer solution as Y_* (the independent variable replacing the stream function ψ) goes to zero:

$$
\begin{aligned}
y &\sim y_s[1 + f_1(t)\varepsilon \ln Y_* + f_2(t)(\varepsilon \ln Y_*)^2 + \cdots] \\
\varrho &\sim (\varrho_\infty/\varepsilon)g(t)Y_*^{1+\nu}[1 + g_2(t)\varepsilon \ln Y_* + \cdots]
\end{aligned}
\tag{16}
$$

This behavior is precisely what will be needed for the eventual matching
with the solution of the inner region (entropy wake).

For the inner region, a variable is introduced (the standard affine
transform employing constant scales does not work)

$$\zeta \equiv Y_*^{2(1+\nu)\varepsilon}/\sigma^{2\varepsilon} \tag{17}$$

where σ is a constant related to γ and the initial energy release*. Since Y_*
is exponentially small where $\varepsilon \ln Y_* = O(1)$, so too is the transverse pressure
gradient. The density and the particle trajectory (the streamline) are then
evaluated, with an exponentially-small error, as

$$
\begin{aligned}
\tilde{\varrho} &= \frac{(\gamma + 1)}{2\gamma}\left(\frac{\gamma + 1}{2}\right)^{1/\gamma}\tilde{p}_b^{1/\gamma}/\sigma\zeta \\
\tilde{y}^{1+\nu} &= \tilde{y}^{-+\nu} + \left[\sigma\zeta \Big/ \left(\frac{\gamma + 1}{\gamma}\right)\left(\frac{\gamma + 1}{2}\right)^{1/\gamma}\tilde{p}_b^{1/\gamma}\right]
\end{aligned}
\tag{18}
$$

where $\tilde{p} \equiv p/\varrho_\infty u_\infty^2(b/\tau)$, $\tilde{\varrho} \equiv \varepsilon\varrho/\varrho_\infty Y_*^{1+\nu}$, $\tilde{y} \equiv y/b$, with τ and b as reference
time and length scales.

There is a common region of validity for the above solution and that
for the shock layer at $\varepsilon |\ln \varepsilon| \ll \varepsilon |\ln Y_*| \ll 1$, where matching of the two
solutions readily recovers the pressure–area relation of Cheng and co-
workers (^{8,12}):

$$\tilde{p}_0(t)[Y_0^{1+\nu}(t) - Y_b^{1+\nu}(t)] = \sigma_0/2 \tag{19}$$

---

* Of course, one may also choose the variable as Y_*^ε or any positive power of it, or
$\varepsilon \ln Y_*$, so long as the variable chosen remains of order unity in the range where
$\varepsilon \ln Y_* = O(1)$.

where $Y \equiv y_s/b$, the subscript 0 refers to the leading approximation, and, without loss of generality, we can set $\sigma_0 = 2$. The higher-order terms in the matching give a linear relation (not written out) between the first-order pressure correction, say $\tilde{p}_1$, and the shock correction $Y_1(t)$. It is more interesting for the present purpose to combine the two equations involving Y_0, Y_1, $\tilde{p}_0$, and $\tilde{p}_1$, and recast it into a (composite) pressure–area relation valid to the order ε:

$$2\left[\frac{\tilde{p}(t)}{\sigma}\right]^{1/\gamma}[Y_e^{1+\nu}(t) - Y_b^{1+\nu}(t)] = 1 + O(\varepsilon^2)$$
$$Y_e \equiv Y(t) - \varepsilon\text{F.P.}\delta(t)$$

(20)

where F.P.δ stand for the finite part of the shock-layer thickness (divided by b) calculated by the standard theory, which could not have existed without taking the F.P. This relation could have been anticipated from a consideration of the particle isentropy of the whole entropy-wake region; although, in such a consideration, the outer edge for the wake could not be unambiguously defined.

With $\tilde{p}$ expressed in terms of Y through the Busemann formula for $\tilde{p}_0$ and the corresponding equation for $\tilde{p}_1$, these pressure–area relations lead to a second-order ordinary differential equation for Y, or Y_0 and Y_1. In the limit $t \to 0$, the differential equations admit the singular solution

$$Y = At^{2/(3+\nu)}\left[1 + \frac{2\ln 2}{3 + \nu}\varepsilon + O(\varepsilon^2)\right]$$
$$\tilde{p} = \frac{1}{2}\left(\frac{2}{3 + \nu}\right)^2 At^{-2(1+\nu)/(3+\nu)}\left[1 - \left(4\frac{2 + \nu}{3 + \nu}\ln 2\right)\varepsilon + O(\varepsilon^2)\right]$$

(21)

The results agree well with similar solutions for γ as large as 1.40.

The behavior of the shock at large t, i.e., far downstream, will depend on whether asymptotically the aftbody grows faster than a pure blast wave $Y \propto t^{2/(3+\nu)}$. If $Y_b/t^{2/(3+\nu)} \neq \infty$ as $t \to \infty$, then the above equations yield $Y \propto t^{2/(3+\nu)}$. On the other hand, if $Y_b/t^{2/(3+\nu)} \to \infty$ as $t \to \infty$, then the equations give

$$Y \sim Y_b + \text{F.P.}\delta$$

(22)

That is, the shock layer will reattach far downstream if the aftbody grows faster than a pure blast wave (e.g., a blunted wedge and cone). In fact, this bears out the relation between the shock and body exponents given by Fig. 2 as proposed by Freeman.

It is of interest to examine more closely the manner in which the shock layer approaches the aftbody surface. For simplicity, consider an aftbody

$$Y_b \sim t^\omega \tag{23}$$

with $\omega > 2/(3 + \nu)$. Equation (19) then *admits* a large-time solution

$$Y_0 \sim Y_b + F(t, \omega, \nu) + \varphi \tag{24}$$

where $F(t, \omega, \nu)$ is an algebraic function of t, ω, and ν,

$$\varphi = \frac{A_0}{t^{[(7+5\nu)/4]\omega-1}} \cos\left[\frac{2\omega[(2 + \nu)\omega - 1]}{(3 + \nu)\omega - 2} t^{[(3+\nu)\omega-2]/2} + \text{const} \right] \tag{25}$$

and A_0 is a constant of integration. Hence, except when $A_0 = 0$, there will be a damped oscillation whenever reattachment occurs—a generalization of the results obtained in ([8,12]) for the blunted wedge and cone.

Since the frequency of oscillation shown above does not decrease with increasing time as rapidly as the rate of the amplitude decay, the higher derivatives of Y_0 will soon be dominated by the oscillation, and the foregoing expansion in ε (for finite time) will eventually break down at large time. It may be shown that there is in general a range of nonuniformity at large time,

$$t = O(\varepsilon^{-1/[(3+\nu)\omega-2]}) \qquad \text{for} \quad \omega > 2/(3 + \nu) \tag{26}$$

where the *procedure* based on a finite time will fail. The leading approximation for all flow variables, as well as the shock-layer thickness, remains, however, valid, with relative errors of the order $\varepsilon^{3/4}$ for a blunted wedge and $\varepsilon^{1/2}$ for a blunted cone. Meanwhile, there is a general class of aftbodies of practical interest with $\omega < 12/(13 + 3\nu)$, for which the foregoing solution remains valid to the order ε for all t.

The treatment of the problem of transition to reattachment involving the above range of time will not be elaborated here except to note that the proper scales for all flow variables can be determined from Eqs. (24) and (26).*

Figure 5 illustrates results of the zeroth- and the first-order surface pressure in the problem corresponding to a blunted wedge, where a set of experimental data obtained by Henderson *et al.* ([23]) is also included for comparison.

---

* There are two characteristic time scales for the flow during the period of Eq. (26). One is given by Eq. (26) and the other by the frequency of oscillation indicated by Eq. (25). A solution to this two-time problem is discussed in ([6]).

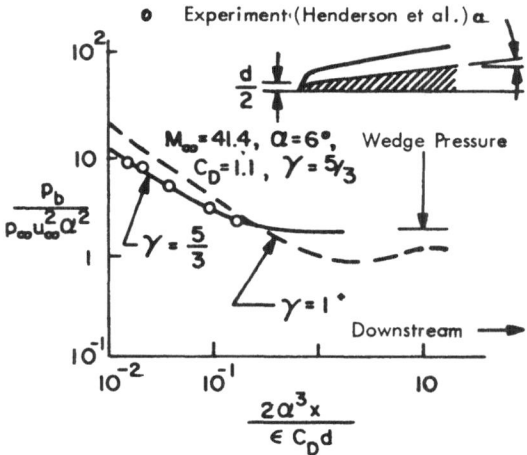

Fig. 5. Comparison of the zeroth and the first-order solutions in pressure. Experimental data ([23]) are also included.

The comparatively low density level in the entropy wake (as well as the entropy layer) suggests that the (two) transverse pressure gradients in the low-density region may be neglected for a blunt-nosed slender body with asymmetrical shape, or with (small) angle of attack, since the dynamic pressure is low. It follows that the pressure and the outer field must be (approximately) axisymmetrical, and are determined primarily by the cross-sectional area of the aftbody. This is Ladyzhenskii's ([24]) "hypersonic area rule."

While a complete theory based on this idea has not yet been firmly formulated, an interesting experiment related to the study by Krasovskii ([25]) may be noted. The experiment carried out for $M_\infty = 18$ and $Re \approx 10^6$ attempts to correlate the drag of equivalent bodies with the same cross-sectional area, i.e., circular cones, bodies of triangular, rectangular, and other cross sections. The result is *not* conclusive (although the author appears to claim a confirmation), owing primarily to the viscous phenomena. On the other hand, the blunted-circular-cone data show convincingly that blunting can indeed lower the drag (slightly) from the sharp-cone value, and that the optimum body length turns out to be rather close to that predicted by Chernyi's integral method noted previously. Yawed-cone data ([25]) show that reductions in lift from the pure-cone value can be as large as 60%. This is to be anticipated, since, according strictly to Ladyzhenskii's model, there would have been no lift.

The knowledge of the field structure in the entropy wake discussed previously provides, however, a *stronger* form of the area rule, as follows.

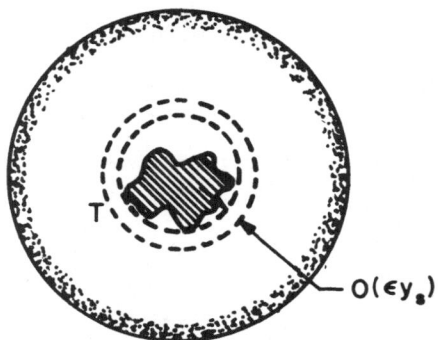

Fig. 6. Illustration of a stronger form of the "hypersonic area rule." Only a limited part of the entropy wake will be affected by the aftbody asymmetry.

From Eqs. (18), we can write

$$\varrho/\varrho_\infty \propto (\tilde{p})^{1/\gamma} Y_*^{2\varepsilon(1+\nu)}$$
$$\propto [\tilde{p}(\tilde{y}^{1+\nu} - \tilde{y}_b^{1+\nu})]^{(1/2\varepsilon)-1} \tag{27}$$

For small ε, this gives a *highly-stratified, symmetrical* density distribution. Thus the field at a point at a finite distance farther from the axis than another will be much stiffer, dynamically speaking. It follows that the disturbance resulting from asymmetrical change of body cross section or yaw can only affect a *limited* portion of the entropy wake. The boundary of this disturbed region is given by the surface of constant Y_*, with Y_* corresponding to the most outward symmetrical stream surface that touches the aftbody. Figure 6 illustrates such a boundary, although the point of contact T need not exist.

1.3. Blow-Hard

Hypersonic slender bodies with strong blowing from the surface provide a topic of considerable theoretical interest, e.g., ([26],[27]). In their blow-hard problem, Cole and Aroesty ([26]) assume a thin layer of inviscid, incompressible injectant, and obtain solutions corresponding to certain nonuniform blowing rates. The possibility of solution for a uniform blowing rate is not, however, obvious from this model. Meanwhile, the experimental results of Hartunian and Spencer ([28]) show wedge-like flow regions of the blown gas; however, so far, theoretical confirmation is incomplete*.

---

* It may not be inappropriate to mention that Stewartson ([29]) has found boundary layer separation (vanishing shear) on a flat plate with a (rather-weak) uniform blowing rate.

1.4. Minimizing Drag

The monograph edited by Miele [30] provides an interesting array of problems of minimizing drag and their solutions. The contributions in [30] by J. D. Cole and by W. D. Hayes contain some results in the shock-layer theory not available elsewhere.

A class of bodies of interest from the minimum (inviscid) drag viewpoint is that of bodies with star-shaped cross sections [[7], for example]. There was, until recently, an apparent discrepancy on the subject between the works of Gonor [31] and of Maikapar [32,33]. The reader may refer to Chernyi's article [34] for a reconciliation of the two sets of results. When a uniform skin friction is assumed, a cross section resembling a triangle appears to be more advantageous [35].

In the past, most work on drag was concerned with nonlifting bodies; a refreshing change in direction is found in Cole and Zien's recent study [36], where a problem of drag due to lift is studied within the full framework of the hypersonic small-disturbance theory. The idea is to construct the compression side of a lifting wing from the known streamlines in the flow behind an axisymmetrical power-law shock $R \propto x^n$ ($n \geq \frac{1}{2}$). They have obtained the optimum wing shape for $\gamma = 1.40$ and several values of n. The smallest value of drag is that generated from the cylindrical blast-wave solution ($n = \frac{1}{2}$), with a wing surface being rather close to the shock. It appears that a nonsimilar explicit solution to the blast-wave interaction problem (discussed in Section 1.2) should provide one more degree of freedom to this approach.

2. THE OUTER-EDGE PROBLEM OF THE HYPERSONIC BOUNDARY LAYER ON A SLENDER BODY

2.1. The Outer-Edge Problem

In the classical theory of the hypersonic boundary layer on a slender body [consult [7], 1st ed.; [37,38]], the field is analyzed as one composed of two layers—a viscous boundary layer (with the self-induced pressure and the transverse curvature fully taken into account) and an inviscid region governed by the HSDT. The model leads to a Mach-number-independent similitude for the viscous flow.

Underlying this similitude is the assumption of a vanishing temperature at the outer edge, which may be justified in view of the relatively high temperature level attained in the hypersonic boundary layer. On the other

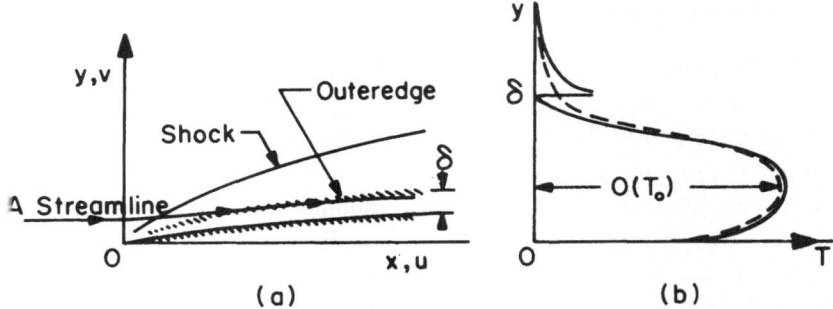

Fig. 7. Sketch for the study of hypersonic viscous Flow over a thin body: (a) the shock, the streamline, and the boundary layer outer edge. (b) the singular behavior in the solutions for the temperature.

hand, this, coupled with the assumption that viscosity vanishes with a vanishing temperature, leads to a sharply defined outer edge, with an accompanying singularity in the boundary layer solution. Meanwhile, the outer flow in various hypersonic problems of interest also develops a singularity as the boundary layer edge is approached (from outside) (see Fig. 7). The theory is obviously invalid near the boundary-layer edge, and the successive approximations in the conventional boundary layer theory ([16]) are evidently inapplicable.

It is of interest to mention in this connection the analogy between the hypersonic outer-edge problem and the one of unsteady diffusion. If the diffusion coefficient depends on the concentration, say, according to a power law, one may anticipate the existence of a singular front which separates a mixed and an unmixed region.*

The behavior of the boundary layer near its outer edge has been studied quite extensively by Freeman and Lam ([39]) for the zero-pressure-gradient case[†]. The singularity in the inviscid solution is attributable to the uneven heating of the inviscid flow through the singularly-curved bow shock; the effect due to the uneven shock heating, and the associated vorticity, were recognized and studied earlier by Lees ([40]) for the flat plate with strong self-induced pressure. Subsequently, Oguchi ([41]) treated the problem more systematically as one of a second-order hypersonic boundary-layer; however, the result is incomplete in that the pressure and outer flow are not determined consistently, and the boundary layer solution contains another

---

* This fact was called to the author's attention by Dr. J. D. Cole, who also pointed out that experimentally-observed singular fronts have been reported.
[†] The results reported by Freeman and Lam ([39]) for the tangential velocity and for the temperature (with a linear viscosity law) appear to be in error.

error. In fact, it is not clear that a consistently-determined higher approximation will always lead to an increase in skin friction and surface heat transfer, as a simple consideration would suggest. Recently, Matveeva and Sychev ([42]) analyzed the inverse problem of determining the viscous-flow field and the body shape that support a $\frac{3}{4}$-power-law shock. Underlying these analyses are the classical two-layer model and a linear viscosity–temperature law.

The study to be made below is concerned with the more recent development on the outer-edge problem, beginning with the analysis by Bush ([43]), who has made a study of the hypersonic strong-interaction problem for the flat plate with a viscosity law $\mu \propto T^{\omega}$, applying limit-process expansions to the Navier–Stokes equations (for the limit $M_{\infty}(\delta/L) \to \infty$, $\delta/L \to 0$, with δ standing here for a boundary layer thickness). He treats the viscous transition layer as being distinct from the HSDT and the boundary layer regions, and, interestingly, finds it possible to demonstrate the existence of a self-similar solution for that region. The analysis indicates that solutions based on the two-layer model cannot be matched, in the Kaplan–Lagerstrom–Cole sense ([16,44]), unless an intermediate layer is introduced. Although Bush's analysis fails to give results for $\omega \geq 1$, two questions may now be raised: (1) Are previous analyses of the strong-interaction problem, based on the two-layer model with $\omega = 1$, correct asymptotically? (2) Are second-order approximations required for a uniformly valid description of the transition layer? The second question arises because in Bush's formulation, the temperature in the transition layer appears to be completely specified by only the leading approximations in the neighboring regions.

In answering the first, as well as the second, question, it is essential to know the various scales associated with the transition layer and their dependence on the viscosity-law exponent ω. This will be studied in Section 2.2. Bush's analysis of the transition layer, generalized to other power-law shocks, will be discussed in Section 2.3, where the relation of the outer-edge problem and the higher-order approximation will also be examined for $\omega < 1$. The outer-edge problem raised for the linear viscosity law ($\omega = 1$) is delineated and resolved in Section 2.4.

2.2. The Scales Associated with the Viscous-Transition Layer

We denote by "viscous-transition layer" a thin region where the molecular transport becomes comparable to the convective effect, while the velocity field is still close to the inviscid one. To obtain proper estimates

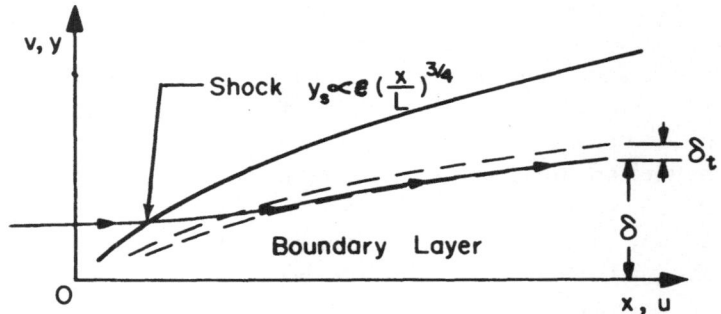

Fig. 8. Division and notation for the study of the outer-edge problem for a flat plate in the hypersonic strong-interaction regime.

for the various scales of this region, we may follow Lees' earlier work,*
without having to rely on the existence of self-similarity and the result of
matching as in ([43]). For simplicity, we consider the strong interaction
regime

$$M_\infty(\delta/L) \gg 1, \qquad \delta/L \ll 1 \qquad (28a)$$

and that the upstream portion of the shock is dominated by (for small x)

$$y_s \propto x^{3/4} \qquad (28b)$$

The resulting set of scales will be applicable to plane and axisymmetrical
flows with bodies no blunter than $x^{3/4}$, i.e., $y_b/x^{3/4} \neq \infty$ as $x \to 0$. We
shall concern ourselves mainly with cases $\omega \leq 1$.

Comparing the molecular-transport and the advective terms in the momentum equation leads to an estimate for the thickness of the layer δ_t

$$\delta_t/L \propto (\mu/\varrho u L)^{1/2}_{\mathrm{HSDT}} \qquad (29)$$

where the subscript HSDT refers to the condition based on the HSDT
solution. Since the boundary-layer thickness δ may be estimated in a similar
manner but based on boundary layer properties, we may obtain

$$\delta_t/\delta \propto (T_t/T_0)^{(1+\omega)/2} \qquad (30)$$

where T_t and T_0 refer, respectively, to the temperature in the transition layer

---

* The results arrived at below do not completely agree with Lees' results.

and that in the boundary layer; the latter is at the same level as the stagnation temperature.

The temperature can be expressed in terms of δ and a dimensionless stream function $\bar{\psi}_t$ in the layer; this is done by making use of Eq. (28b) and tracing isentropically a typical streamline in the layer back to the point where it leaves the shock (see Fig. 8):

$$\frac{T_t}{T_0} \propto (\bar{\psi}_t)^{-2/3\gamma(1+\nu)} \left(\frac{\delta}{L}\right)^{(8/3\gamma)+2(\gamma-1/\gamma)} \tag{31}$$

On the other hand, mass fluxes $\Delta\bar{\psi}_t$ and $\bar{\psi}_{\rm BL}$ in the transition and boundary layers, respectively, can be estimated in terms of the temperature ratio as

$$\Delta\bar{\psi}_t/\bar{\psi}_{\rm BL} \propto (T_t/T_0)^{(\omega-1)/2} \tag{32a}$$

where

$$\Delta\bar{\psi}_t \equiv \bar{\psi}_t - \bar{\psi}_{\rm BL} \tag{32b}$$

Equations (31) and (32a,b) may be combined to yield under $\omega \leq 1$

$$\frac{\delta_t}{\delta} \propto \left(\frac{\delta}{L}\right)^{[(1+\omega)/2]\varkappa}, \qquad \frac{T_t}{T_0} \propto \left(\frac{\delta}{L}\right)^{\varkappa}$$

$$\frac{\Delta\psi_t}{\psi_{\rm BL}} \propto \left(\frac{\delta}{L}\right)^{[(\omega-1)/2]\varkappa}, \qquad \frac{\Delta p_t}{p_{\rm BL}} \propto \left(\frac{\delta}{L}\right)^{2+[(\omega-1)/2]\varkappa} \tag{33}$$

where

$$\varkappa = \frac{2[3\gamma(1+\nu) - 2]}{3\gamma(1+\nu) - 1 + \omega} \tag{33'}$$

Note that the range of interest for $\varkappa$ is $0 < \varkappa < 2$, and that $\Delta\psi_t \gg \psi_{\rm BL}$ for $\omega < 1$. The above results are consistent with Bush's ([43]) for $\omega < 1$ and agree with Lees' ([10]) for $\omega = 1$. The above shows that for a linear viscosity law ($\omega = 1$), the mass fluxes through the transition layer and the boundary layer are of the same order. This would suggest that for $\omega = 1$, a two-layer model in a Von Mises variable may suffice; this point will be clarified in Section 2.4. In any case, the physical thickness of the transition layer is much smaller than that of the boundary layer, $\delta_t \ll \delta$, and a three-layer construction will be needed in a formulation based on physical coordinates (x, y).

For $\omega > 1$, Eq. (29), hence Eqs. (30), (32), and (33), do not hold, because $\Delta\psi_t \ll \psi_t$, and the convection is no longer dominated by the advective term. The proper scaling of the transition layer for $\omega > 1$ is given and discussed in ([6]).

2 3. The Transition Layer in the Strong- and Weak-Interaction Regimes Involving Power-Law Shocks

In setting up a framework for the studies below, we assume a field model consisting of three distinct regions: the HSDT region, the boundary-layer layer, and the viscous-transition layer. For convenience in the subsequent studies, we shall write down the differential equations (in Von Mises variables) governing a boundary layer in hypersonic plane or axisymmetrical flow over a slender body with a constant Prandtl number, a viscosity-temperature law $\mu \propto T^\omega$, and also a strong bow shock.

In order to write the equations for the strong and weak interactions into one system, we shall use three typical length scales: L, characterizing the distance in x; δ, the boundary layer thickness; and εL the transverse dimension of the disturbed field. The governing equations may then be written, in dimensionless form:

$$\tilde{r}^\nu \frac{\partial \tilde{y}}{\partial \tilde{\psi}} = \frac{\tilde{T}}{\tilde{p}\tilde{u}}, \qquad \tilde{r} = \tilde{r}_b + \frac{\delta}{\varepsilon L} \tilde{y}$$

$$\tilde{u} \frac{\partial \tilde{u}}{\partial \xi} + \frac{\gamma-1}{2\gamma} \frac{\tilde{T}}{\tilde{p}} \frac{d\tilde{p}}{d\xi} = \tilde{u} \frac{\partial}{\partial \tilde{\varphi}} \left(\tilde{T}^{\omega-1}\tilde{r}^{2\nu}\tilde{p}\tilde{u} \frac{\partial \tilde{u}}{\partial \tilde{\psi}} \right) \qquad (34)$$

$$\frac{\partial \tilde{T}}{\partial \xi} - \frac{\gamma-1}{\gamma} \frac{\tilde{T}}{\tilde{p}} \frac{d\tilde{p}}{d\xi} = \tilde{p} \frac{\partial}{\partial \tilde{\varphi}} \left(\tilde{T}^{\omega-1}\tilde{r}^{2\nu} \frac{\tilde{u}}{\mathrm{Pr}} \frac{\partial \tilde{T}}{\partial \tilde{\psi}} \right) + \tilde{p}\tilde{u}\tilde{T}^{\omega-1}\tilde{r}^{2\nu} \left(\frac{\partial \tilde{u}}{\partial \tilde{\psi}} \right)^2$$

where

$$\tilde{u} \equiv u/u_\infty, \qquad \tilde{v} \equiv v/\varepsilon u_\infty, \qquad \tilde{p} \equiv p/\varrho_\infty u_\infty^2 \varepsilon^2, \qquad \tilde{T} \equiv T/T_0$$
$$\tag{34'}$$

$$\tilde{r} \equiv y/\varepsilon L, \qquad \xi \equiv x/L, \qquad \tilde{\psi} \equiv \psi/\varrho_\infty u_\infty \delta L^\nu \varepsilon^{2+\nu} \left(\frac{2\gamma}{\gamma-1} \right), \qquad \tilde{y} \equiv \frac{(r-r_b)}{\delta}$$

$$\delta^2\varepsilon^2 \equiv \left(\frac{\gamma-1}{2} \right)^{1+\omega} M_\infty^{2\omega} \Big/ \gamma \mathrm{Re}_L, \qquad \mathrm{Re}_L \equiv \varrho_\infty u_\infty L/\mu_\infty \qquad (34'')$$

and the subscript ∞ refers to the uniform free-stream condition.* We shall omit the boundary conditions and equations for the outer flows, except to mention that they contain, in their dimensionless form, the two products $M_\infty^2 \varepsilon^2$ and τ/ε, where τ is a body thickness ratio. We may assign

---

* If a reference temperature is used in conjunction with a power viscosity law, then Re_L should be replaced by Re_L/C_*, with $C_* = \mu_* T_\infty^\omega/\mu_\infty T_*^\omega$.

ε in the above as

$$\varepsilon = \delta/L \qquad \text{for } O\!\left(\frac{1}{M_\infty} + \tau\right) \le O(\delta)$$

$$= \tau \qquad \text{for } O\!\left(\frac{1}{M_\infty} + \delta\right) \le O(\tau)$$

$$= 1/M_\infty \quad \text{for } O(\delta + \tau) \le O\!\left(\frac{1}{M_\infty}\right)$$

This permits complete elimination of the parameters M_∞ and Re_L from the momentum and energy equations.

Note that in the strong-interaction regime, we have $M_\infty^2 \varepsilon^2 = M_\infty^2 \delta^2/L^2 \propto M_\infty^{2+\omega}/\mathrm{Re}_L^{1/2}$, which may be identified with the standard interaction parameter χ; ε or δ/L in this case corresponds to the "vee bar" in the rarefied gas dynamics literature.

It is essential to recognize that the system of Eqs. (34) includes *all* terms in the equations relevant to the transition layer, because the pressure variation across the transition layer can be neglected, and the layer is thinner than the boundary layer (also because u and v retain the same orders of magnitude as in the boundary layer). Equations (34) are subject to an error of order ε^2, and are therefore sufficient for analyzing higher-order effects (slip, shock-heating, etc.) so long as the corrections remain of lower order than ε^2.

Before going into the study of the transition layer for the flat plate, it is essential to observe a singularity of Eqs. (34) corresponding to a sharp outer edge—namely, with $\tilde{u} \ne 0$ and $\tilde{\psi} > 0$, the system of partial differential equations has a singular behavior near the outer edge where $\tilde{T} \to 0$

$$\tilde{T} \sim f(\xi)\tilde{\psi}^{-2/(1-\omega)} \propto F(x)[\delta(x) - y]^{2/(1+\omega)}, \; \omega < 1$$
$$\sim g(\xi)\tilde{\psi}^{(2/\gamma-3)}/\exp\big|\,h(\xi)\tilde{\psi}^2\,\big| \propto G(x)(\delta - y)\,\big|\ln(\delta - y)\big|^{1/2}, \; \omega = 1 \quad (35)$$

which may be shown to be the only (saddle-point) singularity at $\tilde{T} = 0$, $\tilde{u} \ne 0$, in the self-similar cases. This behavior of the boundary layer temperature does not directly match that of the HSDT temperature. For example, for a strong power-law shock $y_s \propto x^m$, the HSDT gives a temperature near the boundary layer edge (written in terms of $\tilde{T}$ and $\tilde{\psi}$)

$$\tilde{T} \sim J(\xi)\tilde{\psi}^{-2(1-m)/m(1+\nu)\gamma} \qquad (36)$$

which is obviously not comparable to that in Eq. (35).

For the flat plate in the strong-interaction regime ($m = \frac{3}{4}$) with $\omega < 1$

and a constant wall temperature, Bush ([43]) introduces a transition layer and shows that a self-similar transition solution exists and matches Eqs. (35) and (36) at the two ends. The strong interaction is not crucial to the study of the outer-edge problem. Below we will arrive at the transition solution for temperature with a strong power-law shock for an arbitrary m.*

The equations governing the transition layer can be directly obtained from Eqs. (34) making use of the assumptions

$$\frac{u}{u_\infty} \sim 1, \qquad \frac{v}{u_\infty} \sim \frac{d\delta}{dx}, \qquad y \sim \delta(x) \qquad (37)$$

The equation for the temperature can be decoupled and can be written, subject to an error of the order ε_T, as

$$\frac{\partial \hat{T}}{\partial \xi} - \frac{\gamma - 1}{\gamma} \hat{T} \frac{d}{d\xi} \ln \tilde{p} = \tilde{r}_\delta^{2\nu} \tilde{p} \frac{\partial}{\partial \hat{\psi}} \frac{\hat{T}^{\omega-1}}{\text{Pr}} \frac{\partial \hat{T}}{\partial \hat{\psi}} \qquad (38)$$

where

$$\hat{T} \equiv \frac{T}{T_0 \varepsilon_T}, \qquad \tilde{r}_\delta \equiv \frac{\delta(x)}{\varepsilon L}, \qquad \hat{\psi} \equiv \psi / \varrho_\infty u_\infty \left(\frac{2\gamma}{\gamma - 1}\right) L^\nu \delta \varepsilon^{2+\nu} \varepsilon_T^{-(1-\omega)/2} \qquad (38')$$

Here, ε_T is a small parameter characterizing T_t/T_0. In the case where $m = \frac{3}{4}$, T_t/T_0 was estimated as $\varepsilon^\varkappa$ in Section 2.2.

In cases for which self-similar solutions exist in both (the leading approximation of) the HSDT and boundary layer regions (which require a constant wall temperature and power-law shock),[†] we seek a self-similar solution for $\tilde{T}$ in a manner similar to that used by Bush ([43]):

$$\tilde{p} = P_0 \xi^{2(m-1)}, \qquad \hat{T} = \xi^{C_1} \theta, \qquad \hat{\psi} = \lambda \xi^{C_2} \zeta, \qquad \tilde{y}_b = E \xi^m \qquad (39)$$

Equation (38) then reduces to

$$\frac{E^{\xi \nu} P_0}{\lambda^2 C_2 \text{Pr}} \frac{d}{d\zeta} \theta^{\omega-1} \frac{d\theta}{d\zeta} + \zeta \frac{d\theta}{d\zeta} - \frac{1}{C_2}\left[C_1 + 2(1 - m) \frac{(\gamma - 1)}{\gamma}\right]\theta = 0 \qquad (40)$$

provided

$$C_2 = [2m(1 + \nu) - 1 - (1 - \omega)C_1]/2 \qquad (41)$$

---

* For $m \neq \frac{3}{4}$, the self-similar solutions require a weak interaction. There is, however, a critical value of m (less that $\frac{3}{4}$) below which the speed and entropy defects discussed in Section 1.1 must be taken fully into account. It turns out that Eqs. (37)–(45) can be generalized to analyze effects of entropy and speed defects in the external flow ([6]).

† Recall that for $m \neq \frac{3}{4}$, the self-similarity requires χ or $M_\infty^{2+\omega}/\text{Re}_L^{1/2} \to 0$.

As $\zeta \to \infty$, Eq. (40) is dominated by the last two terms, which bears out the HSDT behavior, Eq. (36). Hence the matching is possible and requires that*

$$-\frac{2(1-m)}{m(1+v)\gamma} = \frac{1}{C_2}\left[C_1 + 2\frac{(1-m)(\gamma-1)}{\gamma}\right] \qquad (42)$$

Equations (41) and (42) completely determine C_1 and C_2,

$$C_1 = -2(1-m)\frac{2m(1+v)\gamma - 1}{m(1+v)\gamma - (1-\omega)(1-m)}$$
$$C_2 = m(1+v) - \tfrac{1}{2} - \tfrac{1}{2}(1-\omega)C_1 \qquad (43)$$

Putting

$$\lambda^2 \equiv E^{2v}P_0/C_2\mathrm{Pr}$$

we arrive at

$$\frac{d}{d\zeta}\theta^{\omega-1}\frac{d\theta}{d\zeta} + \zeta\frac{d\theta}{d\zeta} + \frac{2(1-m)}{m(1+v)\gamma}\theta = 0 \qquad (44)$$

For $\zeta \to 0$, Eq. (44) gives

$$\theta = \left[\frac{1+\omega}{1-\omega}\frac{m(1+v)\gamma}{m(1+v)\gamma - (1-m)(1-\omega)}\right]^{1/(1-\omega)}\zeta^{-2/(1-\omega)} \qquad (45)$$

which is in fact a particular solution to Eq. (44), and gives a $\hat{T}$ or $\tilde{T}$ that indeed matches with the self-similar form of Eq. (35) from the boundary layer. The second-order differential equation (44) is invariant with respect to a multiplicative transform, and therefore is reducible to a first-order one. It is then possible to study the existence and uniqueness of the desired solution from the topology of the integral curves, as Bush does for $m = \tfrac{3}{4}$ [43].

After θ or $\hat{T}$ are obtained, other equations of (34) can be used to determine the speed and streamline deflections,

$$\hat{u} \equiv (u - u_\infty)/u_\infty\varepsilon_T$$
$$\hat{y} \equiv [y - \delta(x)]/\delta(\varepsilon_T)^{(1+\omega)/2} \qquad (46)$$

The above analysis along Bush's lines ceases to be valid as $\omega \to 1$. This is evident from Eq. (45), and may be anticipated from the study of

---

* The matching will also determine ε_T in terms of ε, although ε_T can also be determined for a general m in the manner of Section 2.2.

Section 2.2. The outer-edge problem for the linear viscosity law ($\omega = 1$) will be studied specifically in Section 2.4. Even for $\omega < 1$, it is not altogether clear from the above analysis how the presence of a transition layer may affect the boundary layer and the external flow, and in what manner these effects are related to the external vorticity. This question has been studied by Lee and Cheng ([45]) for the special case of a flat plate near the strong interaction limit.

Whereas the basic analytical frameworks of Bush ([43]) and of Lee and Cheng ([45]) are equivalent, the latter analysis serves to clarify two points of theoretical interest. First, there are effects of the transition layer on the boundary layer which do *not* result *directly* from the shock heating nor from vorticity (i.e., not from matching in T and u), but chiefly through matching in v, or the streamlines. Of interest in this connection is the small parameter that controls and represents the order of magnitude of the effect, which is not ε_T, characterizing the shock heating, but ε_V, characterizing the transition layer thickness.* For example, for $m = \frac{3}{4}$ and $\nu = 0$

$$\tilde{T} = \tilde{T}_0 + \left(\frac{\varepsilon}{x^{1/4}}\right)^{[(1+\omega)/2]\varkappa} \tilde{T}_1 + \cdots$$

$$y_s \propto \varepsilon x^{3/4}\left[1 + \text{const}\left(\frac{\varepsilon}{x^{1/4}}\right)^{[(1+\omega)/2]\varkappa} + \cdots\right]$$

(47)

Thus for $\omega < 1$, it is the *streamline displacement* resulting from the *physical presence* of the transition layer that is responsible for the effect.

A second point brought out in the results of Lee and Cheng is that the flow field of the transition layer, in terms of physical coordinates (x, y), cannot be *uniquely* determined without going beyond the leading approximation for the boundary layer solution. The reason should have been obvious; since $\delta_t \ll \delta$, the flow properties in the transition layer cannot be unambiguously determined without having determined the displacement effect to an order comparable to δ_t/δ. From the viewpoint of the von Mises (x, ψ) formulation, used in Bush's work ([43]), this degree of arbitrariness could also have been observed if one were to write out the equation of the streamline $y(x, \psi)$ in the transition layer.

Following the same vein as ([43]), the classical problem of a flat plate in the weak-interaction regime ($M_\infty\delta \sim \chi \to 0$, $\delta \to 0$) has also been examined recently by Bush and Cross ([46]). Leading approximations are obtained for the flow in the boundary layer, the transition layer, and

* Note that for $\omega < 1$, $\varepsilon_V \gg \varepsilon_T$.

the perturbed outer inviscid field, as well as in the region near the shock front.* From a fluid-mechanical viewpoint, Bush and Cross' analysis of the transition layer is not fundamentally different from that given earlier by Freeman and Lam ([39]). It may be observed that, for the flat-plate problem in the weak-interaction limit, existence of a self-similar solution does not require $M_\infty \to \infty$. One could therefore study the region of interest directly from the self-similar solution of the boundary layer [as in ([39])]. The treatment by Bush and Cross ([46]) on nonuniformity near the shock may be of some interest; they give a weak shock displacement at large distances as $M_\infty y_s \sim x + \text{const}(M_\infty \delta)^{2/3} x^{2/3}$ which confirms the result of Kuo ([47]) obtained by Lighthill's technique. We shall discuss the far-field, weak-shock analysis in Section 3 along with the axisymmetrical problem.

It must be said that the results obtained for the hypersonic weak-interaction problem concerning self-induced pressure, skin friction, and heat transfer are quite extensive [see ([7]), 1st ed., and ([37])]. Based on integral methods, Kubota and Ko ([48]) have recently extended the hypersonic, insulated, flat-plate results for unit Prandtl number to the finite Mach-number range. The analysis is developed to the order of χ^3.[†]

2.4. The Outer-Edge Problems for the Flat Plate in the Strong-Interaction Regime: $\omega = 1$

While the analysis based on a three-layer model fails to describe the neighborhood of the outer edge for $\omega \to 1$, a treatment of the problem for $\omega = 1$ providing the desired matched temperature behavior at the outer edge has been given by Oguchi ([41]). Aside from the incomplete solution and several algebraic errors in his second approximation, the validity of the two-layer model underlying Oguchi's analysis requires a critical examination. This will be made in the following, where the uniform validity of the two-layer model will be confirmed by comparison with the strict asymptotic solutions ([49]). In the process, we will find that there are some interesting differences between cases $\omega < 1$ and the case $\omega = 1$ regarding the nature of interaction of the boundary layer and the region around the outer edge.

In order to see clearly the problem encountered by the two-layer model at the outer edge, we shall briefly outline a development equivalent to that

---

* For the same reason stated in the preceding paragraph, a unique solution to the flow field cannot be claimed here.
† However, the question of uniqueness concerning terms of order χ^3 [as noted by Stewartson ([37])] has not been discussed.

of Oguchi ([41]). For the outer HSDT region, one may assume, following Oguchi, the form

$$\bar{y}_s \equiv \frac{y_s}{\delta} = A\xi^{3/4}\left[\bar{V}_0(1) + a_n \frac{\varepsilon_T}{\xi^n} + \cdots \right]$$

$$\bar{p} \equiv \frac{p}{\gamma M_\infty^2 p_\infty \varepsilon^2} = \xi^{-1/2}\left[\bar{P}_0(\Omega) + \frac{\varepsilon_T}{\xi^n} \bar{P}_1(\Omega) + \cdots \right]$$

$$\bar{T} \equiv \frac{T}{T_0} = \xi^{-1/2}\left[\Theta_0(\Omega) + \frac{\varepsilon_T}{\xi^n} \Theta_1(\Omega) + \cdots \right] \qquad (48)$$

$$\bar{v} \equiv \frac{v}{\delta u_\infty} = \xi^{-1/4}\left[\bar{V}_0(\Omega) + \frac{\varepsilon_T}{\xi^n} \bar{V}_1(\Omega) + \cdots \right]$$

$$\bar{u} \equiv 1 - \frac{u}{u_\infty} = \xi^{-1/2}\left[\bar{U}_0(\Omega) + \frac{\varepsilon_T}{\xi^n} \bar{U}_1(\Omega) + \cdots \right]$$

where a_n is an unknown constant, $\xi \equiv x/L$, $\Omega \equiv \psi/\psi_{\text{shock}}$ and

$$\varepsilon_T = \varepsilon_V = \varepsilon^{\varkappa}, \qquad \varkappa \equiv 2[1 - (2/3\gamma)], \qquad n = \tfrac{1}{2}[1 - (2/3\gamma)] \qquad (49)$$

These values for ε_T and n had in fact been indicated by the discussion in Section 2.2. The boundary conditions are obtained from the Rankine–Hugoniot relation at $\Omega = 1$. In approaching the boundary-layer edge, i.e., $\Omega \to 0$, the self-similar HSDT solution and its higher-order approximation behave like

$$\bar{P}_0 \sim P_0 + P_1\Omega + \cdots, \qquad \bar{V}_0 \sim V_0 + V_1\Omega^{1-(2/3\gamma)} + \cdots$$
$$\bar{T}_0 \sim \Omega^{-2/3\gamma}[\Theta_0 + \Theta_1\Omega + \cdots], \qquad \bar{U}_0 \sim \frac{\gamma}{\gamma - 1}\frac{P_0}{D_0}\Omega^{-2/3\gamma} + \cdots \qquad (50a)$$

$$\bar{P}_1 \sim a_n P_{00} + \cdots, \qquad \bar{V}_1 \sim a_n V_{00} + \cdots$$
$$\bar{T}_1 \sim \text{const } \Omega^{-(2/3\gamma)-(4n/3)} + \cdots, \qquad \bar{U}_1 \sim \text{const } \Omega^{-(2/3\gamma)-(4n/3)} + \cdots \qquad (50b)$$

where P_0, P_{00}, V_0, V_{00}, Θ_0, and Θ_1 are known constants. This is to be matched with solution from the next inner region.* In ([41]) the higher-order terms $\bar{P}_1$, $\bar{V}_1$, etc. are not obtained.

Meanwhile, for the solution to the second-order boundary-layer

---

* From Eqs. (50a,b), the range of nonuniformity for the HSDT expansion is found in $\zeta = O(\varepsilon) \ll 1$.

problem, Eqs. (34) admit the form

$$\tilde{u} = \tilde{U}_0(\zeta) + (\varepsilon_T/\xi^n)\tilde{u}_1(\zeta) + \cdots$$

$$\tilde{v} = (1/A)\xi^{-1/4}[\tilde{V}_0(\zeta) + (\varepsilon_T/\xi^n)\tilde{V}_1(\zeta) + \cdots]$$

$$\tilde{p} = (1/A^2)\xi^{-1/2}[P_0 + (\varepsilon_T/\xi^n)a_n P_{00} + \cdots] \qquad (51)$$

$$\tilde{T} = \tilde{T}_0(\zeta) + (\varepsilon_T/\xi^n)\tilde{T}_1(\zeta) + \cdots$$

where

$$\zeta \equiv \tilde{\psi}/A\xi^{1/4} = \xi^{1/2}\Omega/\varepsilon^2 \qquad (51')$$

The differential equations governing $\tilde{U}_0(\zeta)$, $\tilde{V}_0(\zeta)$, and $\tilde{T}_0(\zeta)$ are the same as in Stewartson's classical work [see exposition in (^{37,41})]. The next-order terms $\tilde{U}_1(\zeta)$, $\tilde{V}_1(\zeta)$, and $\tilde{T}_1(\zeta)$ obey a system of linear ordinary differential equations. For the expressed purpose of studying the uniform validity of the solutions around the outer edge, we need only to examine the behavior of each of the terms in Eqs. (51) as $\zeta \to \infty$ (where the outer edge supposedly lies).

The differential equations for the energy conservation in the leading approximation can be reduced to a nondegenerate form as $\zeta \to \infty$ (with $\tilde{U}_0 \to 1$),

$$\tilde{T}_0'' + \frac{\text{Pr}}{4P_0}\zeta\tilde{T}_0' - \frac{\text{Pr}}{2P_0}\left(\frac{\gamma-1}{\gamma}\right)\tilde{T}_0 = 0 \qquad (52)$$

from which the asymptotic solution for $\tilde{T}_0$ can be found. Similarly, one can reduce the momentum, conservation, and continuity equations for the study of asymptotic properties of $\tilde{U}_0$ and $\tilde{V}_0$. The asymptotic solutions, which are bounded at infinity, are

$$\tilde{T}_0 \sim C\zeta^{-(3\gamma-2)/\gamma} \exp[-(\text{Pr}/8P_0)\zeta^2][1 + O(1/\zeta^2)]$$

$$\tilde{U}_0 \sim 1 \qquad (53)$$

$$\tilde{V}_0 \sim V_0$$

where C is a known constant and the remainders in $\tilde{U}_0$ and $\tilde{V}_0$ are exponentially small like $\tilde{T}_0$. The energy equation in the next order can be reduced, for $\zeta \to \infty$ (with $\tilde{U}_0 \to 1$), to

$$\frac{\zeta}{4}\frac{\tilde{T}_1}{\tilde{T}_0}\tilde{T}_0' + \left(\frac{1}{3\gamma} - \frac{1}{2}\right)\tilde{T}_1 - \frac{\zeta}{4}\tilde{T}_1' \sim -\frac{P_0}{\text{Pr}}\frac{\tilde{T}_1}{\tilde{T}_0}(\tilde{U}_0\tilde{T}_0')' + P_0\tilde{T}_0'' \qquad (54)$$

this, together with the corresponding equations from momentum and continuity, gives

$$\tilde{T}_1 \sim E_1 \zeta^{-2/3\gamma}\left[1 + O\left(\frac{1}{\zeta^2}\right)\right]$$

$$\tilde{U}_1 \sim \frac{\gamma}{\gamma - 1} E_1 \zeta^{-2/3\gamma}\left[1 + O\left(\frac{1}{\zeta^2}\right)\right] + E_2 \zeta^{-4[(1/2)-(1/3\gamma)]} \tag{55}$$

$$\tilde{V}_1 \sim \frac{3E_1}{2(3\gamma - 2)P_0} \zeta^{1-(2/3\gamma)}\left[1 + O\left(\frac{1}{\zeta^2}\right)\right] + E_3$$

where E_1 is an undetermined constant and E_2 and E_3 are determined from integration and are fixed if E_1 and the inner boundary condition for $\tilde{U}_1$, $\tilde{V}_1$, and $\tilde{T}_1$ are specified. The remainders are again exponentially small like $\tilde{T}_0$. In view of the behavior shown in Eqs. (53) and (55), through the relation between Ω and ζ, the leading terms of the boundary layer solution, Eqs. (51), can be made to match the leading terms of the HSDT solution, Eqs. (48), except for the temperature. On the other hand, if second-order approximations are included in the boundary layer solution, a formal matching with the HSDT solution then appears possible for the temperature. This follows (if one can ignore the exponential function of ζ) from the comparison

$$\tilde{T} \sim C\zeta^{-(3\gamma-2)/\gamma}\left[\exp - \frac{\text{Pr}}{8P_0}\zeta^2\right] + \left(\frac{\varepsilon}{\xi^{1/4}}\right)^{2[1-(2/3\gamma)]} E_1 \zeta^{-2/3\gamma} + \cdots \tag{56a}$$

$$\varepsilon^2\tilde{T} \sim \varepsilon^2\xi^{-1/2}\Theta_0\Omega^{-2/3\gamma} + \cdots \sim \left(\frac{\varepsilon}{\xi^{1/4}}\right)^{2[1-(2/3\gamma)]}\Theta_0\zeta^{-2/3\gamma} + \cdots \tag{56b}$$

Thus, matching in the temperature determines E_1 as $E_1 = \Theta_0$, while matching in normal velocity determines the shock displacement parameter a_n as $a_n = E_3/V_{00}$. This in essence completes the development based on the two-layer model.

The basic difficulty in establishing the foregoing development as valid asymptotic theory under $(\varepsilon/\xi^{1/4})^{2[1-(2/3\gamma)]} \ll 1$ lies in the fact that in the range of ζ where $\tilde{T}$ and $\varepsilon^2\tilde{T}$ are formally matched, Eq. (56a) is *dominated* by the second-order terms proportional to $\varepsilon_T\zeta^{-2/3\gamma}$. It would appear that as ζ increases, a region of nonuniformity for the boundary layer expansion may be encountered *before* the matching can take place. Generally speaking, the uniform validity of an n-term expansion will depend on whether or not the remainder belongs to an order higher than the nth term. Hence one may settle the question at hand by carrying out the problem to the *third* order. However, in the absence of the third-order results, the observation that the second term in Eq. (56a) is no longer of higher order than the first, and the fact that a third region is actually required in the case of $\omega < 1$, constitute a strong enough argument for a closer scrutiny of the two-layer model for this problem ([49]).

In the following, we analyze the problem by allowing a third intermediate layer in the suspected region of nonuniformity, and demonstrate matchings of leading approximations of the three regions; finally, we show that in the intermediate layer, the intermediate-layer solution and the second-order boundary-layer solution are equivalent.

For this purpose, we introduce a new variable for the third region

$$z \equiv \varepsilon_T \zeta^\alpha \exp(k\zeta^2) \tag{57}$$

with $\alpha = (q\gamma - 8)/3\gamma$ and $k = \mathrm{Pr}/8P_0$. In the region of nonuniformity, z is of unit order. The boundary layer region is identified with a small $z = O(\varepsilon_T)$ and the HSDT region with an exponentially large $z = O[\varepsilon_T \varepsilon^{-2\alpha} \times \exp(k/\varepsilon^4)]$. Note that

$$\zeta^2 = \zeta_*^2 + \frac{\ln z}{k}\left(1 - \frac{\alpha}{2k\zeta_*}\right) + O\left(\frac{1}{\zeta_*^4}\right) \tag{58}$$

where ζ_* is the value of ζ at $z = 1$:

$$\zeta_*^2 = \frac{|\ln \varepsilon_T|}{k}\left[1 - \frac{\alpha}{2}\frac{\ln|\ln \varepsilon_T|}{|\ln \varepsilon_T|} + \cdots\right] \gg 1$$

It is essential to observe here that the range of $z = O(1)$ corresponds to a location in ζ which is logarithmically far from the boundary layer [where $\zeta = O(1)$], yet a thickness in ζ smaller than that of the boundary layer, i.e., $z = O(1)$, corresponds to $\zeta - \zeta_* = O(1/\zeta_*)$. The outer limit of the second-order boundary layer solution Eq. (56a) now reads

$$\tilde{T} \sim \varepsilon_T \zeta_*^{-2/3\gamma}[(C/z) + E_1 \xi^{[(1/3\gamma)-(1/2)]} + \cdots][1 + O(1/\zeta_*^2)] \tag{59}$$

where C is the known constant. This form suggests that we introduce

$$\vartheta \equiv T/\varepsilon_T \zeta_*^{-2/3\gamma} T_0 = \zeta_*^{2/3\gamma} \tilde{T}/\varepsilon_T \tag{60}$$

Notice that the temperature scale for this region belongs to an order (slightly) higher than $\varepsilon_T T_0$ by a logarithmic factor $\zeta_*^{2/3\gamma}$. In terms of ζ, z, and ϑ, with the simplification following Eqs. (37), the energy equation of the

Navier–Stokes system yields

$$z^2 \frac{\partial^2 \vartheta}{\partial z^2} + 2z \frac{\partial \vartheta}{\partial z} = 0 \qquad (61)$$

with a remainder of the order $1/\zeta_*^2$.

Its solution is*

$$\vartheta = f_1(\xi) + [f_2(\xi)/z] \qquad (62)$$

Since the HSDT expansion remains uniform so long as $\zeta \gg \varepsilon$,[†] we may write the inner limit of the HSDT solution, Eq. (56b), as

$$\varepsilon^2 \bar{T} \sim \frac{\varepsilon_T}{\xi^{[1-(2/3\gamma)]/2}} \, \Theta_0 \zeta_*^{-2/3\gamma} \left[1 + O\!\left(\frac{\ln z}{\zeta_*^2} \right) \right] \qquad (63)$$

It is then possible to match $\tilde{T}$ from Eq. (62) over the region

$$1 \ll |\ln z| \ll \zeta_*^2 \sim |\ln \varepsilon_T|/k \qquad (64)$$

This results in

$$f_1(\xi) = \Theta_0/\xi^{(1/2)-(1/3\gamma)} \qquad (65)$$

Now, ϑ of Eq. (62) is identified with the outer limit of the boundary layer solution Eq. (59), provided we set

$$f_2(\xi) = C \qquad \text{and} \qquad E_1 = \Theta_0 \qquad (66)$$

Hence the temperature in the range of $z = O(1)$ is determined as

$$\vartheta = \frac{\Theta_0}{\xi^{(1/2)-(1/3\gamma)}} + \frac{C}{z} \qquad (67)$$

We conclude that the uniformly valid solution in the range $z = O(1)$, where nonuniformity in the boundary layer is suspected, is no more or less than the outer limit of the second-order boundary layer solution, Eq. (56a), and that the results based on the two-layer model (in the von Mises variables) are verified.

Unlike the case with $\omega < 1$, the shock-heating effect is chiefly responsible for the interaction and is directly carried into the boundary layer, i.e., the constant E_1 of the boundary layer is determined by Θ_0 in the outer flow

---

* The next-order solution, accurate to the order of ζ_*^{-2}[(49)], will not be given here.
† Recall footnote on p. 156.

[refer to Eq. (66)]. Contrary to common belief ([7,40,41]), its effect on heat transfer turns out to be numerically small. For the examples studied in ([49]), it is in fact slightly negative.

3. TRANSVERSE CURVATURE AND CROSS FLOW IN THE STRONG-INTERACTION REGIME

Before embarking on the study of the transverse-curvature and cross-flow effects, it is expedient to identify four limits of the inviscid/viscous interaction in hypersonic flow around thin and slender bodies:

I. $M_\infty(\delta/L) \to \infty, \ \delta/\tau \to \infty$;

II. $M_\infty(\delta/L) \to \infty, \ \delta/\tau \to O$;

III. $M_\infty(\delta/L) \to O, \ \delta/\tau \to \infty$;

IV. $M_\infty(\delta/L) \to O, \ \delta/\tau \to O$.

In the axisymmetrical case, the transverse-curvature effect is strong in I and III, weak in II and IV. Only in the first limit does a strong interaction imply a strong transverse-curvature effect. Thus the early work of Probstein and Elliot [see ([7]), 1st ed.] on the transverse-curvature effect on slender cones corresponds to Limit IV. The self-similar solutions of Yasuhara ([50]) and of Stewartson ([37]) for $\frac{3}{4}$-power-law body in plane and axisymmetrical flows apply in a regime spanning I and II. The theory of Stewartson on viscous hypersonic flow past a very slender cone ([51]) and its extension to a more general model gas by Ellinwood and Mirels ([52]) belong to Limit I. The method proposed recently by Mirels and Ellinwood ([53]) applies to the regime between Limits I and II. The study of Shen and Sun ([54]) analyzes a boundary layer problem corresponding to Limit III, and also treats the problem over a range spanning Limits III and IV. An analysis of the flow field in Limit III has also been made recently by Cross ([55]).

For a slender body with cross flow, such as that around a yawed slender cone, Ladyzhenskii ([56]) formulated the hypersonic strong-interaction problem under the assumption $\tau/\delta = O(1)$ (with τ a thickness ratio characterizing the largest transverse body dimension). This work falls into a regime between I and II. The introduction of a cross flow in the strong-interaction regime makes possible a number of changes in the basic flow model. This formulation will be sketched and discussed in Section 3.1. Developments in the solutions to the axisymmetrical hypersonic needle problem are discussed in Sections 3.2 and 3.3.

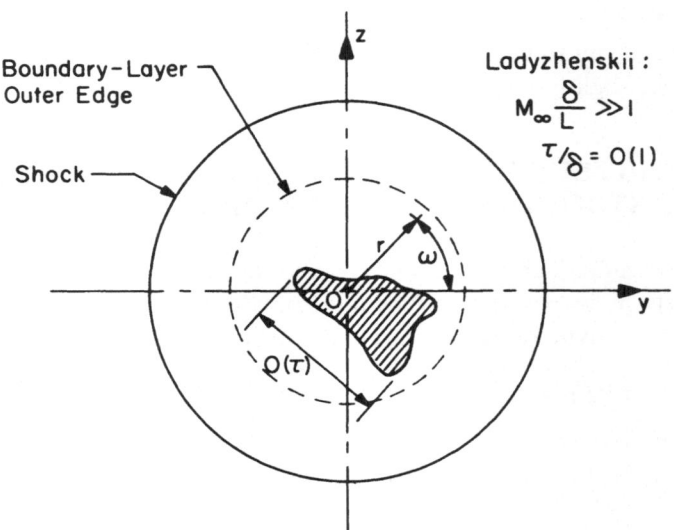

Fig. 9. A sketch in the transverse plane illustrating the persistence of axial symmetry in a viscous hypersonic flow according to Ladyzhenskii ([56]).

3.1. A Strong-Interaction Formulation of Hypersonic Viscous Flow around Asymmetrical Slender Bodies

In the hypersonic boundary layer on a slender body, the density is lower than the outer flow by a factor of δ^2/L^2. For this reason, the boundary layer cannot support a normal pressure gradient as large as that in the outer flow. From the low density, it also follows that the three-dimensional boundary layer on a slender body is characterized in the weak-interaction regime by a strong secondary flow ([57]); namely, if one lets α denote a parameter controlling flow asymmetry, such as a yaw angle, the circumferential velocity in the layer will be of an order much lower than αu_∞. This follows because, with a circumferential pressure gradient being given, the fluid accelerates and decelerates (in the circumferential direction) at a rate in inverse proportion to the local density.

The strong-interaction regime offers, however, another possibility regarding the secondary flow. Let us consider a hypersonic viscous flow around a general slender body of which the maximum transverse dimension of the configuration, measured from the wind axis through the apex (nose), does not exceed the viscous core (see Fig. 9), i.e.,

$$\max.(\tau, \alpha) = O(\delta/L) \tag{68}$$

For simplicity, we require that the nose bluntness is not important.* By relieving the transverse pressure gradient through interaction, cross flow can now remain small at the order αu_∞, in spite of the low density, which in fact helps to reinforce a uniform pressure. The situation at hand is very similar to that associated with the entropy wake and the hypersonic-area rule for the inviscid flow (see Section 1.1).

It is obvious that if this alternative of a small cross flow is to be stipulated, as it was by Ladyzhenskii ([56]), the flow model consists of a nearly axisymmetrical outer (HSDT) flow and an inner viscous core characterized by a nearly uniform pressure.† It is essential to study the proper scales and variables assumed in the formulation for the two regions. In terms of cylindrical polar coordinates (x, r, ω), as specified in Fig. 9, Ladyzhenskii defines a set of dimensionless variables, distinguished by a tilde, for the viscous region:

$$x = \tilde{x}, \qquad r = \delta r, \qquad \omega = \omega$$

$$u = u_\infty \tilde{u}, \qquad v = \delta u_\infty v, \qquad w = \delta u_\infty \tilde{w}$$

$$\varrho = \delta^2 \varrho_\infty \tilde{\varrho}, \qquad H = \tfrac{1}{2} u_\infty^2 \tilde{H} \tag{69a}$$

$$p = \varrho_\infty u_\infty^2 \delta^2 [\tilde{p}_0(\tilde{x}) + \delta^2 \tilde{p}_1(\tilde{x}, r, \omega)]$$

$$r_\delta = \delta [\tilde{R}_\delta(\tilde{x}) + \delta^2 \tilde{y}_\delta(\tilde{x})]$$

and for the outer flow (with the dimensionless quantities denoted by a bar),

$$x = \bar{x}, \qquad r = \delta \bar{r}, \qquad \omega = \omega$$

$$u = u_\infty [1 + O(\delta^2)], \qquad v = \delta \bar{v}, \qquad w = \delta^3 \bar{w}$$

$$\varrho = \varrho_\infty \bar{\varrho}, \qquad H = \tfrac{1}{2} u_\infty^2 \tag{69b}$$

$$p = \varrho_\infty u_\infty^2 \delta^2 [\bar{p}_0(\bar{x}) + O(\delta^2)]$$

$$r_s = \delta [\bar{R}_s(\bar{x}) + O(\delta^2)]$$

In the above, the reference length for the x has been taken to be unity, i.e., $L = 1$, and v and w are the velocity components associated with the variables r and ω. The subscripts δ and s refer to the viscous core outer edge and the shock, respectively. All dimensionless variables will be assumed to be of

---

* A sufficient condition for fulfilling this requirement is $r_b/x^{2/(3+\nu)} \neq \infty$ as $x \to 0$.
† For a complete description, there is of course a viscous transition layer in the sense of Section 2.4 which is not essential, however, for the flows in the two main regions studied here.

order unity in the regions analyzed. It is essential to observe that the second term in the pressure $\delta^2 \tilde{p}_1$ accounts solely for the transverse pressure variation, but does *not* represent the entire second-order correction to the pressure.

The dependence of these scales on δ is made consistent, under Eq. (68), with the model of a small cross flow and a viscous core of nearly uniform pressure discussed earlier. The particular exponents of δ appearing in $\delta^2 \tilde{p}_-$, $\delta^3 \tilde{y}_\delta$ and $\delta^3 \bar{w}$ may become evident by observing the fact that if the pressure were not nearly uniform, these terms would appear as $(\alpha/\delta)\tilde{p}_1$, $\delta(\alpha/\delta)\tilde{y}_\delta$, and $\alpha \bar{w}$, or as $\tilde{p}_1$, $\delta \tilde{y}_\delta$, and $\delta \bar{w}$ under the conditions of Eq. (68). With the low density now taken into account, each of these scales has to be multiplied by δ^2. It may be pointed out that the forms of Eqs. (69) and the formulations that follow are also applicable to a flat wing of very low aspect ratio under the condition $\tau \ll \alpha = O(\delta)$.*

Based on the set of scales shown in Eqs. (69a), the Navier–Stokes equations can be reduced to a system of parabolic differential equations, subject to a relative error of the order δ^2. It suffices to point out that the resulting equations corresponding to the two transverse momenta involve the transverse gradients of $\tilde{p}_1$, which, except for the absence of the second-order partial x-derivative, contains all terms of the Navier–Stokes equations. This means that the transverse pressure variation $\delta^2 \tilde{p}_1(\tilde{x}, \tilde{y}, \omega)$, though belonging to an order δ^2 higher than $\tilde{p}_0$, has to be solved simultaneously along with the leading approximations of the other variables. This is so because with the density of order $\delta^2 \varrho_\infty$, a transverse pressure change of order $\delta^2 \varrho_\infty u_\infty^2$ is sufficient to cause appreciable change in the transverse velocities, which are of order δu_∞.

According to this formulation, the magnitudes of pressure and of the viscous stress components are in the proportion, in the case where $\alpha/\tau = O(1)$,

$$p : p_1 : |\sigma_{xy}| : |\sigma_{yz}| \approx \delta^2 : \delta^4 : \delta^3 : \delta^4 \tag{70}$$

It follows that the drag X (contributed equally by the pressure $\tilde{p}$ and skin friction), and the transverse forces Y and Z (shared equally by the pressure $\tilde{p}_1$ and the viscous stresses) may be estimated, respectively, as

$$\frac{X}{\varrho_\infty u_\infty^2 L^2} = O(\delta^4), \qquad \frac{Y}{\varrho_\infty u_\infty^2 L^2}, \frac{Z}{\varrho_\infty u_\infty^2 L^2} = O(\delta^5) \tag{71}^\dagger$$

---

* The immediate vicinity of the two sharp leading edges must of course be excluded.

$\dagger$ Then, the corresponding lift/drag ratio $\sim N/D = O(\delta) = O(\alpha)$, which may be compared to $O(\alpha/\tau^2)$ for a slender yawed cone in an inviscid hypersonic flow.

Interestingly, for a uniform body temperature, a viscosity law $\mu \propto T^{\omega}$, and a body shape

$$r = \chi_b(\omega)x^{3/4} \tag{72}$$

Ladyshenskii's formulation admits a self-similar solution which is a generalization of Stewartson's [37] and Yasuhara's [50] self-similar solutions for the interaction problem. Ladyzhenskii [56] has presented a study of this self-similar system including a detailed examination of the singular behavior of the solution at the outer edge. Inasmuch as the existence of a solution to the reduced two-point boundary-value problem has not yet been demonstrated, the validity of the foregoing formulation (based on the alternative of a small cross flow) cannot be considered as being theoretically established.

3.2. The Needle in the Strong-Interaction Regime

For study of the transverse curvature effect, let us introduce the parameter

$$\mathcal{S} \equiv \tau L/\delta \tag{73}$$

where τ and δ signify, respectively, the body thickness and the boundary layer thickness as before. To make the subsequent analysis more specific, we shall define τ through the equation of the body

$$y_b/L = \tau F(x/L) \tag{73'}$$

where F is regular function of x/L independent of τ; we also choose δ according to Eq. (34″) with $\varepsilon = \delta/L$, i.e.,

$$\delta = \left\{ \frac{\gamma - 1}{2\gamma} \frac{\mu_0}{\varrho_\infty u_\infty L} \right\}^{1/4} L \tag{73''}$$

Yasuhara's self-similar solution [50] for the $\frac{3}{4}$-power-law body of revolution is valid for a strong shock and a nonslip surface at a uniform temperature over the entire range of $\mathcal{S}$, namely, $0 < \mathcal{S} < \infty$. The system of ordinary differential equations develops a singularity in the limit $\mathcal{S} \to 0$ (corresponding to a vanishingly thin needle), but is amenable to a treatment similar to that used in the asymptotic theory of a boundary layer on a circular cylinder in axial incompressible flow [[58,59]; also see [37]]. In fact, as noted by Stewartson [51], the self-similar solution, except in a region near the body, remains a valid asymptotic description of the strong-interaction problem for a needle of general shape. In view of its relevance to a number of recent studies [52-55,60,61], as well as to unpublished current investigations, Stewartson's solution [51] and its extension [52] will be developed below. The approach and procedure employed here, which avoids treating $\ln \mathscr{S}$ as being of order unity for small $\mathscr{S}$, as was done in [51,52], displays the salient features of the basic flow field more clearly, and is more in keeping with the viewpoint of limit-process expansions [16,44].* For the more general problems, such as the one involving a yawed needle, we find the more systematic approach used here indispensable.

Let us begin with the special example of a $\frac{3}{4}$-power-law body and write Eq. (73′) for this case as

$$y_b/\tau L = F(x/L) = R_0(x/L)^{3/4} \tag{74a}$$

$$y_s/\delta = A(x/L)^{3/4} \tag{74b}$$

where R_0 is a specified constant of order unity and A an unknown parameter which may depend on both δ and τ. The self-similar solution may be expressed in terms of the similarity variable for this case ($m = \frac{3}{4}$),

$$\zeta \equiv \frac{\tilde{\psi}}{\frac{3}{4}\sqrt{2P(0)/(\gamma + 1)}(A^2\tilde{x})} \tag{75}$$

where $P(0)$ is the ratio p_b/p_s and is 0.846 for $\gamma = \frac{5}{3}$ and 0.835 for $\gamma = \frac{7}{5}$. For the present purpose, it suffices to study the case of a heat-insulated slender body and a unit Prandtl number. For this particular case, the equations for the self-similar boundary layer under the strong shock of Eq. (74b) are

$$-\zeta\tilde{u}\tilde{u}' = \tilde{u}[(1 - \tilde{u}^2)^{\omega-1}Q\tilde{u}\tilde{u}']' + \frac{\gamma - 1}{4\gamma}(1 - \tilde{u}^2)$$

$$A^2Q' = k(1 - \tilde{u}^2)/\tilde{u} \tag{76}$$

$$\tilde{u}(\infty) = 1, \qquad Q(\infty) = Y^2(0)$$

$$\tilde{u}(0) = 0, \qquad Q(0) = \mathscr{S}R_0^2/A^2$$

$$Q \equiv \left(\frac{r}{r_s}\right)^2, \qquad k \equiv \frac{4}{3}\sqrt{\frac{\gamma + 1}{2P(0)}}$$

where the prime stands for derivative with respect to ζ, and $Y(0) = y_b/y_s$ and is 0.819 for $\gamma = \frac{5}{3}$ and 0.875 for $\gamma = \frac{7}{5}$.

---

* A similar work has recently been reported by Bush [62]. A more general formulation is presented in [6].

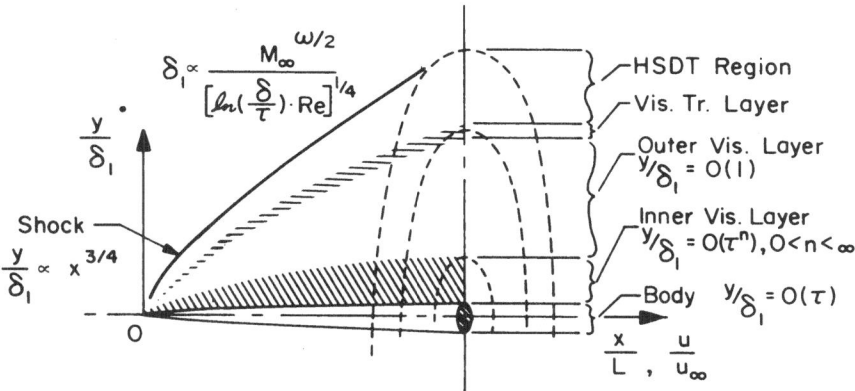

Fig. 10. Classification of flow regions in the viscous hypersonic needle problem: $M_\infty\delta/L \gg 1$, $\mathscr{S} = \tau/\delta \to 0$.

In view of the outer boundary condition, Q is seen to be of order unity. However, if $\mathscr{S} \to 0$, Q is expected to vanish with $\mathscr{S}$ near the body; it is apparent that the boundary layer solution involves two distinct regions.* In the following, we shall first analyze on the basis of Eqs. (76) the outer viscous region which supports a $\frac{3}{4}$-power-layer shock; we subsequently analyze an inner viscous region which does not require the similarity assumption. We may then establish, through matching of the two solutions, the validity of the $\frac{3}{4}$-power-law, self-similar solution for the outer viscous and inviscid fields for a general axisymmetrical needle (Fig. 10).

Following Stewartson ([51]), we assume that A^2 vanishes with $\mathscr{S}$, and that the flow speed in the outer viscous region is near that of the free stream. Equations (76) then admit a solution (asymptotic in $\mathscr{S}$) of the form

$$\tilde{u} = 1 - \varepsilon_1\breve{u}, \qquad \zeta = \varepsilon_1^{(\omega-1)/2}\breve{\zeta}, \qquad (77)$$

with $\varepsilon_1 \equiv (A^2/k)^{2/(1+\omega)}$, where $\breve{u}$, $\breve{\zeta}$, and Q are of order unity and the proper relation between A^2 (hence ε_1) and $\mathscr{S}$ is to be determined later. The differential equations for $\breve{u}$ and $\breve{\zeta}$ reduced from Eq. (76) are subject to an error of order ε_1,

$$2^\omega \frac{d}{dQ}\left[Q\breve{u}^\omega \frac{d\breve{u}}{dQ}\right] + \breve{\zeta}\frac{d\breve{u}}{dQ} - \frac{\gamma-1}{4\gamma} = 0$$

$$\frac{d\breve{\zeta}}{dQ} = \frac{1}{2\breve{u}} \qquad (78)$$

---

* There may, of course, be a viscous transition layer in the sense of Section 2.

Integrating once, eliminating $\breve{u}$, and making use of one of the outer boundary conditions $Q(\infty) = Y^2(0)$, one has a second-order differential equation for Q as function of ξ

$$\tfrac{1}{2}Q(Q')^{\omega-1}Q'' + \tfrac{1}{2}\xi Q' - [(3\gamma - 1)/4\gamma]Q = -\breve{C}_0$$

$$\breve{C}_0 \equiv [(3\gamma - 1)/4\gamma]Y^2(0)$$

(79)

from which $\breve{u}$ can be obtained as $\breve{u} = Q'/2$. Equation (79) admits an outer limit which satisfies the remaining boundary condition $\breve{u}(\infty) = 0$ with a behavior basically the same as that for a plane flow, i.e., Eq. (35) of Section 2.3. In the inner limit, $\xi \to 0$ or $Q \to 0$, and Eq. (79) gives

$$\xi \sim \left[\frac{1}{2\breve{C}_0(1 + \omega)}\right]^{1/(1+\omega)} \int_0 \frac{dQ}{[\ln(C'/Q)]^{1/(1+\omega)}} + C''$$

$$\breve{u} \sim \frac{1}{2}\,[2(1 + \omega)\breve{C}_0]^{1/(1+\omega)}\left[\ln\!\left(\frac{C'}{Q}\right)\right]^{1/(1+\omega)}$$

(80)

It may be inferred from Eqs. (79) and (80) that with $\breve{u}(\infty)$ or $Q'(\infty)/2 = 0$ as an outer boundary condition, the constant C' appearing in Eq. (80) is determined by integration of Eq. (79) from infinity.

Now, the second of Eqs. (80) indicates a nonuniformity in the expansion for small ε_1, namely, as $Q \to 0$,

$$\breve{u} = 1 - \varepsilon\breve{u} \approx 1 - \frac{\varepsilon_1}{2}\,[2(1 + \omega)\breve{C}_0]^{1/(1+\omega)}\,|\ln Q|^{1/(1+\omega)} + \cdots \quad (81)$$

which gives a region of nonuniformity at

$$\varepsilon_1^{1+\omega}\,|\ln Q| = O(1) \qquad \text{or} \qquad Q^{\varepsilon_1^{1+\omega}} = O(1) \quad (81')$$

This is a situation similar to that of the analysis of the entropy wake discussed in Section 1.2. We shall accordingly introduce a new variable for this inner region*:

$$\hat{Q} \equiv (Q/k_1)^{\varepsilon_1^{1+\omega}} \quad (82)$$

where k_1 is a constant to be chosen later.

One remarkable feature of such a region of Q as represented by Eqs. (81'), or $\hat{Q} = O(1)$, is its extensive range for $Q \ll 1$. As will be seen later, $\varepsilon_1^{1+\omega} = O(1/|\ln\mathscr{S}|)$; thus any range of Q representable as $Q = O(\mathscr{S}^n)$, $0 < n$

---

* One may, of course, also use $\hat{Q} = \varepsilon_1^{1+\omega}\ln Q$.

$< \infty$, falls into this region. Assuming that u/u_∞ is of order unity, Eqs. (76), as well as the Navier–Stokes equations (without the similarity assumption) give a Couette-type equation for $\hat{u} \equiv u/u_\infty$ in the inner region,

$$\frac{\partial}{\partial \hat{Q}} \left[(1 - \hat{u}^2)^\omega \hat{Q} \frac{\partial \hat{u}}{\partial \hat{Q}} \right] = 0 \tag{83}$$

with a remainder being exponentially small in ε_1. One may also evaluate ζ or the corresponding stream function in the general situation, which is also found to be exponentially small in ε_1.

The solution to Eq. (83) is

$$\int_1 (1 - \hat{u}^2)^\omega \, d\hat{u} = \hat{C}_0 \ln \hat{Q} + \hat{C}_1 \tag{84}$$

In studying the outer limit of $\hat{u}$, it is convenient to work with the quantity $(1 - \hat{u})$. Considering $(1 - \hat{u}) \to 0$, Eq. (84) yields

$$\frac{2^\omega}{1 + \omega} (1 - \hat{u})^{1+\omega} \sim - \hat{C}_1 + \varepsilon_1^{1+\omega} \hat{C}_0 [\ln k - \ln Q] + O[\varepsilon_1^{2+\omega} f(Q)] \tag{85a}$$

The corresponding inner limit of the outer viscous solution, Eq. (81), can be arranged as

$$\frac{2^\omega}{1 + \omega} (1 - \hat{u})^{1+\omega} \sim \varepsilon_1^{1+\omega} \check{C}_0 [\ln C' - \ln Q] + O[\varepsilon_1^{2+\omega} f(Q)] \tag{85b}$$

Matching the two forms, noting $\hat{u} = \check{u}$, we determine all three constants,

$$\hat{C}_1 = 0, \qquad \hat{C}_0 = \check{C}_0 \equiv \frac{3\gamma - 1}{4\gamma} Y^2(0), \qquad k_1 = C' \tag{86}$$

Hence, in terms of the variable $\hat{Q}$, the inner solution is completely determined, subject to an error of order ε_1^2, as

$$\int_1 (1 - \hat{u}^2)^\omega \, d\hat{u} = \frac{3\gamma - 1}{4\gamma} Y^2(0) \ln \hat{Q} \tag{87}$$

The boundary condition for $\hat{u}$ on the non-slip surface of a general slender body is

$$\hat{u} = 0 \qquad \text{at} \quad Q = \mathscr{S} R^2 / A^2 \tag{88}$$

where $R(\tilde{x}) \equiv f(\tilde{x})/(\tilde{x})^{3/4}$. With Eq. (87) and the transformation Eq. (82), this remaining boundary condition furnishes an equation determining ε_1 and the shock amplitude A in terms of $\mathscr{S}$, subject to a relative error of order ε_1^2,

$$\varepsilon_1^{1+\omega} \equiv \frac{A^4}{k^2} = \frac{\int_0^1 (1-t^2)^\omega \, dt}{[(3\gamma-1)/4\gamma]Y^2(0)\ln(A^2C'/\mathscr{S}R^2)} \tag{89}$$

A valid explicit result for A^2 is

$$A^2 \sim k \left| \frac{K}{\ln\mathscr{S}} \right|^{1/2} \left[1 - \frac{1}{2} \frac{\ln(KkC'/|R^2\ln\mathscr{S}|)}{|\ln\mathscr{S}|} + O\left| \frac{\ln|\ln\mathscr{S}|}{\ln\mathscr{S}} \right|^2 \right] \tag{90}$$

with

$$K \equiv \frac{\int_0^1 (1-t^2)^\omega \, dt}{[(3\gamma-1)/4\gamma]Y^2(0)}$$

Combining Eqs. (87) and (89) and expressing $\hat{Q}$ in terms of Q, we recover the results of ([51,52]) for the inner region,

$$\int_0 (1-\hat{u}^2)^\omega \, d\hat{u} = \frac{3\gamma-1}{4\gamma} Y^2(0)\varepsilon_1^{1+\omega} \ln\left(\frac{QA^2}{\mathscr{S}k_1}\right) \tag{91}$$

From A, one determines the shock radius $y_s(x)$, the boundary layer outer edge $y_\delta(x)$, and the surface pressure $p_b(x)$,

$$y_s(x) = \delta A x^{3/4}, \qquad y_\delta(x) = \delta A Y(0) x^{3/4} \tag{92}$$

$$p_b(x) = \frac{2\gamma}{\gamma+1} \frac{9}{16} P(0)p_\infty M_\infty^2 A^2 \delta^2 x^{-1/2}$$

According to Eq. (90), the leading term in A^2 is determined by the order of magnitude (the "ball-park value") of τ/δ, and is *independent* of the body shape, so long as $R(\tilde{x})$ is bounded. Thus in the leading approximation Eq. (92) gives a shock, boundary-layer thickness, and pressure the same as for a $\frac{3}{4}$-power-law body, except that the magnitudes of y_s and y_δ are reduced by a factor of $|\ln\mathscr{S}|^{1/4}$, and p_δ by $|\ln\mathscr{S}|^{1/2}$. Furthermore, the leading terms are independent of the constant of integration C', as noted by Ellinwood and Mirels ([52]). In view of Eq. (90), the next approximation depends on C' and also on the departure from a $\frac{3}{4}$-power-law body shape, i.e., $R(\tilde{x})$. However, there is nonuniformity in v near the body (discussed on p. 173) which will affect the tangential velocity to the same order as C' does, i.e., $\varepsilon_1^{1+\omega}$. Consequently, it does not warrant further discussion.

On the basis of Eq. (87), we have

$$r(1 - \hat{u}^2)^{\omega} \frac{\partial \hat{u}}{\partial r} = \frac{3\gamma - 1}{4\gamma} Y^2(0)$$

It follows that the friction drag per unit length is independent of x and of the body shape, and that the magnitude of the slip (and that of the temperature jump) is increased from their plane-flow values by a factor of δ/τ. In fact, by allowing a low body temperature and a small $(\gamma - 1)/2\gamma$, it is not difficult to show that the slip effect belongs to

$$\frac{\Delta u}{u_\infty} \sim \frac{\delta}{\tau} \sqrt{\frac{2\gamma}{\gamma - 1} \frac{T_w}{T_0}} \qquad \frac{\delta}{L} \sim \frac{1}{\mathscr{S}} \left(\frac{\Delta u}{u_\infty} \right)_{r=0} \tag{93}$$

Hence the foregoing conclusion of slip corrections is independent of the specific-heat ratio and the wall temperature.

The needle solutions to both the insulated and the noninsulated cases are given in Stewartson's original paper [51] for a linear viscosity law and unit Prandtl number.* Its extension to a viscosity law $\mu \propto T^{\omega}$ with $\mathrm{Pr} \neq 1$ has been made by Ellinwood and Mirels in [52]. In the same paper, they also give an extensive tabulation of numerical data for the self-similar $\frac{3}{4}$-power case over a wide range of $\mathscr{S}$. Mirels and Ellinwood, in a separate investigation [53], proposed a procedure to treat viscous hypersonic flow past axisymmetrical bodies of general shape under a strong shock, encompassing both the strong- and weak-interaction regimes. The procedure employs a local-similarity method similar to that of Dewey [63] to handle boundary layer equations with transverse curvature. The analysis reproduces the correct solutions for cones in both limits $\mathscr{S} \to 0$ and $\mathscr{S} \to \infty$.

Mirels and Ellinwood [53] have correlated recent experimental data on slender-cone drag from [64,65] in terms of a parameter Λ which is essentially $1/\mathscr{S}$, and compared them with their local-similarity calculation. Their results are reproduced in Fig. 11. Included is the solution for the limit $\mathscr{S} \to 0$, which, in spite of the fact that $1/|\ln \mathscr{S}|$ is not a very small number in practice, shows reasonable agreement in the data of Kussoy [64]. Experimental confirmation of Stewartson's cone theory has also been reported by Solomon [60] and recently by Horstman and Kussoy [61]. Solomon [60] also includes test results for a $3°$ half-cone at $10°$ yaw. Horstman and Kussoy's [61] cone-pressure data are reproduced in Fig. 12. Considering

---

* Professor K. Stewartson pointed out to me a few algebraic errors in his results for $T_w/T_0 \neq 1$, Eqs. (4.23)-(4.27) of [51], which are also noted in [52,60].

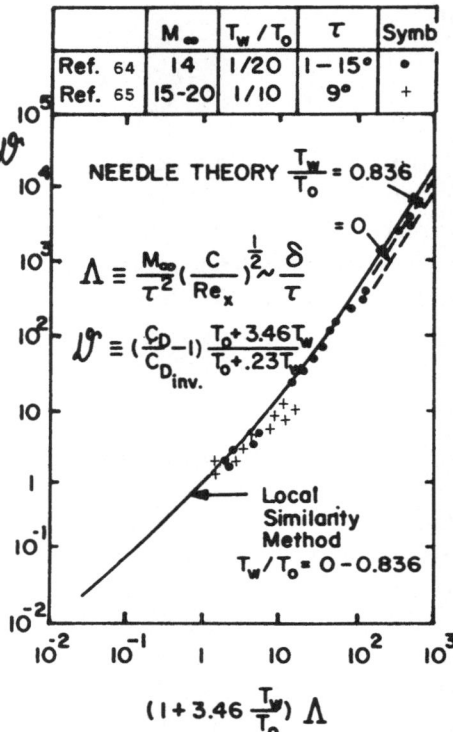

Fig. 11. Correlation of the slender-cone drag data according to Mirels and Ellinwood [53].

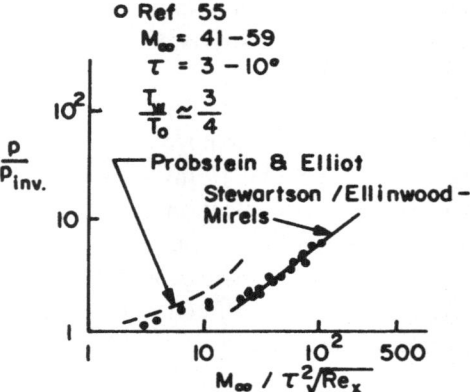

Fig. 12. Comparison of experimental and theoretical cone-pressure data for helium flow [61].

the rather low orders of magnitude of the remainder in the asymptotic formula and of the slip effects, the close agreement reported seems remarkable.

For hypersonic viscous flow around a yawed slender cone, or an asymmetrical needle of arbitrary shape, a similar approach to the strong-interaction problem applies, provided $M_\infty^2 \delta^2 \to \infty$ and $\tau_1/\delta \equiv \mathscr{S} \to 0$, where τ_1 now characterizes the largest transverse distance from the wind axis. The analysis of the asymmetrical aspect of the needle problem is valuable in furnishing examples with which Ladyzhenskii's formulation and the flow-symmetry property discussed in Section 3.1 may be tested theoretically. In the following, we discuss the basic ideas and features of a current study of this topic being conducted by H. T. Iida in collaboration with the author([66]).

To apply the asymptotic approach to the problem, we need an additional region with a transverse dimension of $O(\tau_1)$. This is found necessary not only to accommodate the asymmetrical inner boundary, but as a consequence of a nonuniformity in the transverse velocity v according to the development based on Eq. (83) for the inner viscous region (even for the axisymmetrical case).* To distinguish it from the inner viscous region where $\hat{Q} = O(1)$, we refer to this region where $r/L = O(\tau_1)$ as the Stokes-like region for reasons to become obvious later.

Physically, the outer flow field shall remain axisymmetrical so long as $\tau_1/\delta \to 0$. Indeed, it is possible to formulate the problem in the three viscous regions[†] in a consistent manner so that the u and v fields in the two outer regions are nearly axisymmetrical.

The formulation depends critically on the scales and forms of the variables used. For the innermost region, we find the following are admissible by the equations:

$$x = L\check{x}, \qquad y = \tau_1 L\check{y}, \qquad z = \tau_1 L\check{z}$$

$$u = \varepsilon_1^{1+\omega} u_\infty \check{u}, \qquad v = \varepsilon_1^{1+\omega} \tau_1 u_\infty \check{v}, \qquad w = \varepsilon_1^{1+\omega} \tau_1 u_\infty \check{w} \qquad (94)$$

$$p = \varrho_\infty u_\infty^2 \delta_1^{\ 2} [\check{p}(\check{x}) + \delta_1^2 \check{p}_1(\check{x}, \check{y}, \check{z})], \qquad T = T_w \check{T}$$

where $\delta_1 \equiv \delta/|\ln \delta/\tau_1|^{1/4}$. The overall scales for u, v, and p are the same as indicated by the axisymmetrical solution at $Q = O(\mathscr{S}/A^2)$ [see Eq. (91)]. Of interest is the manner in which the asymmetry enters into the pressure field—only through the term $\delta_1^2 \check{p}_1(\check{x}, \check{y}, \check{z})$.

---

* This nonuniformity in v does not, however, affect the leading solution for u ([66]).
† We again omit the viscous transition layer from the present discussion, since it does not play a major role in this study.

In the leading approximation, $\check{T} = 1 + o$, and the Navier–Stokes equation for the x-momentum yields a Laplace equation for $\check{u}$:

$$\nabla_1{}^2 \check{u} = 0 \qquad (95)$$

where $\nabla_1{}^2 \equiv \partial^2/\partial \check{y}^2 + \partial^2/\partial \check{z}^2$. With $\check{u} = 0$ on the body, $\check{u}$ is completely determined by matching with the nearly-symmetrical u-field of the neighboring region. The leading approximation of the latter is given by Eq. (91). The matching gives an *equivalent* body of revolution for the outer fields and also determines the ε_1, and hence the shock amplitude A and other flow variables (except the circumferential velocity) in the leading approximation of the two outer regions.

With the scales and variables of Eqs. (94), the continuity equation and the Navier–Stokes equations for the transverse momenta yield a system for the leading approximations for $\check{v}$, $\check{w}$, and $\check{p}_1$:

$$\frac{\partial \check{v}}{\partial \check{y}} + \frac{\partial \check{w}}{\partial \check{z}} = -\check{u}\,\frac{\partial}{\partial x}\ln(\check{p}\check{u})$$

$$\left(\frac{T_w}{T_0}\right)^{\omega} \nabla_1{}^2 \begin{pmatrix} \check{v} \\ \check{w} \end{pmatrix} = \begin{pmatrix} \partial/\partial \check{y} \\ \partial/\partial \check{z} \end{pmatrix}\left(\check{p}_1 + \frac{\check{u}}{3}\,\frac{d}{dx}\ln \check{p}\right) \qquad (96)$$

Since $\check{u}$ and $\check{p}$ are known, Eqs. (96) may be compared with the system for a Stokes flow with distributed sources and body forces. At the body surface, $\check{v}$ and $\check{w}$ are specified; e.g., $\check{v} = 0$ and $\check{w} = 0$ on a nonslip, impermeable surface.

It is easy to find a stream function $\check{\psi}$ that satisfies the first of Eqs. (96) and reduce the remaining equations after eliminating $\check{p}_1$ to a form

$$\nabla_1{}^2\nabla_1{}^2\check{\psi} = \check{R}(\check{x}, \check{y}, \check{z}) \qquad (96')$$

where $\check{R}$ is a function of $\check{x}$, $\check{y}$, and $\check{z}$ depending on the definition of $\check{\psi}$. The reduced boundary-value problem is therefore the same as for the classical two-dimensional Stokes flow. In the outer limit $\check{r} \equiv (\check{y}^2 + \check{z}^2)^{1/2} \to \infty$, Eqs. (96) give a radial velocity which is dominated by $\check{r}\ln \check{r}$ and matches its counterparts in the Couette-like region. However, logarithmic singularities identified with those in the Stokes paradox [21,44] arise, and their strengths are determined after the solutions for the three viscous regions are matched. The order of the perturbation in the HSDT region is established as order $\alpha\delta_1/\varepsilon_1$.

It is of interest to compare the orders of magnitude of certain flow variables of interest near the body. Consider the case of a body of revolution of thickness τ at an angle of attack $\alpha = O(\tau)$. The two pressures p and p_1, the skin frictions in the axial and in the transverse directions, are in the proportion

$$p : p_1 : |\sigma_{xy}| : |\sigma_{yz}| \approx \tau | \ln \mathscr{S} |^{1/2} : (T_w/T_0)^\omega \delta^2 \alpha : \delta^2 : (T_w/T_0)^\omega \delta^2 \alpha \quad (97)$$

If we further confine the case to a T_w/T_0 of order unity and $\delta^2 \ll \tau \ll \delta$ [which also ensures a small slip; see Eq. (93)], we find the ratio of the Burnett terms to the Chapman–Enskog terms in both the Stokes-like and Couette-like regions to be small*:

$$\sigma_{xx}/p = (\mu \, \partial u/\partial x)/p \sim \delta^3/\tau \, | \ln \mathscr{S} |^{3/4} \ll 1 \quad (97a)$$

the drag coefficient to be

$$C_D = D/\varrho_\infty u_\infty^2 L^2 \tau^2 \sim \delta^2/ |\ln \mathscr{S} |^{1/2} \ll 1 \quad (97b)$$

and the lateral force/drag ratio to be[†]

$$N/D \sim \alpha \delta^2/\tau^2 \, | \ln \mathscr{S} |^{1/2} \ll 1 \quad (97c)$$

3.3. The Needle Problem in the Weak-Shock Regimes

In the weak-interaction regimes involving weak shock, the range of Mach number is not of basic importance, and one may study the transverse-curvature effect without considering the special limit $M_\infty \to \infty$. In a relatively recent paper, Shen and Sun [54] analyze the strong transverse-curvature limit, $\mathscr{S} \equiv \tau/\delta \to 0$, for an axisymmetrical body with a unit Prandtl number and a linear viscosity law. The inner viscous region discussed earlier is treated in their analysis by the PLK method. Their results constitute a generalization of the earlier works on incompressible flow [see [58,59]), for example] and include higher-order corrections in $\mathscr{S}$ for a power-law body. The displacement effects are not evaluated, however. Therefore the theory is subject to an error of the order $M_\infty^2 \delta/| M_\infty^2 - 1 |^{1/2}$. In the same paper, a momentum integral method employing Crocco's variables is proposed for treating the weak-shock regime for the entire range $0 < \mathscr{S} < \infty$.

The problem in the limit $\mathscr{S} \to 0$, $M_\infty \delta_1 \to 0$ was analyzed recently by Cross [55]; this analysis was carried out as an axisymmetrical counterpart of an earlier development for a flat plate in hypersonic plane flow [46].

---

* The ratio $(\mu \, \partial u/\partial y)/p$ does not occur in the equations of change.
† The ratio appears to be much lower than the inviscid value.

An asymptotic solution to the entire field under the condition $\mathscr{S} \to 0$, $M_\infty \delta_1 \to 0$ is sought, including the regions near the shock and near the boundary layer edge. The analysis is restricted to a viscosity law $\mu \propto T^\omega$ with $\omega < 1$, and a needle of paraboloidal shape, i.e., $y_b \propto \sqrt{x}$.*

The nonuniformity of the linearized solution in the vicinity of the bow shock of a slender body is well known ([16,44,67,68]). The problem arising in the "exterior inviscid layer" of Cross and Bush's model [in both ([46]) and ([55])] is of the same nature, although some differences in viewpoint may be noted. Shock decays in problems of supersonic bangs ([44,67]) are analyzed as for a far field, i.e., $y \to \infty$ and $x - \beta y = O(1)$; in the problem at hand, it is analyzed with $y = O(1)$ and $x - \beta y \to 0$. However, the two cases equivalent, by virtue of the scale invariance of an inviscid flow.

Following the method of Cole ([44]), the shocks in Bush and Cross' studies are analyzed as those produced by a parabola in plane flow and by a paraboloid in axisymmetrical flow, say,

$$y_b/L = \varepsilon R(x/L) = \varepsilon (x/L)^{1/2} \tag{98}$$

Below, the principal results of the shock problem will be recovered from the potential theory for both axisymmetrical and plane flow.

Near the Mach envelope from the apex, the linearized potential solution for a parabolic body behaves as (taking $L = 1$)

$$\varphi = -\frac{\varepsilon}{B}(x - By)^{1/2}, \qquad \nu = 0 \tag{99}$$
$$= -\varepsilon^2 u_\infty \int_0^{x-By} \frac{R(x_1)R'(x_1)\,dx_1}{[(x-x_1)^2 - B^2 y^2]^{1/2}} \sim -\varepsilon u_\infty \left(\frac{x-By}{2x}\right)^{1/2}, \qquad \nu = 1$$

where $B \equiv (M_\infty^2 - 1)^{1/2}$, and the perturbation velocities φ_x and ψ_y tend to infinity like $1/(x - By)^{1/2}$. The region of nonuniformity may best be studied in terms of the variables

$$\xi = x - By, \qquad \eta = By, \qquad \bar{\varphi} = B^{1-\nu}\varphi/\varepsilon^{1+\nu}u_\infty \tag{100}$$

Then, Eq. (99) becomes

$$\bar{\varphi} \sim -[\xi/(2\eta)^\nu]^{1/2}, \qquad \xi \to 0 \tag{99'}$$

The nonlinear equation governing the perturbation potential becomes

$$2\bar{\varphi}_{\xi\eta} + (\nu/\eta)\bar{\varphi}_\xi + k\omega\bar{\varphi}_\xi\bar{\varphi}_{\xi\xi} = \bar{\varphi}_{\eta\eta} + (\nu/\eta)\bar{\varphi}_\eta + O[\omega B^{\nu-2}\bar{\varphi}_\xi^2] \tag{101}$$

* It appears that the case of $\omega \leq 1$ with $y_b \propto x^n$, $n < \frac{1}{2}$, may be treated in a similar manner.

with $k \equiv (\gamma + 1)M_\infty^2/(M_\infty^2 - 1)$, $\omega \equiv \varepsilon^{1+\nu}B^{\nu-1} \ll 1$ $(^{16,44})$. As $\xi \to 0$, according to Eq. (99'), the second and the third terms on the left, and the first, second, and the remainder on the right of Eq. (101) behave, respectively, as $\xi^{-1/2}$, $\omega\xi^{-3/2}$, $\xi^{-1/2}$, $\xi^{1/2}$, and $\omega B^{\nu-2}\xi^{-1}$. Evidently, the third nonlinear term on the left-hand side of Eq. (101) involving the product of $\bar{\varphi}_\xi$ and $\bar{\varphi}_{\xi\xi}$ is crucial in bringing about nonuniformity of the expansion

$$\bar{\varphi}_\xi = \bar{\varphi}_\xi^{(0)} + \omega\bar{\varphi}_\xi^{(1)} + \omega^2\bar{\varphi}_\xi^{(2)} + \cdots \tag{102}$$

Indeed, with the help of the foregoing estimates, one may integrate Eq. (101) with respect to η to arrive at an estimate

$$\bar{\varphi}_\xi^{(1)} \sim \int \bar{\varphi}_\xi^{(0)}\bar{\varphi}_{\xi\xi}^{(0)} \, d\eta \sim F(\tilde{x})/\xi^2 \tag{103}$$

A region of nonuniformity in ξ is therefore found where

$$\omega\bar{\varphi}_\xi^{(1)}/\bar{\varphi}_\xi^{(0)} = O(1), \qquad \text{i.e.,} \qquad \omega F(\eta)/\xi^{3/2} = O(1)$$

Hence

$$\xi = O[\omega^{2/3}] \tag{104}$$

To analyze this region, we introduce the variables

$$\zeta \equiv \xi/\omega^{2/3}, \qquad \hat{u} \equiv \omega^{1/3}\bar{\varphi}_\xi = u'/\omega^{2/3}u_\infty \tag{105}$$

and transform Eq. (101) to

$$2\hat{u}_\eta + \nu(\hat{u}/\eta) + k\hat{u}\hat{u}_\zeta = 0 \tag{106}$$

where all variables are to be treated as of order unity, and the remainder is of order $\omega^{2/3}$. The general solution to this quasilinear P.D.E. is

$$\zeta = k\hat{u}\eta + g(\hat{u}^{1+\nu}\eta^\nu) \tag{107}$$

where g is a function of the argument $(\hat{u}^{1+\nu}\eta^\nu)$ and can be determined through matching with the "outer limit" of the linearized solution Eq. (99').* The general solution in this form has been given by Van Dyke $(^{16})$ for plane flow ($\nu = 0$) considering the far field $\eta \to \infty$.

---

* Equation (106) and its solution, Eq. (107), are valid for the general case in which Eq. (99') is replaced by $\bar{\varphi} \approx F(\eta)\xi^\sigma$, $\xi \to 0$, with $\sigma < \frac{3}{2}$. The transformations Eq. (105) will have to change, of course.

For a parabolic contact surface, the outer limit mentioned is

$$\hat{u} \sim - \frac{1}{2(2\eta)^{\nu/2}} \frac{1}{\sqrt{\zeta}} \tag{108}$$

This is to be compared with Eq. (107). The identification of $\hat{u}$ in the two forms requires that $\hat{u} \to 0$ as $\zeta \to \infty$; comparing Eq. (108) with $\zeta \sim g(\hat{u}^{1+\nu}\eta^{\nu})$, we identify

$$g(\hat{u}^{1+\nu}\eta^{\nu}) = 1/4(2\eta)^{\nu}\hat{u}^2$$

The particular solution near the shock for a prabolic contact surface is therefore

$$\zeta = k\hat{u}\eta + [1/4(2\eta)^{\nu}\hat{u}^2] \tag{109}$$

To locate the shock, we need the relation between the perturbation in local Mach angle μ and axial perturbation velocity u' right behind the shock; we will also make use of the fact that a weak shock with shock angle β bisects the two local Mach angles immediately upstream and downstream of it. We may then relate $\Delta\beta \equiv \beta - \mu_{\infty}$ and u' right behind the shock. In terms of $\hat{u}$, ζ, and η, the shock relation in question is

$$\hat{u} = \frac{4}{\gamma + 1} \frac{B^2}{M_{\infty}^2} \frac{d\zeta}{d\eta} \tag{110}$$

Application of Eq. (109) at the shock (eliminating $\hat{u}$) then gives a differential equation for the shock geometry

$$4k\eta\left(\frac{4}{\gamma+1} \frac{B^2}{M_{\infty}^4} \frac{d\zeta}{d\eta}\right)^3 - 4(2\eta)^{\nu}\zeta\left(\frac{4}{\gamma+1} \frac{B^2}{M_{\infty}^4} \frac{d\zeta}{d\eta}\right)^2 + 1 = 0$$

which has a solution

$$\zeta = A\eta^{(2-\nu)/3}$$

$$A \equiv \frac{(\gamma+1)^2}{2^{6+\nu}} \left(\frac{3}{2-\nu}\right)^2 \frac{M_{\infty}^8}{B^4} \bigg/ \left[1 - \frac{4}{\gamma-1} \frac{2-\nu}{3} k \frac{B^2}{M_{\infty}^4}\right] \tag{111}$$

In terms of x and y, and simplified for high M_{∞}, the above equation yields a shock shape

$$x - (M_{\infty}^2 - 1)^{1/2}y = F(\gamma, \nu)(M_{\infty}\varepsilon)^{2(1+\nu)/3}x^{2/3(1+\nu)} \tag{112}$$

in agreement with results of ([47,55]) for the "weak-interaction boom."

ACKNOWLEDGMENT

The author would like to thank Drs. J. Aroesty, W. Bush, J. D. Cole, J. W. Ellinwood, H. T. Iida, J. W. Kirsch, R. S. Lee, and H. Mirels, also A. K. Cross for helpful discussions on some of the studies. Partial support from NASA through research grant NGL-05-018-044 to the University of Southern California is acknowledged.

REFERENCES

1. G. G. Chernyi, *Introduction to Hypersonic Flow*, Fizmatgiz, Moscow (1959); English translation by R. F. Probstein, Academic Press, New York (1961).
2. C. Gazley, Jr., "Atmospheric Entry," in: P-2052 *Handbook of Astronautical Engineering*, McGraw-Hill Book Co., New York (1961).
3. W. H. T. Loh, *Dynamics and Thermodynamics of Re-Entry*, Prentice-Hall, Englewood Cliffs, New Jersey (1962).
4. F. S. Nyland, "Hypersonic Turning with Constant Bank Angle Control," The RAND Corporation, RM-4483-PR (March 1965).
5. H. Hidalgo, R. Vaglio-Laurin, and R. G. Finke, "High Altitude on Lifting Re-Entry Performance," paper presented at the 18th International Astronautical Congress, Belgrade, Yugoslavia, September 24–30, 1967.
6. H. K. Cheng, "Notes on Hypersonic Flow Past Slender Bodies," Univ. Southern California Dept. Aerospace Eng. Report USCAE 108 (1969).
7. W. D. Hayes and R. F. Probstein, *Hypersonic Flow Theory*, Academic Press, New York, 1st ed. (1959), 2nd ed. (1966).
8. H. K. Cheng, G. J. Hall, T. C. Golian, and A. Hertzberg, "Boundary Layer Displacement and Leading-Edge Bluntness Effects in High Temperature Hypersonic Flow," *J. Aerospace Sci.* **28**, 353–381 (1961).
9. V. V. Sychev, "On the Theory of Hypersonic Gas Flow with a Power-Law Shock Wave," *I Meckhan. Prikl. Matem.* **24**, 756–764 (1960).
10. J. K. Yakura, "Theory of Entropy Layers and Nose Bluntness in Hypersonic Flow," in: *Hypersonic Flow Research*, F. R. Riddell, ed., Academic Press, New York (1962), pp. 421–470.
11. N. C. Freeman, "Newtonian Theory of Hypersonic Flow at Large Distances from Bluff Axially Symmetric Bodies," in: *Hypersonic Flow Research*, F. R. Riddel, ed., Academic Press, New York (1962), pp. 345–377.
12. H. K. Cheng, "Hypersonic Flow with Combined Leading-Edge Bluntness and Boundary-Layer Displacement Effect," Cornell Aero. Lab. Report AF-1285-A-4 (1960).
13. R. Vaglio-Laurin, "Asymptotic Flow Pattern of a Hypersonic Body," Dept. Aero. Eng. and Appl. Mech., Poly. Inst. Brooklyn, PIBAL Report 805 (1964).
14. J. P. Giraud, D. Vallee, and R. Zolver, "Bluntness Effects in Hypersonic Small Disturbance Theory," in: *Basic Developments in Fluid Dynamics*, Vol. 1, M. Holt, ed., Academic Press, New York (1965), pp. 127–247.
15. J. P. Guiraud, "Asymptotic Theory in Hypersonic Flow," in: *Fundamental Phenomena in Hypersonic Flow*, G. J. Hall, ed., Cornell University Press, Ithaca, New York (1965), pp. 70–84.
16. M. D. Van Dyke, *Perturbation Methods in Fluid Mechanics*, Academic Press, New York (1964).

17. A. F. Messiter, "Asymptotic Theory of Inviscid Hypersonic Flow at Large Distance from a Blunt-Nosed Body," Univ. Mich. Inst. Sci. and Tech. BOMIRAC Report ⊿63-81-1 (1964).

18. H. Mirels, "Hypersonic Flow over Slender Bodies Associated with Power-Law Shocks," in: *Advances in Applied Mechanics*, Vol. 7, Academic Press, New York (1962), pp. 1–54, 317–319.

19. N. C. Freeman, "Asymptotic Solutions in Hypersonic Flow, An Approach to Second Order Solutions of Hypersonic Small-Disturbance Theory," in: *Research Frontier in Fluid Dynamics*, R. J. Seeger and G. Temple, eds., Interscience Publishers, New York (1965), pp. 284–307.

20. K. Stewartson and B. W. Thompson, "On One-dimensional Unsteady Flow at Infinite Mach Number," *Proc. Roy. Soc. London, Ser. A.* **304**, 255–273 (1968).

21. H. K. Cheng and J. Kirsch, "Nonsimilar Structure of Blast Wave with an Expanding Interface, and Application to Hypersonic Flow," paper read at the 18th International Astronautical Congress, Belgrade, Yugoslavia, September 25–29, 1967.

22. J. D. Cole, "Newtonian Flow Theory for Slender Bodies," *J. Aeron. Sci.* **24**, 448–455 (1957).

23. A. Henderson, R. D. Watson, and R. D. Wagner, "Fluid Dynamic Studies to $M = 41$ in Helium," *AIAA J.* **4** (12) 2117–2124 (1966).

24. M. D. Ladyzhenskii, "The Hypersonic Area Rule," *Inzh. Zh.* **1** (1) 159–196 (1961); *AIAA J.* **1**, 2696–2698 (1963).

25. V. M. Krasovskii, "Experimental Study of Hypersonic Helium Flow over Blunt-Nosed Bodies," *Inzh. Zh.* **5** (2), 249–253 (1965).

26. J. D. Cole, and J. Aroesty, "The Blowhard Problem—Inviscid Flows with Surface Injection," The RAND Corporation, RM-5196-ARPA (1967); to be published in the *International Journal of Heat and Mass Transfer*.

27. M. Holt and T. D. Taylor, "High Speed Flow Past a Cone with Large Wall Injection Velocities," paper presented at the 18th International Astronautical Congress, Belgrade, Yugoslavia, September 24–30, 1967.

28. R. A. Hartunian and D. J. Spencer, "Experimental Results for Massive Blowing Studies," *AIAA J.* **5**, 1397 (1967).

29. K. Stewartson, "Some Recent Developments in Boundary Layer Theory," paper presented at the 7th Symposium on Advanced Problems and Methods in Fluid Mechanics, Gdansk, Poland, September 1965.

30. A. Miele, ed., *Theory of Optimum Aerodynamic Shapes*, Academic Press, New York (1965).

31. A. L. Gonor, "On the Form of Three-Dimensional Bodies of Minimum Drag at Hypersonic Speeds," *Prikl. Matem. i Mekhan.* **27** (1) (1963).

32. G. L. Maikapar, "On the Wave Drag of Non-Axisymmetric Bodies in Supersonic Flow," *Prikl. Matem. i Mekhan.* **23** (2) (1959).

33. G. L. Maikapar, "On the Form of a Supersonic Airplane," *Tr. Tsent. Aero-Gidrodin. Inst. Mosk. (TsAGI)*, No. 841 (1961).

34. G. G. Chernyi, "On the Analysis of Bodies of Minimum Drag at Hypersonic Speed," *Prikl. Matem. i Mekhan.* **28** (2), 387–389 (1964).

35. A. Miele, "Optimum Transversal Contour of a Lifting Body at Hypersonic Speeds," paper presented at the 18th International Astronautical Congress, Belgrade, Yygoslavia, September 24–30, 1967.

36. J. D. Cole and T. F. Zien, "A Class of Three-Dimensional Optimum Hypersonic Wings," AIAA paper 68–158 (1968).
37. K. Stewartson, *The Theory of Laminar Boundary Layers in Compressible Fluids*, Oxford University Press (Clarendon) (1964), pp. 161–183.
38. F. K. Moore, "Hypersonic Boundary Layer," in: *Theory of Laminar Flows*, F. K. Moore, ed., Princeton University Press, Princeton, New Jersey (1963).
39. N. C. Feeman and S. H. Lam, Princeton University, Dept. Aero. Eng. Reports 468 and 471 (1959).
40. L. Lees, "Influence of the Leading-Edge Shock Wave on the Laminar Boundary Layer at Hypersonic Speeds," *J. Aeron. Sci.* **23** (6), 594–600, 612 (1956).
41. H. Oguchi, "First-Order Approach to a Strong Interaction Problem in Hypersonic Flow over an Insulated Flat Plate," Univ. Tokyo, Aero. Res. Inst. Report 330 (1958).
42. N. S. Matveeva and V. V. Sychev, "On the Theory of Strong Interaction of the Boundary Layer with an Inviscid Hypersonic Flow," *Prikl. Matem. i Mekhan.* **29** (4), 644–657 (1965).
43. W. B. Bush, "Hypersonic Strong-Interaction Similarity Solutions for Flow Past a Flat Plate," *J. Fluid Mech.* **25** (1), 51–64 (1966).
44. J. D. Cole, *Perturbation Methods in Applied Mathematics*, Random House (Blaisdell), New York (1968).
45. R. S. Lee and H. K. Cheng, "On the Higher-Order Asymptotic Theory of Hypersonic Boundary Layers on Slender Bodies in the Strong-Interaction Regime," paper read at the AGARD Seminar on Numerical Methods for Viscous Flows, Teddington, England, September 18–21, 1967; to be submitted for publication.
46. W. B. Bush and A. K. Cross, "Hypersonic Weak-Interaction Similarity Solutions for Flow Past a Flat Plate," *J. Fluid Mech.* **29** (2), 349–359 (1967).
47. Y. H. Kuo, *J. Aero. Sci.* **23**, 125 (1956).
48. T. Kubota and D. R. S. Ko, "A Second-Order Weak Interaction Expansion for Moderately Hypersonic Flow Past a Flat Plate," *AIAA J.* **5** (10) (1967).
49. R. S. Lee and H. K. Cheng, "On the Outer-Edge Problem of a Hypersonic Boundary Layer" (accepted for publication in the *J. Fluid Mech.*).
50. M. Yasuhara, "Axisymmetric Viscous Flow Past a Very Slender Body of Revolution," *J. Aerospace Sci.*, **29**, 667–679 (1962).
51. K. Stewartson, "Viscous Hypersonic Flow Past a Slender Cone," *Phys. Fluids* **7** (5), 667–675 (1964).
52. J. W. Ellinwood and H. Mirels, "Axisymmetric Hypersonic Flow with Strong Viscous Interaction," Aerospace Corp. Report TR 0 158(3240-10)-1 (1967).
53. H. Mirels and J. W. Ellinwood, "Hypersonic Viscous Interaction Theory for Slender Axisymmetric Bodies," AIAA paper 68-1 (1968).
54. J. T. Shen and T. F. Sun, "Effect of Transverse Curvature on Axisymmetric Compressible Laminar Boundary Layer," (in Chinese) *ACTA Mechanica Sinica* **9** (2), 150–171 (1966).
55. A. K. Cross, private communication.
56. M. D. Ladyzhenskii, "Hypersonic Viscous Flow over Slender Bodies," *Prikl. i Matem. Mekhan.* **27** (5), 667–675 (1963).
57. H. K. Cheng, "The Shock-Layer Concept and Three-Dimensional Hypersonic Boundary Layer," Cornell Aero. Lab. Report AF-1285-A-3 (1960).
58. M. B. Glauert, and M. J. Lighthill, *Proc. Roy. Soc.* **A230**, 188 (1950).

59. K. Stewartson, "The Asymptotic Boundary Layer on a Circular Cylinder in Axial Incompressible Flow," *Quart. Appl. Math.* **13**, 113–122 (1955).

60. J. M. Solomon, U. S. Navy Ord. Lab., White Oak, Maryland NOLTR 66-225 (1967).

61. C. C. Horstman, and M. I. Kussoy, "Hypersonic Viscous Interaction on Slender Cones," AIAA paper 68-2 (1968).

62. W. B. Bush, "Hypersonic Strong-Interaction Similarity Solutions for Flow Past a Very Slender Axisymmetric Body," RAND Corp. Report RM-5699 (1968).

63. C. F. Dewey, Jr., *AIAA J.* **1**, 20–33 (1963).

64. M. Kussoy, "Hypersonic Viscous Drag on Cones in Rarefied Flow," NASA TND 4036 (1967).

65. J. D. Whitfield and B. J. Griffith, *AIAA J.* **3**, 1165 (1965).

66. H. T. Iida, "On the Three-Dimensional Hypersonic Needle Problem in the Viscous Strong-Interaction Regime," Univ. Southern California Dept. Aerospace Eng. Thesis (1969).

67. G. B. Whitham, "The Flow Pattern of Supersonic Projectile," *Comm. Pure Appl. Math* **5**, 301–348 (1952).

68. M. J. Lighthill, "A Technique for Rendering Approximate Solutions to Physical Problems Uniformly Valid," *Phil. Mag.* **40** (7), 1179–1201 (1949).

Chapter 5

HYPERSONIC BLUNT-BODY GAS DYNAMICS

J. F. McCarthy, Jr.

Vice President, Research, Engineering, and Test
Space Division
North American Rockwell Corporation
Downey, California

INTRODUCTION *

The intent in this chapter is to present some of the methods and techniques being used in the aerospace industry to determine the flow field around a blunt body moving at hypersonic speed. In this context, attention is given primarily to presenting the more significant results obtained in recent years, and, where possible, experimental results are compared with theoretical predictions. Before discussing these methods and results, however, it is worthwhile to view the role that flow-field analysis plays in the design of a hypersonic vehicle.

Flow-field analyses are performed primarily to obtain data for use in structural and thermal analyses for vehicle design and to develop vehicle trajectories. As shown in Fig. 1, analysis of the flow field is usually initiated on the basis of an assumed vehicle configuration, from which atmospheric trajectories, along with preliminary values for the flow-field parameters at the stagnation point, have been predicted. From these preliminary data, pressure, temperature, and velocity distributions around the body are computed for a flow field assumed to be inviscid. These results are then used to compute detailed temperature (enthalpy) and heat flux data for thermal analyses, shear–stress distribution for refining the aerodynamic parameters, and distribution of boundary layer displacement thickness for determining the effective body shape, which is used in related analyses. In addition, the detailed chemical composition of the flow field is determined

---

* Symbols used in this chapter are defined on pp. 228–231.

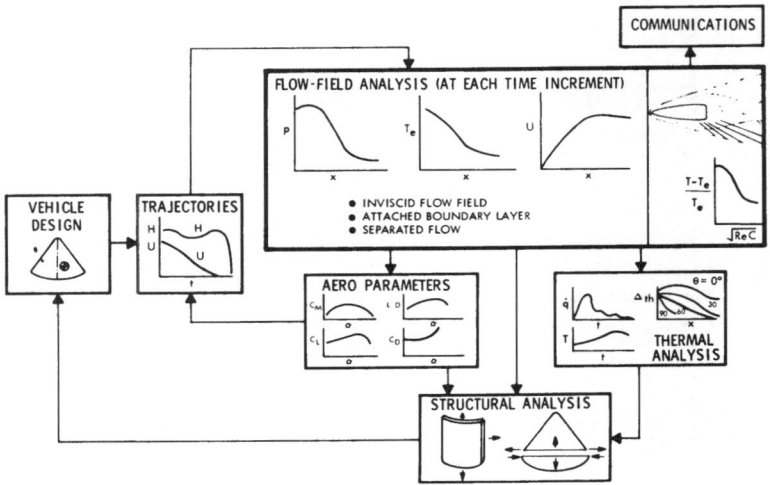

Fig. 1. Interrelationships among technical disciplines applied to the design of aerospace vehicles.

for use in the analysis and for evaluating the electrical-disturbance effects on communications between the vehicle and ground. The results of this preliminary analysis of the flow field thus permit the configuration to be modified and the trajectories to be redefined; the complete process is then iterated. In addition, as part of the design process, the results are used to determine the response of the vehicle structure to the aerodynamic and thermal loads.

Because of the vast breadth of subject matter encompassed in such flow-field analyses, the material in this chapter is intentionally limited to continuum flow over hypersonic blunt bodies, with consideration given only to adiabatic (nonradiating) and chemically-inactive flow. Body shapes and flow regimes are defined, and, within these definitions, the discussion covers selected analytical details for the inviscid flow field, the viscous layers (attached and separated), and some of the details of the region in which these two distinctly separate flow fields interact.

1. CLASSIFICATION OF BODY SHAPES

Techniques used to analyze the hypersonic flow field around a body depend to a considerable degree on the shape of the body. Thus the shape of a hypersonic vehicle is frequently classified as slender, intermediate, or blunt. Each of these classifications can be described by three primary

TABLE I

Classification of Hypersonic-Body Shapes

Body shape	Configuration	Parametric description		
		Body angle	Velocity change	Local mach No.
Slender	Fig. 2a	$\theta_{\text{geo}} \ll 1$ (Attached shock)	$\lvert \Delta U/U_\infty \rvert \ll 1$	$M_e \gg 1$
Intermediate	Fig. 2b	$\theta_{\text{geo}} < 1$ (Attached shock)	$\lvert \Delta U/U_\infty \rvert < 1$	$M_e > 1$
Blunt	Fig. 2c	$0 \le \theta_{\text{geo}} \le \pi/2$ (Detached shock, stagnation point, and highly rotational flow)	$\lvert \Delta U/U_\infty \rvert \approx 1$	$0 < M_e < M_\infty$

parameters (Table I; Fig. 2): ratio of the velocity change across the shock wave (ΔU) to the velocity of the free stream (U_∞); local Mach number in the shock layer (M_e); and the angle made by the body surface relative to the free-stream velocity (θ_{geo}).

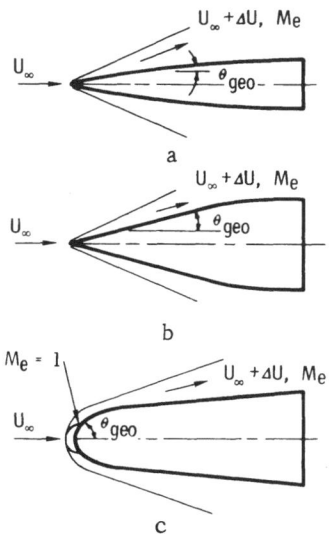

Fig. 2. (a) Slender, (b) intermediate, and (c) blunt body configurations.

The slender-body classification is based mainly on the fact that the local surface slope is everywhere very small. For such a body, the flow velocity in the shock layer is nearly the same as the free-stream value despite the large changes in pressure and density that occur across the shock wave. Thus the local Mach numbers are still hypersonic. Moreover, the shock wave is attached to the pointed nose of a slender body.

The intermediate classification describes a body shape on which some blunt features are evident, but for which these bluntness characteristics are not yet in limiting form. For example, the body-surface slope and the velocity perturbations in the flow field are larger for an intermediate than for a slender body, and the local Mach numbers may be even slightly subsonic despite the fact the shock wave is attached to the nose of the body.

The blunt-body classification is associated with a detached shock wave over a body having a surface slope at the nose (with respect to the free stream) approaching $\pi/2$ rad. For this body, a large subsonic flow region is always present in the vicinity of a clearly-defined stagnation point. Also, as the fluid passes through the shock wave, the change in its velocity is the same order of magnitude as the free-stream velocity, and the local Mach number varies from zero (at the stagnation point) to a supersonic value. Furthermore, the flow is generally highly rotational. This rotation invalidates the conventional assumption that a velocity potential exists, and, although the aft portion of these bodies may revert to a slender shape, the presence of the blunt nose influences the downstream flow of the viscous boundary layer.

2. BLUNT-BODY FLOW FIELDS

The blunt-body shape considered here is typified by the Apollo command module (Fig. 3). This shape encompasses most of the problems that arise from both viscous and inviscid flow-field considerations.

As it is shown in Fig. 3, the body is at an arbitrary angle of attack (achieved by an offset center of gravity) relative to the free-stream flow. Thus, though the body is axisymmetrical, the geometry of the flow field is not. The only plane of symmetry is the pitch plane, which contains the free-stream velocity vector and the body axis. The shock wave in front of this configuration is fully detached, and a subsonic region and a transonic region exist behind the shock wave on the front side. As the flow expands around the body, a sonic surface is created ($M = 1$); since the flow continues to increase in speed, it becomes supersonic downstream of this surface.

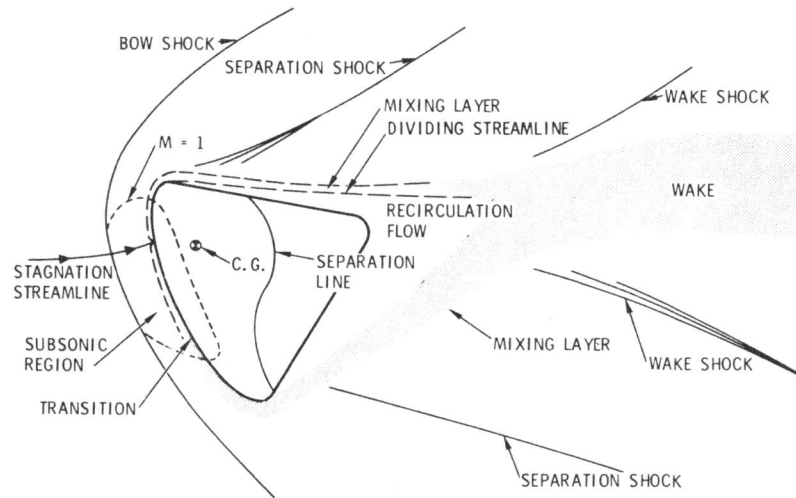

Fig. 3. Flow regions around a blunt body traveling at hypersonic speed.

The viscous boundary layer is attached in the front region of the body,
but downstream (because of the build up of the boundary layer thickness
and the possible adverse pressure gradients), the boundary layer tends to
separate. Viscous-flow phenomena increase after the flow separates and
goes downstream.

Emanating from the separation region is the dividing-stream surface,
which defines the boundary of the recirculation region—an area in which
the flow is "trapped." The flow field outside the dividing streamline proceeds
from the separation point in a viscous mixing layer that converges to a
single flow stream at the neck of the wake. Oblique shocks are generated
near the neck because the flow external to the wake is turned supersonically.
The surrounding inviscid flow and the viscous mixing-layer flow build up
in the wake much like a boundary layer and develop into the far wake
downstream of the base flow field.

3. FLOW REGIMES

A hypersonic vehicle entering the earth's atmosphere encounters a
continuously-varying flow regime. The definition of the specific altitude at
which atmospheric entry begins is somewhat arbitrary. It has frequently
been defined as 400,000 ft because at this altitude, a hypersonic vehicle
normally will not have encountered a measurable decelerating force due to

the atmosphere; when the Apollo spacecraft returned from the moon, e.g., the atmospheric deceleration built up to a force of only 0.05 g at an altitude of about 300,000 f, and at 400,000 f, this decelerating force was essentially zero.

At altitudes greater than 400,000 ft, the free-stream density is so low that the mean-free-path is much greater than any of the characteristic dimensions of the vehicle; flow at these altitudes (for vehicles whose characteristic dimension is of the order of feet) is called the "free-molecular-flow regime." At altitudes from 400,000 to 300,000 ft, the local mean free path is the same order of magnitude as the characteristic dimension of the vehicle; flow at these altitudes is called the "transition regime." At altitudes between 300,000 and 200,000 ft, the flow is moderately rarefied, and is called the "slip-flow regime." This name derives from the fact that under these conditions, the gas layer immediately adjacent to the vehicle does not adhere to the surface, but slips along with a velocity proportional to the mean free path. At altitudes below 200,000 ft, the flow is considered to be a continuum to which usual gas-dynamic equations can be applied.

Below 400,000 ft, the flow regime can be characterized by the Knudsen number, which is defined as the ratio of the local mean free path of the flowing fluid to a physical dimension of either the flow field or the vehicle. Through its relationship to the velocity of sound and the fluid viscosity, the mean free path can be related to both the Mach number and the Reynolds number; consequently, the flow regimes defined above can be redefined as follows [1]:

1. The free-molecular-flow regime is that for which $K \geq 10$ or $(M/Re) \geq 10$.
2. The slip-flow regime is that for which $0.01 \leq (M/\sqrt{Re}) \leq 0.1$ for $Re > 1$, and $0.01 \leq (M/Re) \leq 0.1$ for $Re < 1$.
3. The transition regime is that which lies between the free-molecular-flow and slip-flow regimes.
4. The continuum-flow regime is that for which $(M/\sqrt{Re}) < 10^{-2}$.

Shown in Fig. 4 is a plot of these regimes and a corresponding characterization of three typical trajectories. As stated previously, this chapter is concerned with only the continuum-flow regime. Nevertheless, for a blunt body at a given flight condition, the free-stream properties may not correspond to a continuum-flow regime. However, the density increase across the detached shock wave reduces the local mean free path, making the flow a continuum adjacent to the body. Thus several regimes may be required to define the flow field for a given flight condition. For example,

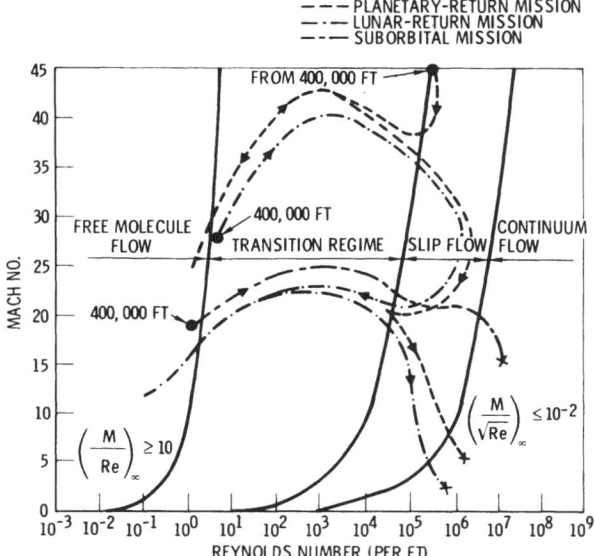

Fig. 4. Flow regimes based on the free-stream condition
for three typical trajectories.

if a flow-field analysis is made along a normal to a hypersonic blunt body
at an altitude of 300,000 ft, the free-stream flow will be in the free-molecular-
flow regime but approach the continum-flow regime at the wall. The classification of flow regimes for a blunt body developed by Probstein and Kemp
([2]) accounts for this effect. In this latter classification, the ratio of the mean
free path to the body radius is compared with the ratio of the gas densities
before and after the vehicle shock wave.

4. FLOW-FIELD GAS PROPERTIES

Due to high gas temperatures, chemical changes occur in the flow
field of a hypersonic vehicle traveling through the atmosphere. These
changes play a significant role in the flow-field analysis and also in the
evaluation of the potential problem of communications blackout during
hypersonic flight. Worthy of particular note are the three principal chemical
changes that occur at the stagnation point. The first change, which occurs
at velocities of about 9000 to 15,000 ft/sec, is the dissociation of molecular
oxygen into atomic oxygen. The second change occurs when the nitrogen
dissociates at velocities in excess of 15,000 ft/sec. The third change occurs

at velocities higher than about 30,000 ft/sec, when both oxygen and nitrogen ionize. All three of these regions of chemical change are encountered by most entry vehicles at some time during atmospheric flight.

5. INVISCID-FLOW ANALYSIS

The inviscid flow field is analyzed by solving the equations for conservation of mass, momentum, and energy. The equation of state, which may be complex for a gas that undergoes chemical reactions, is also needed to relate flow-field properties to each other. In vector notation, these basic equations are as follows:

Conservation of mass:

$$\nabla \cdot (\varrho\mathbf{U}) = -\partial\varrho/\partial t \tag{1}$$

Conservation of momentum:

$$D\mathbf{U}/Dt - (1/\varrho)\,\nabla p \tag{2}$$

Conservation of energy:

$$\frac{D(h + \frac{1}{2}\mathbf{U} \cdot \mathbf{U})}{Dt} = -\frac{1}{\varrho}\frac{\partial\varrho}{\partial t} \tag{3}$$

Equation of state:

$$h = h(p, \varrho) \tag{4}$$

5.1. Methods of Analysis—Subsonic–Transonic Region

Methods of analyzing the subsonic–transonic region can be categorized as analytical or numerical. The analytical methods are based on limiting assumptions that restrict the application of results. However, these methods do provide rapid approximations of the pressure, temperature, and velocity distribution in the flow field, from which considered judgments can be made. Numerical methods, although accurate, have proven quite expensive because of the computation time required to obtain solutions.

Analytical methods for determining the properties of a hypersonic flow field are based primarily on the assumption that the density ratio ε across the shock wave approaches zero. These methods, discussed in detail by Hayes and Probstein [3] include the Newton–Busemann approximation, the thin-shock-layer theory, and asymptotic-expansion theories for a small density ratio across the shock wave.

5.1.1. *Inherent Characteristics*

Some inherent difficulties arise from the basic mathematics used in integrating the flow-field equations by numerical methods. These difficulties can be examined in some detail by an investigation of the equation for the velocity potential in the somewhat simplified case of unsteady, two-dimensional, compressible, linearized, irrotational flow. In this case, the expression for the velocity potential is

$$\nabla^2\phi - \frac{1}{\bar{a}^2}\left(U\frac{\partial}{\partial x} + \frac{\partial}{\partial t}\right)^2\phi = 0 \tag{5}$$

where $\bar{a}$ is the local speed of sound and U the local fluid velocity. In expanded form, this equation becomes

$$(1 - M^2)\phi_{xx} - \frac{2U}{\bar{a}^2}\phi_{xt} - \frac{1}{\bar{a}^2}\phi_{tt} + \phi_{yy} = 0 \tag{6}$$

where M is the local Mach number and the subscripts x, y, and t indicate that the velocity potential ϕ has been differentiated with respect to the indicated subscript. For the case of steady flow, Eq. (6) reduces to

$$(1 - M^2)\phi_{xx} + \phi_{yy} = 0 \tag{7}$$

which, when compared with the more general differential equation in the form

$$A\phi_{xx} + B\phi_x\phi_y + C\phi_{yy} = 0 \tag{8}$$

shows that the term (discriminant) $B^2 - 4AC$ [$= -4(1 - M^2)$ in this case] can take on positive or negative values depending on whether the Mach number is greater or less than one.

For subsonic conditions (M < 1), the discriminant $(B^2 - 4AC)$ is negative, and the equation for the velocity potential is elliptical in character. The techniques for obtaining solutions to elliptic equations with initial-value conditions [i.e., where $\phi(x, 0) = f(x)$ and ϕ_x is known initially] have very stringent requirements imposed on them. For example, if one considers the simple case where the velocity potential is assumed to be periodic in the y direction (imaginary exponential) and grows with x in the form

$$\phi = \phi_0 e^{\alpha x}e^{i\beta y} \tag{9}$$

where α and β are real and positive, the derivatives of the potential function are

$$\phi_{xx} = \alpha^2 e^{\alpha x}\phi_0 e^{i\beta y}, \qquad \phi_{yy} = -\beta^2 e^{\alpha x}\phi_0 e^{i\beta y} \tag{10}$$

Further, for simplicity, let $M = 0$, so that Eq. (7) becomes

$$\alpha^2 e^{\alpha x}\phi_0 e^{i\beta y} - \beta^2 e^{\alpha x}\phi_0 e^{i\beta y} = 0$$

$$\alpha^2 - \beta^2 = 0, \qquad \alpha = \pm\beta \tag{11}$$

Thus, for subsonic incompressible flow, the form of the equation is elliptical —i.e., $B^2 - 4AC < 0$—and a small disturbance in the y direction during the integration process results in a disturbance in the x direction that is unbounded; consequently, the result is inherently unstable. Because of this inherent instability, therefore, the numerical integration must be performed where $\beta = 0$ (i.e., no disturbance in the y direction) so that the solution is bounded. This technique requires that the initial conditions for the solution be known to a high degree of accuracy to assure that the integration can be done without a disturbance (or inaccuracy) becoming predominant.

Where the flow is supersonic ($M > 1$), the discriminant of Eq. (7) becomes positive, and the equation is of hyperbolic form. In this case, the effect of a small disturbance in the integration process can again be investigated using Eq. (8), where the velocity potential is assumed to be periodic in the y direction and grows with x. In this hyperbolic case, the derivatives ϕ_{xx} and ϕ_{yy} are also the same as those used before. Then, if it is assumed for simplicity that $M = \sqrt{2}$, Eq. (7) becomes

$$-\alpha^2 e^{\alpha x}\phi_0 e^{i\beta y} - \beta^2 e^{\alpha x}\phi_0 e^{i\beta y} = 0$$

$$\alpha^2 = -\beta^2, \qquad \alpha = i\sqrt{\beta^2} \tag{12}$$

Thus, for the hyperbolic-equation form (which is the case for supersonic steady flow), a periodic disturbance in the y direction is also periodic with x. For this case, therefore, the solution is bounded under all conditions, and inaccuracies are not amplified during the numerical integration.

For the case of sonic flow ($M = 1$), the discriminant in Eq. (7) is zero, which is the condition for a parabolic equation, and the solution is the sonic line in the flow field. The parabolic equation is typified by the boundary layer equations discussed below. However, it should be noted here that this class of equations is well understood, and stability is assured when the solution is obtained by an implicit integration technique.

Recent investigations show that the analysis of the subsonic–transonic flow region by numerical techniques can be improved by the use of time-dependent methods. The chief advantage of using time-dependent flow fields is that the solution automatically converges to the correct solution for steady flow without the costly and uncertain iterations required for steady-flow solutions. The time-dependent technique is actually a real relaxation of the solution to the problem, with an arbitrary set of initial conditions.

To examine the stability of the time-dependent flow equations, a bounded disturbance in both the x and y directions can be introduced into the two-dimensional, linearized equation for the velocity potential [Eq. (6)], and the variation of this disturbance with time can be investigated. For this purpose, it can be assumed that the velocity potential $\phi = \phi(x, y, t)$ is periodic (bounded) in both the x and y directions and varies exponentially with time. In this case

$$\phi = \phi_0 e^{i\beta y} e^{i\alpha x} e^{\gamma t} \tag{13}$$

If the appropriate partial derivatives are obtained for this function and substituted into Eq. (6), the resultant equation reduces to

$$(M^2 - 1)\alpha^2 - \frac{2U}{\bar{a}^2} i\alpha\gamma - \frac{\gamma^2}{\bar{a}^2} - \beta^2 = 0 \tag{14}$$

This result requires that

$$\gamma = [-U\alpha \pm \bar{a}(\beta^2 + \alpha^2)^{1/2}]i \tag{15}$$

Thus, since γ is an imaginary number, a periodic diturbance of ϕ in its physical plane (x, y) results in a periodic (bounded) disturbance with time.

For analyzing the subsonic–transonic region, numerical methods have been developed with only minor limitations, other than the slight inaccuracy incurred by the finite-difference techniques. These methods are used to obtain results for steady-state flow fields as well as for those that are time dependent, and the methods cover the complete range of the density ratio across the shock wave, from zero to one. Initial attempts to employ numerical methods to obtain solutions for the inviscid flow field were oriented toward the steady-flow problem because, historically, this problem, as an initial-value problem, has been more readily solved than has the time-dependent problem. Because of the importance of this steady-flow problem, there are many numerical techniques available for its solution. However, the time-dependent problem is actually more amenable to yielding a steady-state (time approaches infinity) solution when numerical techniques are used.

5.1.2. Steady-Flow Solutions

Each numerical method of analyzing the steady inviscid flow field has been developed for either the direct or indirect (inverse) problem. The direct problem is to seek a solution when the body shape, its flight conditions, and the angle of attack are known and one desires to find the corresponding flow field, i.e., the location and shape of the shock wave for which the Rankine–Hugoniot equations are satisfied. Conversely, the indirect problem is the case in which the shock wave is assumed, and the body shape required to yield the assumed shock wave is then determined for the given angle of attack and the flight conditions. The assumed shock-wave shape is altered after each determination of shape until the calculated body shape agrees with the physical shape of the body.

a. Inverse Problem. For a blunt body at zero angle of attack, the inverse method has been used extensively to determine the flow field in the subsonic–transonic region for steady-flow conditions. It was developed by Van Dyke [4] and Lomax and Inouye [5], among others, to solve the problem in the real physical plane. Garabedian and Lieberstein [6] solved the inverse problem in the complex plane, where the equations are hyperbolic (only complex characteristics exist for $M > 1$). However, this procedure requires that an analysis be made of the entire flow field in the transformed plane before results are obtained for a single point in the physical plane. It is therefore too lengthy to be widely used.

For a blunt body at angle of attack, the inverse method has been used by Webb *et al.* [7] and Joss [8] to determine the flow field in the forebody region. For a small angle of attack, Swigart [9] obtained solutions by using the first term in a power-series expansion for the angle of attack. The computerized technique employed by Webb is to determine the inviscid-flow-field properties in the subsonic region for an axisymmetrical shape at angles of attack from 0 to 40°. To avoid the extensive computer time required to solve this problem by the technique of Garabedian and Lieberstein for the complex plane, the solution is obtained for the real physical plane, where the problem is an initial-value problem. The solution is initiated at the shock wave, for which the initial shape has been assumed; usually, this shape is based on some highly simplified solution that provides a reasonable starting condition. The procedure then is to "march" downstream by numerical integration until a body shape is determined that supports the assumed shock wave. If the computed body shape agrees with the actual shape of the vehicle, the solution is complete. However, if the computed and actual body shapes do not agree, the assumed shape of the shock wave must be altered, and then process iterated.

A body-oriented cylindrical coordinate system is used in Webb's solution, since the shapes analyzed are restricted to axisymmetrical bodies (which are three-dimensional in character when considered at a finite angle of attack). This coordinate system offers a relatively simple means of

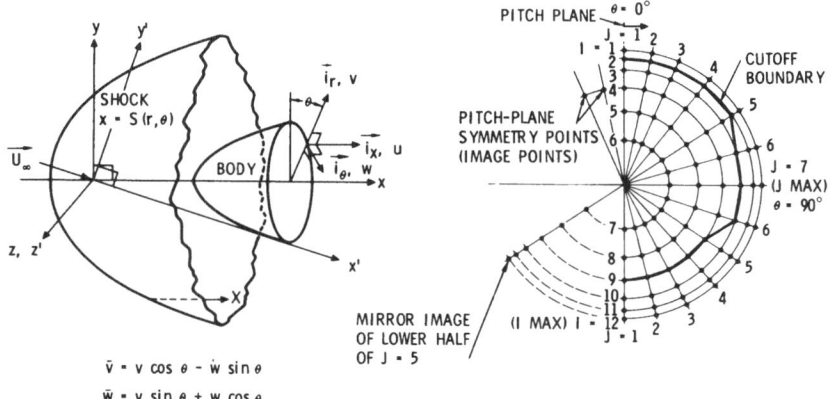

Fig. 5. Coordinate system for the inverse solution of the three-dimensional blunt-body problem.

expressing the shock shape analytically. The basic equations used are the conservation equations for an inviscid, nonconducting medium and the equation of state, which provides the additional relationship required to completely define the flow-field solutions. These equations are then transformed into a shock-oriented, oblique, curvilinear coordinate system (i.e., x, r, θ), as shown in Fig. 5, and "marching" planes are laid off equidistantly in the x direction (constant value of x). To avoid error build up due to numerical inaccuracies, further transformation, to Cartesian lateral velocities, is employed during the data-smoothing phase of the computation for the region about the x axis ($r = 0$). This computational approach permits solutions to be obtained for either perfect gases or for equilibrium real air, with no *a priori* assumption as to the location, shape, or condition of the stagnation streamline.

Figure 5 also illustrates the grid-point mesh in a computational flow-field "marching" plane (one everywhere equidistant from the shock in the x direction). The cutoff boundary shown represents the limit of the flow-field computations. The flow-field symmetry about the pitch plane was used to apply finite-difference principles for the computation of property derivatives in the "marching" plane.

To assure the inclusion of smooth initial data in the computer analysis, a shock shape is used that can be expressed in analytical form as

$$S = \sum_{g=0}^{\infty} \left(\sum_{n=1}^{\infty} a_{n,g} r^n \right) \cos g\theta \tag{16}$$

This expression is based on an extension of the work of Kaattari ([10]).

Flow-field properties behind the shock wave are determined by using the analytically determined slope of the shock surface calculated from the expression for the shock-wave coordinates.

To determine the body shape and streamline pattern for the three-dimensional-flow solution (finite angle of attack), a pair of stream functions —ψ and ϕ (note that ϕ is not the velocity potential in this case)—were introduced to represent stream surfaces. These stream surfaces are defined by the initial values given by

$$\phi = (\varrho_\infty U_\infty)^{1/2} y', \qquad \psi = (\varrho_\infty U_\infty)^{1/2} z' \qquad (17)$$

These two functions must obey the relationship that the cross product of their gradients equal ϱU, i.e., they must conform to

$$\nabla\phi \times \nabla\psi = \varrho U \qquad (18)$$

Both stream functions are computed for every grid point in the flow field, and the body surface is located by a parabolic extrapolation to determine where $\psi = 0$. The stream function ϕ is used to define pitch-plane streamlines, and the body surface is determined by an interpolation for $\phi = \phi_b$, which occurs at the point where $U = 0$ (the stagnation point).

The numerical techniques employed in this solution consist of a fourth-order Runge–Kutta integration procedure applied to the x derivative. The partial derivatives in the lateral direction (i.e., r and θ derivatives in the "marching" plane) are determined by a five-point central-difference scheme supplemented by off-center difference schemes at the edge of the cutoff boundary.

The flow field over the Apollo command module was calculated with this inverse procedure, and the results for the pitch plane are shown in Fig. 6. This solution was obtained for equilibrium real air at an altitude of 150,480 ft, a vehicle speed of 22,754 ft/sec, and a 22° angle of attack. The initial shock shape was estimated from a scaled schlieren photograph in the 90° meridian plane. Subsequent iterations were based on a shock shape modified from the previous shape by a fraction of the local error (as measured between the calculated and the desired body shapes) to obtain local perturbations in all meridian planes of the shock traces, and thereby to get convergence to the true shock wave.

The final converged flow field in the body pitch plane, including the streamline pattern and the sonic-line boundaries, is shown in Fig. 6. Computed body points are shown for comparison with the true body shape. An enlarged view of the stagnation region is included to illustrate that the

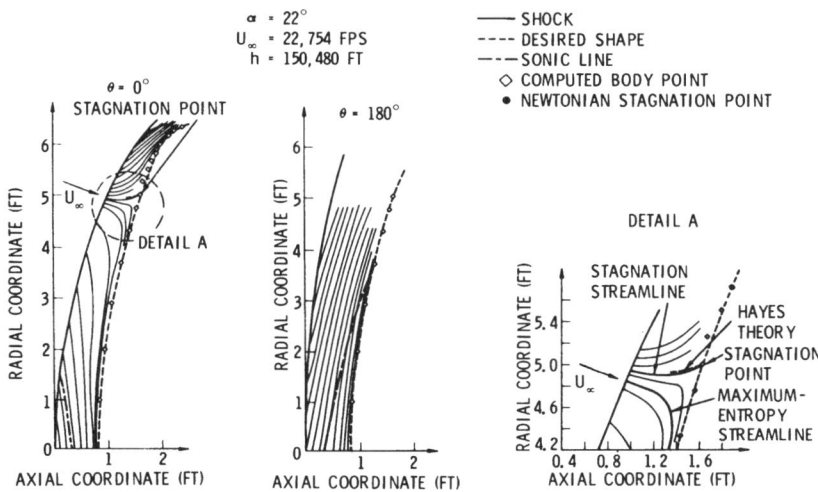

Fig. 6. Pitch-plane flow field determined by the inverse method for the Apollo command module at angle of attack.

stagnation streamline and maximum-entropy streamline are not the same for angle-of-attack flow fields. These results have been substantiated by Muggia, as shown by Hayes and Probstein (^{3}). The enlarged view of the stagnation region also illustrates a good agreement with Hayes' theory for stagnation-point flow over blunt bodies at angle of attack (^{11}). In addition, separate checks of the constancy of total enthalpy and entropy on the body surface were made, and the apparent variation did not exceed 1%.

Shown in Fig. 7 is the pressure distribution at 15° angular intervals in the meridian plane for the same solution shown in Fig. 6. For purposes of comparison, the result based on the modified Newtonian theory is included in Fig. 7. Flight-test data available at the time of this writing are also presented to show the accuracy of the predicted results.

b. Direct Problem. Another technique of analyzing the flow field for zero angle of attack is the method of integral relations. It is used to attack the direct problem (given the body shape, find the shock wave) and was developed in Russia by Belotserkovskii (^{12}) about the time that the inverse problem was solved in the United States. The method requires an initial assumption as to the shock-wave standoff distance, which must be accurate to four or five decimal places to avoid singularities in the solution of flow properties near the sonic line. Other assumptions are required for this technique, and the final solution is obtained by iterating the standoff distance of the shock wave.

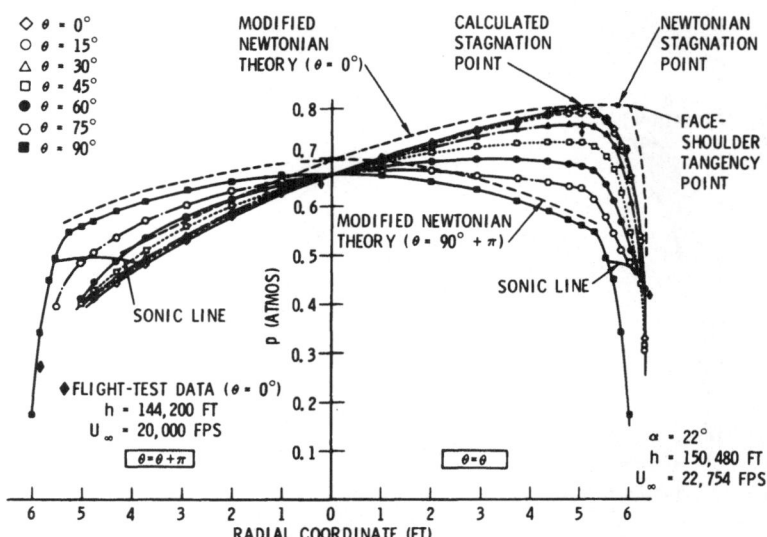

Fig. 7. Pressure distribution obtained by the inverse method for the Apollo command module at angle of attack.

This method was extended by Waldman ([13]) to include a finite angle of attack, but the extension requires that several simplifying assumptions be made. For example, Waldman's method involves the assumption that the stagnation streamline is coincident with the maximum-entropy streamline, and the fluid properties in the flow field are assumed to follow a simple trigonometric variation in the circumferential direction around the body. This technique was applied by Waldman to the Apollo shape, and reasonable results were obtained. However, its use is very limited because of the restrictive assumptions made.

The streamline-curvature method was developed by Gravalos *et al.* ([14]) and Maslen and Moeckel ([15]). This method attacks the direct problem by starting the numerical integration of the body surface, then "marching" outward from the body to the shock wave. The procedure requires that the pressure distribution around the body and entropy distribution in the flow field be assumed at the outset; these assumed distributions must then be refined as the procedure is iterated. The method has not been developed for the case of a blunt body at angle of attack.

Success of the stramline-curvature method for the transonic region of flow is closely related to the vanishingly small dependence of stream-tube area on pressure as the local Mach number approaches unity. The technique of Gravalos is to assume a shock shape which defines the entropy distribution on the streamline. Then, on the basis of an estimated pressure distribution along the body wall, the stream-tube width (Δn) is computed with the continuity equation, expressed in the form

$$\pi\varrho_\infty U_\infty(r_1^2 - r_2^2) = \varrho_{i,j}U_{i,j}2\pi\left(\frac{r_3 + r_4}{2}\right)\Delta n \tag{19}$$

The individual terms are defined in Fig. 8. The normal momentum equation, which is used to calculate the pressure on the outer edge of the stream tube, is given by

$$\Delta p/\Delta n = \varrho_{i,j}U_{i,j}^2/R_{i,j} \tag{20}$$

where R is the radius of curvature of the streamline and the other terms are defined in Fig. 8. With entropy known, all state properties (such as enthalpy) are thus defined. The energy equation is then used to calculate the velocity at the outer edge.

After all streamlines are computed, the intersection of the free-stream and shock-layer streamlines defines the location of a new bow shock. If pressures in the shock layer immediately behind this shock are consistent with the shock equations, the solution is correct. If not, the body-pressure distribution is altered—usually by a fraction of the pressure errors at the shock along the normal from the body point—and the solution is iterated.

Shown in Fig. 9 are two plots of results obtained by the application of the streamline-curvature method to the flow of nonequilibrium real air over a highly blunted cone moving 25,000 ft/sec [16]. The pressure distribution and stramline contours follow directly from the equilibrium solution except very near the shock, where some adjustment has been made to account for the variation of the flow-deflection angle between frozen and equilibrium properties.

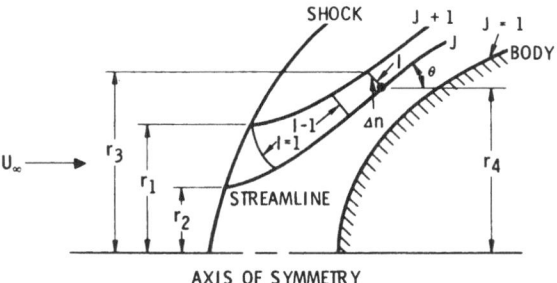

Fig. 8. Coordinate system for streamline-curvature method of analyzing steady-flow field.

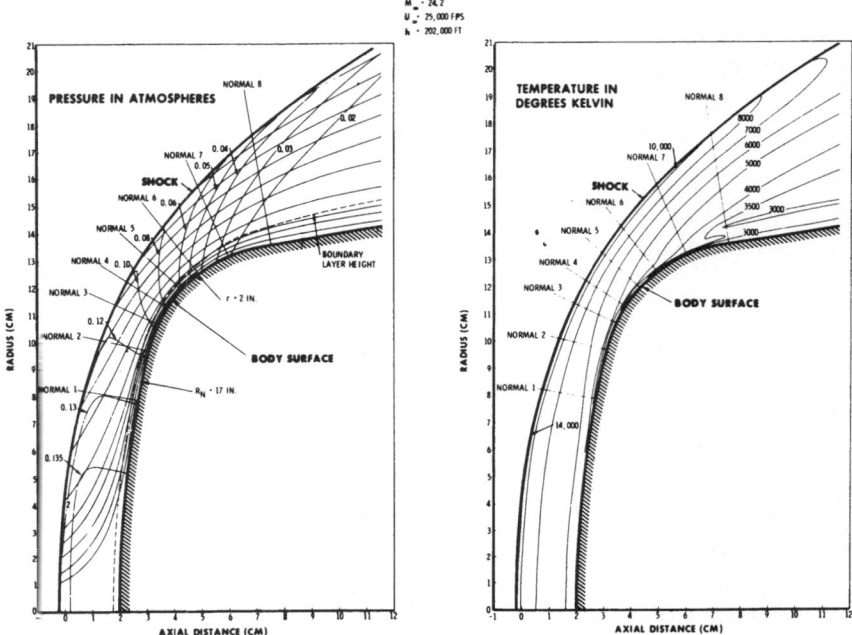

Fig 9. Results obtained by streamline-curvature method for flow of nonequilibrium real air over a highly blunted cone.

Nonequilibrium effects were evaluated by a simple stream-tube analysis; the temperature contours show the effect of the nonequilibrium chemistry. It should be recognized that the use of stream-tube analysis for the boundary layer may lead to significant errors, which result from the neglect of diffusion effects.

This body shape presented the problem of a discontinuous curvature, which occurs where the face and shoulder radii are joined. The result of this discontinuity is that the pressure gradient normal to the body surface is also discontinuous at the body. The problem was overcome by choosing grid points that were not too close to the discontinuity and by paying close attention to the pressure distribution in this region, where a large deviation from the Newtonian pressure occurs.

5.˙.3. Time-Dependent Solutions

The partial differential equations governing unsteady flow are hyperbolic for all Mach numbers. This fact appreciably simplifies numerical integration. The chief advantage of using a time-dependent solution for blunt-body flows is that it is inherently stable and automatically converges to the correct steady-flow solution without the costly and uncertain iterations necessary with other solutions of the direct problem. The time-dependent solution is a real relaxation from an arbitrary initial condition and has become the preferred technique for determining the flow-field properties in the subsonic–transonic region of a hypersonic blunt body.

Recent work, such as that of Bohachevsky and Mates ([17]), is based on the self-generation of shocks using the divergence (or conservative) form of the conservation equations ([18]). For the simple case of one-dimensional unsteady flow, this form of the equations can be written

$$\partial p/\partial t = -\partial(\varrho U)/\partial x \qquad (21)$$

$$\partial(\varrho U)/\partial t = -\partial(\varrho U^2 + p)/\partial x \qquad (22)$$

$$\partial(\varrho E)/\partial t = -\partial(\varrho EU + pU)/\partial x \qquad (23)$$

which express the local conservation of mass, momentum, and energy, respectively. In this form, the corresponding finite-difference equations can be more accurately integrated for regions with internal shock waves. The well-known shock-jump conditions that result from the exact Rankine–Hugoniot equations are more accurately matched by using the divergence form. However, these shocks are distributed over a distance of several grid spaces as a result of the finite-difference approach.

Moretti and Abett ([19]) showed how greatly machine time can be reduced by using a discontinuous bow shock in time-dependent solutions. Since a small grid spacing is not required to accurately locate the distributed shock, only a few points are needed across the shock layer to obtain adequate accuracy of the flow-field structure. No unnecessary points are computed, and the shock wave is accurately computed through use of the Rankine–Hugoniot conditions for a moving shock wave. The increase in grid-point spacing is compounded by the increased integration-time interval, resulting in great savings of machine time.

The possibility of extending Moretti's method has been investigated by Webb and Dresser ([20]). Moretti handled shock and body points through the use of the method of characteristics (in the time dimension). In the extended procedure, this complex approach was successfully replaced with simple end-point formulas in the derivative expressions. The equations were also solved in near-divergence form to make them applicable to flows with secondary shocks. In this procedure, the x, y, coordinates are replaced

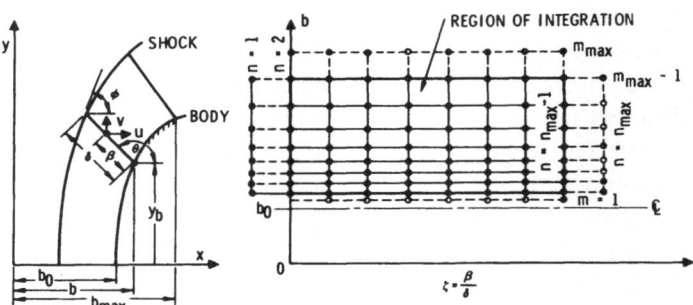

Fig. 10. Generalized coordinate system for time-dependent solution
of flow field at zero angle of attack.

with generalized coordinates to that more of the flow field can be computed,
as shown in Fig. 10. Thus the grid system becomes rectangular. Unequal
spacing permits latitude in spacing along the body surface. This method
may be applied to all parts of the flow field without regard for local Mach
number. Although the method of characteristics is probably more econom-
ical than is the time-dependent solution when $M > 1$, there are situations
when the tradeoff between computation time and the use of two computer
programs suggests continuing the time-dependent solution well into the
supersonic flow region.

In the extended version of Moretti's method, three-point central
differences are used exclusively to compute derivatives for all grid points.
Extrapolated points ahead of the shock and within the body are used in
lieu of end-point formulas to evaluate derivatives at the shock and body.
A similar extrapolation is used for the downstream end of the shock layer,
while symmetry conditions are satisfied on the axis.

Certain integration procedures add "effective viscosity" to the flow,
assuring stability of the solution at the expense of accuracy. Two such
schemes have been widely used: the Lax scheme ([21]) is of first-order accuracy
and introduces a large amount of smoothing for regions of nonlinear variations; the second-order Lax–Wendroff scheme ([22]) is also stable, but adds
very little artificial viscosity (or damping), resulting in a penalty of increased
computation time.

Since a rapid computation procedure is always desirable, a study was
made to evaluate the possibility of improving the accuracy of the simple
Lax scheme without losing integration stability. This improvement can be
realized by removing part of the effective viscosity. The source of the
viscosity (or smoothing) added by the Lax scheme is shown in Fig. 11,

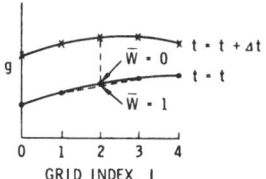

Fig. 11. Coarse grid system for studying the
Lax scheme.

where the valuation of the time derivative, given by

$$\left(\frac{\partial g}{\partial t}\right)_i = \frac{g_{(t+\Delta t)_i} - \frac{1}{2}(g_{t_{i+1}} + g_{t_{i-1}})}{\Delta t} \tag{24}$$

was modified by introducing the factor $\bar{W}$ $(0 < \bar{W} < 1)$ to the expression,

$$\left(\frac{\partial g}{\partial t}\right)_i = \frac{g_{(t+\Delta t)_i} - \frac{1}{2}\bar{W}(g_{t_{i+1}} + g_{t_{i-1}}) - (1 - \bar{W})g_{t_i}}{\Delta t} \tag{25}$$

When $\bar{W} = 1$, the procedure becomes identical to the original Lax procedure. When $\bar{W} = 0$, all smoothing is removed.

It is typical of the solution of hyperbolic equations that regardless of
the integration procedure, the integration interval is always subject to the
limitations of the Courant–Friedrichs–Lewy (CFL) condition [23]. In this
study of the modified Lax scheme, time-dependent solutions were obtained
for several values of the interpolation factor $\bar{W}$ and the CFL term, defined as

$$\text{Value CFL} = (\Delta t/\Delta s)(|u| + \bar{a}) \leq 1 \tag{26}$$

and are shown in Fig. 12 for the flow of a perfect gas over a sphere at
$M_\infty = 4$.

These results show the shock shape and location to be improved as
$\bar{W}$ is reduced (which decreases the artificial viscosity). Reducing the artificial
damping required that the integration time interval be reduced, as reflected
in the CFL value. Consequently, obtaining satisfactory accuracy necessitated
the use of a small time interval for a simple first-order integration scheme.
The shape of the sonic line near the body appears to be particularly sensitive
to the amount of artificial viscosity, and the computed shape is only margin-
ally acceptable for even the most accurate stable computation in which
$\bar{W} = 0.25$.

It can be concluded from this study that for flows with small grid
spacing, as would be experienced when secondary shocks are being sought,

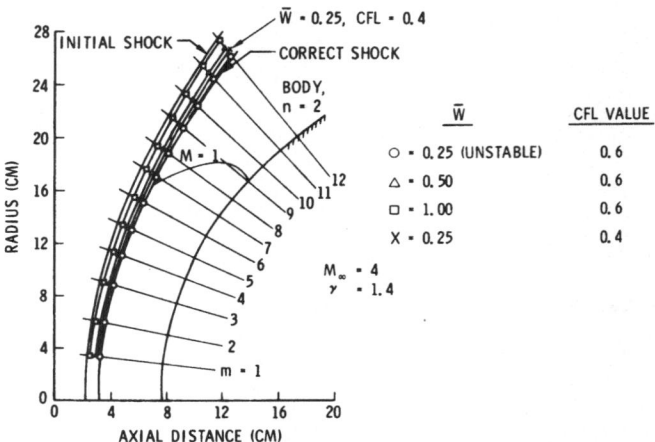

Fig. 12. Time-dependent solutions showin the effect of interpolation constant $\bar{W}$ and CFL (Courant–Friedrichs–Lewy) value for sphere.

the first-order scheme (possible with $\bar{W} \approx 1$) would be acceptably accurate. Fo: the case shown, it is preferable to use the second-order Lax–Wendroff scheme, as did Moretti.

It is interesting to observe the speed (in real time) with which an assumed shock wave relaxes to the correct bow shock in the time-dependent solution. This can be seen in Fig. 13, which shows the x velocity of the shock at three radial locations. The left-hand plot presents the shock velocity

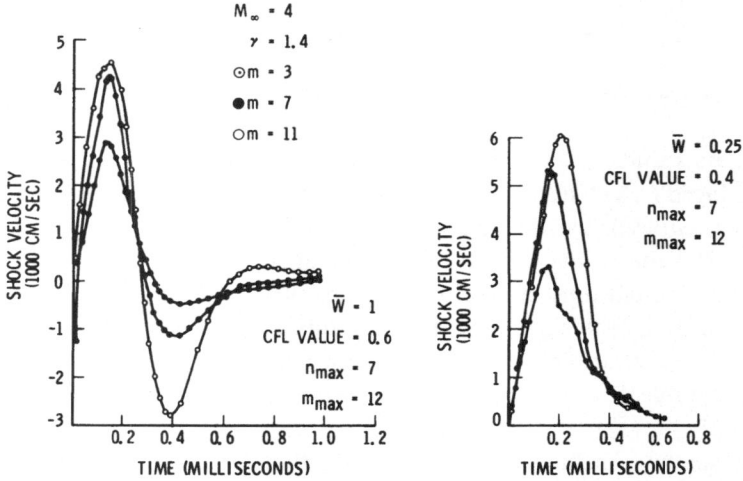

Fig. 13. Speed of shock movement for time-dependent solutions for a sphere.

time for a computer run with the original Lax scheme. Points farthest from the axis are evidently the most oscillatory. The right-hand plot shows a much more highly damped flow after 75% of the artificial viscosity had been removed. The shock moved backward monotonically to its steady-state location (at least to the limit of the calculations). At first, the outer portion of the shock moved aft at a speed greater than that of the portion near the axis of symmetry. Near the end of the calculated transient, the shock shape appears to have settled down, and the shock, as a whole, is asymptotically approaching the steady-state solution.

As a further extension of the time-dependent method, Moretti and Bleich ([24]) obtained the time-dependent convergence of the shock shape over a sphere-come body. In this case (consisting of 594 mesh points), approximately 40 min of IBM 7094 time was required to compute 300 time-integration steps—which is a measure of the computer time required to obtain a solution by the time-dependent method.

5.2. Method of Analysis—Supersonic Region

In a supersonic flow, real characteristics (or Mach lines) exist, one feature of which is to make possible the analysis of discontinuous phenomena. Not only are shock waves present in a supersonic flow field, but velocity discontinuities are propagated along the characteristic lines.

The method of characteristics is the classical method for the numerical analysis of supersonic flow. This method is based on the fact that the partial differential equations that describe the flow reduce to ordinary differential equations that are valid on the streamlines and the Mach lines (the two families of characteristics). Along streamlines, entropy is a constant, whereas along Mach lines, for conventional Cartesian coordinates, the characteristic equations are of the form

$$\frac{dp}{\varrho U^2 \tan \mu} \pm d\theta + \frac{j \sin \theta \sin \mu}{\sin(\theta \pm \mu)} \frac{dy}{y} = 0 \qquad (27)$$

where $\mu = \arcsin(1/M)$; $j = 0$ for two-dimensional flow and $j = 1$ for axisymmetrical flow; and y is the radial coordinate. Geometrical construction of the characteristics and a determination of the flow properties with these equations may involve iterative calculations that require the equation of state, the energy equation, and the constancy of entropy along a streamline as well as the characteristic equations.

Recent applications of this classical method include both perfect and real gas flows, and even nonequilibrium effects ([25]). For a flow with finite chemical-reaction rates, the frozen speed of sound is basic, being the speed of the wavefront [26]. For nearly equilibrium flow, however, the bulk of the disturbances are associated with the Mach wave defined by the equilibrium speed of sound.

Many recently-developed computer programs [27] include three-dimensional-flow capabilities, permitting multiple shocks to be handled with complex shock intersections and downstream slip surfaces. However, development of the logic for such programs does require good understanding of the flow, since instability problems can result from the numerical techniques employed.

A recent result obtained by Rakich [28] for a flow over a blunted cone points out a difficulty, not well understood at this time, with the flow behavior beyond about four nose radii downstream on the lee side, for which the calculations predict zero pressure. This problem is most apparent for slender bodies at large angle of attack. With this method, it is also difficult to keep track of the entropy layer far downstream on slightly-blunted large-angle bodies; however, the problem can be overcome by reducing the grid spacing near the wall [29].

6. VISCOUS-FLOW ANALYSIS

Analysis of the attached boundary layer is the key to the overall success of predicting the performance of a hypersonic vehicle as it traverses the atmosphere. However, the boundary conditions used in the viscous-flow analysis are obtained, in turn, from the results of the inviscid-flow evaluation. Thus there is a great deal of interdependence in the results of the analysis for these two flow-field regions, and the final prediction of vehicle performance is determined from analyses which have been iterated several times.

The viscous flow field is analyzed, as is the inviscid flow field, by solving the equations for the conservation of mass, energy, and momentum, and the equation of state for the flow stream. However, in the viscous analysis. consideration must also be given to the equations for the shear stresses that stem from the fluid viscosity. The basic simplifying assumption for the differential equations for the boundary layer is that for any parameter that changes under the influence of viscosity, the gradient of the parameter taken normal to the body surface must be much greater than its gradient taken parallel to the surface. For the case of a three-dimensional, axisymmetrical body at angle of attack (Fig. 14), the differential equations for the flow field are as follows:

Conservation of mass (continuity):

$$\frac{\partial}{\partial x}(\varrho u) + \frac{\partial}{\partial y}(\varrho v) + \frac{\partial}{\partial z}(\varrho w) = 0 \tag{28}$$

Conservation of momentum:

x-component momentum equation

$$\varrho\left(u\frac{\partial u}{\partial x} + v\frac{\partial u}{\partial y} + w\frac{\partial u}{\partial z}\right) = -\frac{\partial p}{\partial x} + \frac{\partial}{\partial z}\left(\mu\frac{\partial u}{\partial z}\right) \tag{29}$$

y-component momentum equation

$$\varrho\left(u\frac{\partial v}{\partial x} + v\frac{\partial v}{\partial y} + w\frac{\partial v}{\partial z}\right) = -\frac{\partial p}{\partial y} + \frac{\partial}{\partial z}\left(\mu\frac{\partial v}{\partial z}\right) \tag{30}$$

z-component momentum equation

$$\partial p/\partial z = 0 \tag{31}$$

Conservation of energy:

$$\varrho\left(u\frac{\partial H}{\partial x} + v\frac{\partial H}{\partial y} + w\frac{\partial H}{\partial z}\right) = \frac{\partial}{\partial z}\left[\mu\left(\frac{\partial H}{\partial z} + \frac{1-\text{Pr}}{\text{Pr}}C_p\frac{\partial T}{\partial z}\right)\right] \tag{32}$$

where $H = \frac{1}{2}(u^2 + v^2) + h(p, \varrho)$.

Equation of state:

$$h = h(p, \varrho)$$

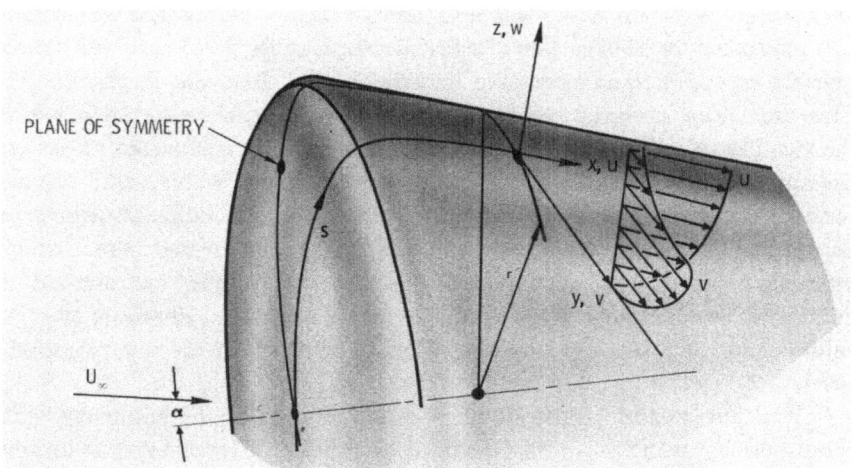

Fig. 14. Coordinate system for analyzing attached boundary layer over an axisymmetrical body at angle of attack.

In addition, the transport properties (μ and either k or the Prandtl number, Pr) must be defined.

In this form, the equations are generalized and valid for laminar, as well as for turbulent, flow. For laminar flow, the transport properties are well known over a wide range of temperatures; in turbulent flow, however, the properties are much more complex, and there is not yet a comprehensive understanding of their characteristics [Patankar ([30]) presents recent work on this problem.]

6.1. Methods of Analysis—Boundary Layer

There are essentially three methods of solving the boundary layer equations. First is the integral technique, which, while not highly accurate in defining the details of the flow field, does provide excellent results for the overall effects of the boundary layer (e.g., surface friction, heat transfer, and boundary-layer displacement thickness). This technique has been widely used to obtain solutions for the turbulent boundary layer because it was, until recently, the only method available. Even though it is limited to the calculation of parameters such as the local skin friction and heat transfer rates, it has been widely applied and has reached a high degree of sophistication [e.g., see ([31])]. Unfortunately, with this method, the local property values inside the viscous layer are required as input data, and details of the flow field are not obtained from the solution.

A second method of solving the boundary layer equations is the similarity solution, which is applicable to laminar flow. This method was originally presented by Blasius [see ([32]) for incompressible flow] and was subsequently extended to compressible flow fields by Cohen and Reshotko ([33]). This technique assumes that a transformed plane can be found in which the profiles of all newly-defined parameters in the transformed plane are similar. Its great advantage is that it provides considerable detail for describing the flow-field composition. It offers the additional advantage of permitting the tabulation of the solution in the transformed plane, which minimizes subsequent calculations. On the other hand, the method is limited to laminar flow fields and does not permit consideration of nonequilibrium chemistry (or mass injection at the wall) except in very specific cases.

The third method is to obtain an exact solution to the boundary layer equations by using a finite-difference technique. Closed-form solutions obtained by quadrature are usually based on simplifying the problem to some degree, but, using the finite-difference approach with a high-speed digital computer to obtain an exact solution, it is not necessary to make such assumptions relative to physical phenomena. In the latter approach, the step size in each direction can be adjusted to obtain as many data points as desired. Exact solutions obtained with machine integration can encompass the effects of nonequilibrium chemistry and wall injection in the basic equations, and considerable detail is obtained in the information about the flow field.

Unfortunately, the finite-difference method requires a great deal of programming and machine time. However, the advent of high-speed computers has opened a whole new era in the solution of physical problems. Mellor ([34]) was the first to use such a computer to solve the equations for the incompressible turbulent boundary layer, employing for the boundary layer the "three-layer" concept, in which three effective viscosity laws are used in the calculations. Each of these laws is used for a separate layer of the viscous flow field as follows: (1) the laminar sublayer adjacent to the body, where the only stresses acting are the viscous stresses of the laminar flow (this layer is several orders of magnitude thinner than the total boundary layer thickness); (2) the overlap layer, in which the velocity fluctuations are so large that they give rise to turbulent shearing stresses, which are comparable to viscous stresses (the introduction of the mixing-length theory by Prandtl in 1925 was the first milestone in the study of this region); and (3) the turbulent core, where the turbulent stresses completely outweigh the viscous stresses. In each of the three layers, the effective kinematic viscosity has a different expression, given by the following relationships:

Velocity-defect layer

$$\nu_{\text{eff}} = KU_e\delta^*(x) \tag{33}$$

where $K = \text{const}$ and $\delta^* = \int_0^\delta [1 - (\varrho u/\varrho_e U_e)] \, dy$;

Overlap layer

$$\nu_{\text{eff}} = k^2 y^2 \,|\, \partial u/\partial y \,| \tag{34}$$

where $k = \text{const}$;

Sublayer

$$\nu_{\text{eff}} = \nu_{\text{mol}}\phi(\zeta) \tag{35}$$

where $\zeta = (k^2 y^2/\nu_{\text{mol}}) \,|\, \partial u/\partial y \,|$ and $\phi(\zeta)$ is an experimental or empirical function.

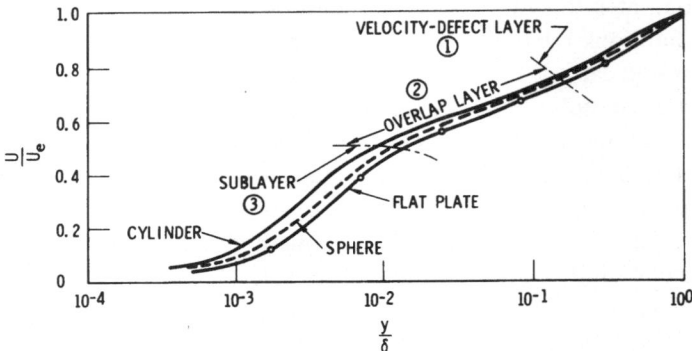

Fig. 15. Typical velocity distribution for a three-layer incompressible boundary layer on a flat plate, sphere, and right-circular cylinder.

For the case of incompressible flow, results obtained from this three-layer concept compare very favorably with experimental data. A typical velocity distribution obtained using this concept is shown in Fig. 15 for a flat plate, cylinder, and sphere. In Mellor's study, unfortunately, the turbulent-boundary-layer equations are transformed before integration, and the results have to be retransformed to the physical plane. As a consequence, the method is applicable only to turbulent attached boundary layers for which the transformation is valid. Moreover, to obtain accuracy for the very thin laminar sublayer, a considerable number of points along each surface normal must be computed.

6.2. Improved Boundary-Layer Analysis

To avoid the difficulties in Mellor's method, a technique of numerical integration has been developed by Waiter and LeBlanc ([35]) whereby the equations are integrated in the physical plane, and a variable step size is also introduced in the direction normal to the wall. This approach starts with a laminar boundary layer and automatically switches to turbulent flow on the basis of any one of a set of "built-in" transition criteria (Re_x, Re/ft, $Re_{\delta*}$, Re_θ, $\partial u/\partial y$). For the turbulent regime, a minimum of ten data points (from a total of 60–80) in the laminar sublayer are valuable, and, as a result, a high degree of accuracy is obtained near the wall (e.g., skin-friction coefficient). The method has been developed for incompressible flow with two-dimensional or axisymmetrical bodies having either positive or negative pressure gradients and with (or without) gas injection or suction

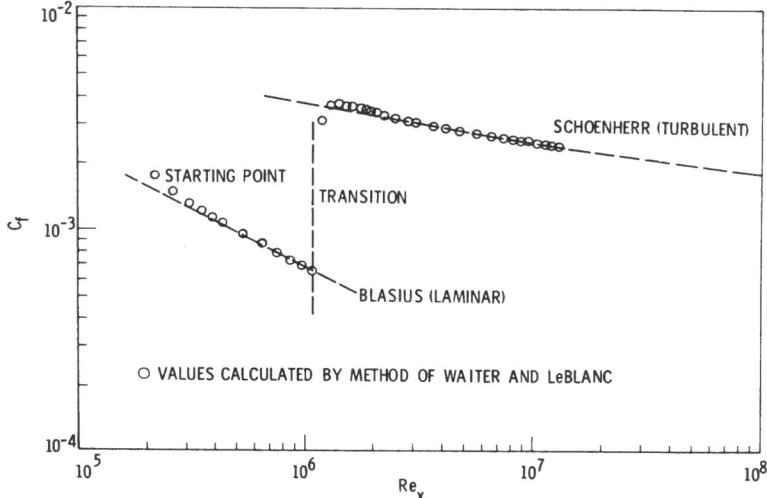

Fig. 16. Skin-friction coefficient for incompressible flow over a flat plate.

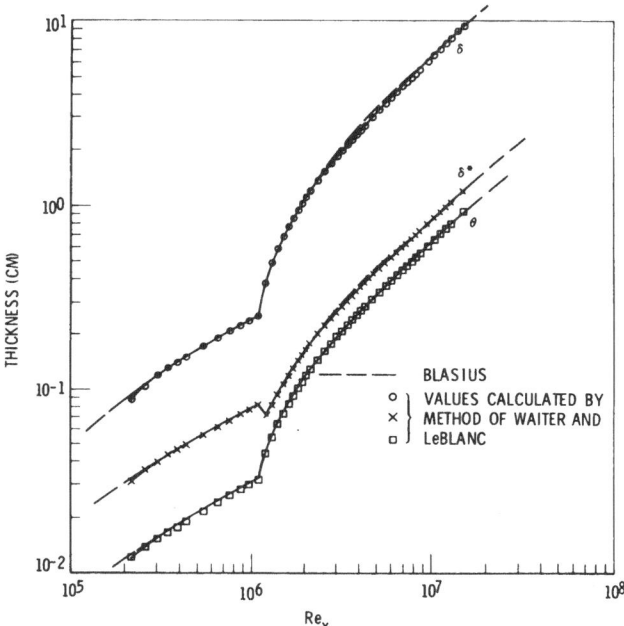

Fig. 17. Boundary-layer thicknesses for incompressible flow
over a flat plate.

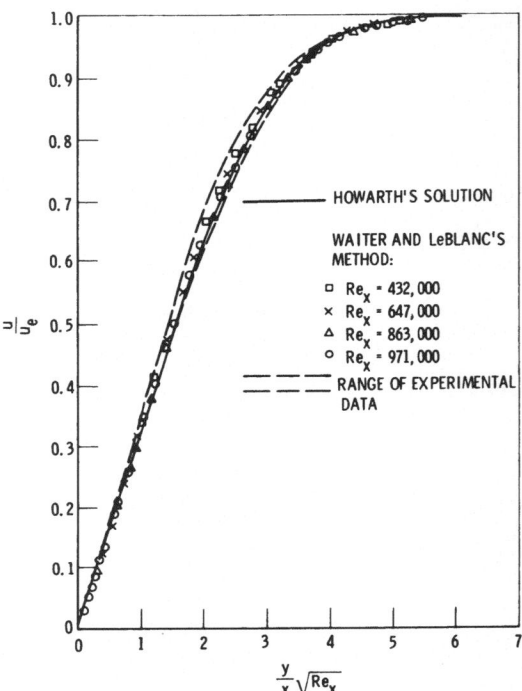

Fig. 18. Comparison of calculated and experimental
incompressible laminar profiles on a flat plate.

at the wall. Typical results obtained using this method are shown for flow
over a flat plate in Figs. 16 and 17. The computation is based on a transition
criterion of $Re_x = 10^6$. The results of closed-form solutions are also included
to facilitate comparison.

To further verify the validity of this technique, the correlation of a
calculated laminar profile in the transformed plane is compared (Fig. 18)
with Howarth's solution and with experimental data. In Fig. 19 a turbulent
profile is compared with recently acquired experimental data [36]; here,
it is impossible to distinguish between the calculated results and the experimental data.

Results for the case of a flat plate, cylinder, and sphere in incompressible
flow with blowing and with suction have been computed using the Waiter–
LeBlanc method [35]. Even with the addition of these complicating parameters, the comparison between analytical results and experimental
data is excellent.

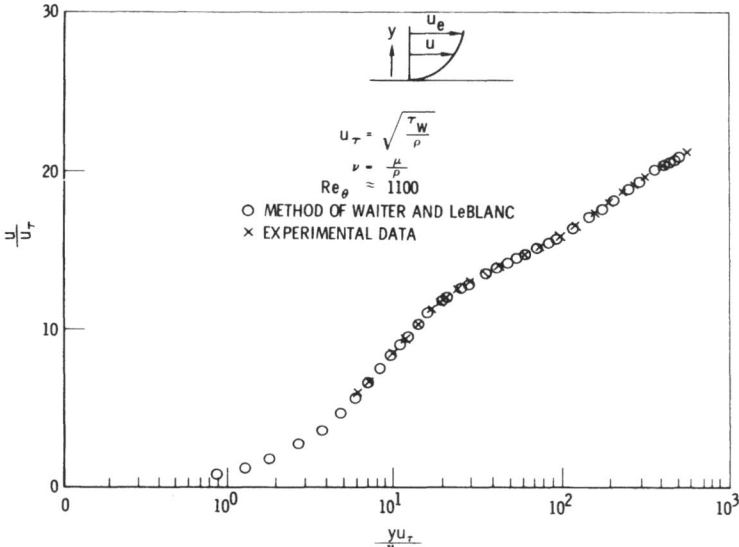

Fig. 19. Comparison of calculated and experimental velocity profiles for the incompressible turbulent boundary layer over a flat plate.

7. INTERACTIONS OF VISCOUS-INVISCID FLOW

One of the assumptions made in boundary layer theory is that the thin viscous layer causes only a very small disturbance in the external flow field. On the basis of this assumption, the pressure field surrounding the body can be calculated from the external inviscid flow for the actual shape of the body, neglecting the displacement effect of the boundary layer. But this assumption breaks down because the boundary layer separates from the wall and creates a noninfinitesimal disturbance. In Fig. 3, the hypersonic flow regions show the boundary layer separating due to the adverse pressure gradient downstream of the shoulder. Location of the separation line depends on the base pressure, nature of the boundary layer (laminar or turbulent), and character of the inviscid flow field. The mixing layer, recirculation flow, and base pressure are determined by coupling the action between the viscous flow and the inviscid flow in the region around, and downstream of, the separation line. This coupling is difficult to accomplish analytically for subsonic speeds, where a disturbance at one point affects the entire flow field. For supersonic speeds, on the other hand, the inviscid-flow equations are of the hyperbolic type, resulting in disturbances being propagated along the characteristics (Mach lines); consequently, interactions with the visous flow can be treated in a relatively simple manner.

7.1. Separated Flow

A simple problem illustrating boundary layer separation in supersonic flow is that of a flat plate on which separation is caused by the impingement of an externally-generated oblique shock wave or by a wedge on the plate (Fig. 20). If the supersonic flow is assumed to be purely inviscid, the pressure rises discontinuously across the impinging and reflecting shock waves or the shock wave attached to a wedge on the plate. In the presence of a boundary layer, however, the pressure rise along the plate does not occur suddenly, because the disturbance is propagated upstream through the subsonic region near the wall, and the boundary layer is then subjected to a rapid but continuous pressure increase. When the pressure rise is small, the boundary layer can negotiate it without separating. There is, however, a certain threshold of pressure rise—called the incipient-separation pressure—which depends on the Mach and Reynolds numbers and above which the boundary layer no longer remains attached.

7.1.1. *Incipient Separation*

Roshko and Thomke [37] have established a correlation for the incipient-separation pressure of a turbulent boundary layer, which can be expressed in the form

$$\frac{[(p_2 - p_1)/p_1]_{\text{incipient}}}{C_{f_1}} = 57(10)^{0.462 M_1} \tag{36}$$

where $C_{f_1} = \tau/\frac{1}{2}\varrho_1 U_1^2$ is the skin-friction coefficient just before the pressure rise is initiated, and the other terms are defined in Fig. 20. This equation can be used to estimate the incipient-separation pressure for a turbulent boundary layer over a wide range of Mach and Reynolds numbers.

For a laminar boundary layer, Popinski and Ehrlich [38] have correlated the incipient-separation pressure in the form

$$[(p_2 - p_1)/p_1]_{\text{incipient}} = \frac{1}{2}\gamma M_1^2 (\text{Re}_{\text{sep}})^{-1/4}[2.03(M_1^2 - 1)^{-0.306}] \tag{37}$$

This correlation essentially relates the incipient-separation pressure to the square root of the friction coefficient at the point of separation (as shown by the exponent on the Reynolds number), whereas the Roshko–Thomke correlation implies that the relationship should be linear in C_{f_1}. This discrepancy between the two sets of results is amplified by the results of Hankey [39] and Needham [40], who correlated the wedge angle β_i for

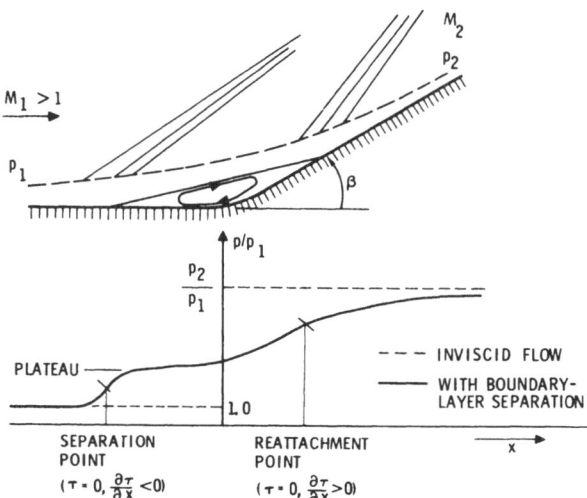

Fig. 20. Flow separation on a flat plate due to a wedge or externally-generated shock wave impinging on the surface.

incipient separation for hypersonic laminar boundary layers in the form

$$M_1\beta_i \sim (M_1^3/Re_{sep}^{1/2})^{1/2} \tag{38}$$

for the range $0.8 \leq M_1\beta_i \leq 8$. If pressure ratios are calculated that correspond to the data collected by Needham for $M_1\beta_i$, the incipient-separation pressure can then be correlated in the form

$$(p_2 - p_1)/p_1 \sim M_1^2/Re_{sep}^{1/2} \tag{39}$$

which is consistent with the results of Roshko and Thomke. These results suggest, therefore, that the laminar-flow correlation needs to be investigated further. However, it should be noted that the incipient pressure rise for a turbulent boundary layer is much greater than for laminar flow because of the differences in the two velocity distributions.

When the wedge angle on the flat plate, or the impinging shock strength, is such that the pressure exceeds the value corresponding to the incipient-separation pressure, the boundary layer separates ahead of the wedge or the shock-impingement point, reattaches downstream of the wedge apex or the impingement point, and forms a recirculation bubble with a shape approximating a small-angle wedge. In this case, the flow separates initially at an angle smaller than the wedge angle (Fig. 20), and the initial pressure rise is less. Thus the compression process takes two steps: first, through the separation region; second, through the reattachment region. If these two regions are sufficiently distant from each other, a region develops between them that has a relatively constant pressure (called the plateau pressure).

It should be noted that the incipient-separation pressure is neither the separation pressure (the pressure at the separation point) nor the plateau pressure. Rather, it is greater than either of these two and is the overall pressure rise through the preseparation and the postreattachment regions.

7.1.2. *Free Separation*

When boundary layer separation occurs far away from the cause, it is referred to as free separation. A simple example of this separation occurs in supersonic flow over a two-dimensional flat plate (Fig. 21). In this illustration, the boundary layer has separated because it has grown to a thickness that can no longer negotiate the continuously increasing pressure.

Chapman *et al.* [41], Gadd *et al.* [42], and Bogdonoff and Kepler [43] observed that the pressure distribution in the region of separated flow at supersonic speed is only indirectly influenced by the cause of separation. Instead, the pressure depends primarily on the local interaction of the flow in the boundary layer with the external inviscid flow. Using this observation, Chapman *et al.* made a simplified analysis that shows the dependence of the pressure rise in the separation region on the Mach and Reynolds numbers. In this analysis, the pressure distribution, which is determined locally by the interaction of the boundary layer, is assumed to be governed by the linearized Prandtl–Meyer relationship given by

$$\frac{p - p_1}{\frac{1}{2}\gamma M_1^2 p_1} \approx \frac{2(\beta - \beta_1)}{(M_1^2 - 1)^{1/2}} \tag{40}$$

The pressure rise for this expression can be determined from a simplified form of the boundary layer equation, where it is assumed that the convective

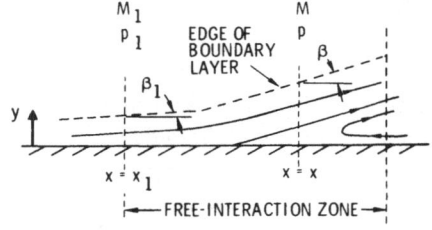

Fig. 21. Free separation in supersonic flow
over a flat plate.

terms can be neglected and the viscous force and the pressure force just balance each other; i.e.,

$$dp/dx \approx \partial\tau/\partial y \tag{41}$$

Substituting this expression into the Prandtl–Meyer expression shows that

$$\frac{p - p_1}{\frac{1}{2}\varrho_1 U_1{}^2} \sim \frac{(C_{f_1})^{1/2}}{(M_1{}^2 - 1)^{1/4}} \tag{42}$$

LeBlanc and Webb ([44]) correlated the data of Chapman *et al.* for many types of separation, basing their work on the assumption that for laminar flow

$$C_{f_1} \sim (\text{Re})^{-1/2} \tag{43}$$

and for turbulent flow

$$C_{f_1} \sim (\text{Re})^{-1/5} \tag{44}$$

Their results can be expressed as follows:

Laminar flow

$$\frac{p_{\text{plateau}} - p_1}{\frac{1}{2}\varrho_1 U_1{}^2} = \frac{1.778}{(\text{Re}_{\text{sep}})^{1/4}(M_1{}^2 - 1)^{1/2}} \tag{45}$$

$$\frac{p_{\text{separation}} - p_1}{\frac{1}{2}\varrho_1 U_1{}^2} = \frac{0.889}{(\text{Re}_{\text{sep}})^{1/4}(M_1{}^2 - 1)^{1/2}} \tag{46}$$

Turbulent flow

$$\frac{p_{\text{plateau}} - p_1}{\frac{1}{2}\varrho_1 U_1{}^2} = \frac{1.77}{(\text{Re}_{\text{sep}})^{1/10}(M_1{}^2 - 1)^{1/2}} \tag{47}$$

$$\frac{p_{\text{separation}} - p_1}{\frac{1}{2}\varrho_1 U_1{}^2} = \frac{1.25}{(\text{Re}_{\text{sep}})^{1/10}(M_1{}^2 - 1)^{1/2}} \tag{48}$$

Lewis *et al.* ([45]) obtained a single curve for the correlation of the pressure distribution in the free-separation region for laminar flow at M_1 in the range from 4 to 6. This correlation was based on a normalized pressure parameter in the form

$$P = \frac{p - p_1}{\frac{1}{2}\varrho_1 U_1{}^2} \cdot \frac{(M_1{}^2 - 1)^{1/4}}{(C_{f_1})^{1/2}} \tag{49}$$

and a normalized length parameter, originally suggested by Curle ([46]), given by

$$X = \frac{x - x_1}{x_1} M_1^{1/2}(\text{Re}_{\text{sep}})^{1/4}\left(\frac{T_0}{T_{\text{wall}}}\right) \tag{50}$$

For the case of a turbulent boundary layer, Zukowski ([47]) reexamined existing data for a wide range of Reynolds numbers, and concluded that the pressure distribution is independent of Reynolds number. He suggests that this apparent independence of Reynolds number may have resulted from the fact that the friction coefficient had not been measured directly in any of the experiments he examined, and that it might have varied little with Reynolds number because of some effect such as wall roughness or changes in the free-stream turbulence level. More importantly, he suggests that this independence results from the relatively short length of the interaction region and from the fact that, due to the increased turbulence level near separation, the wall shear is negligible in the turbulent free interaction.

7 1.3. Methods of Analyzing Separation

From the early 1950's to the present time, the interaction between boundary layer and supersonic external flow has been the subject of numerous investigations. In the method used by Lees and Reeves ([48]), the viscous layer, including the separated region, is assumed to be boundary-layerlike: $\partial/\partial x \ll \partial/\partial y$, and $\partial\varrho/\partial y \approx 0$. The procedure is to transform the equation for the compressible laminar boundary layer into the incompressible form by the Stewartson transformation

$$X = \int_0^x \left(\frac{\bar{a}}{\bar{a}_0} \right)^{(3\gamma-1)/(\gamma-1)} dx \tag{51}$$

$$Y = \frac{\bar{a}}{\bar{a}_0\gamma_0^{1/2}} \int_0^y \frac{\varrho}{\varrho_0} \, dy \tag{52}$$

in order to account for the compressibility effect. Integrating the partial differential equations for momentum, kinetic energy, and total energy across the viscous flow with the assumed profiles provides the ordinary differential equations. The interaction between the boundary layer and the external supersonic flow is obtained by computing the induced flow angle at the edge of the boundary layer. This angle is determined by integrating the mass-continuity equation across the boundary layer; the angle is then related to the local Mach number or to the local pressure by an appropriate supersonic-flow relation. These ordinary differential equations can then be solved by applying appropriate boundary conditions.

Applying the Lees–Reeves method, Ko and Kubota ([49]) considered the flow separation in compression corners. The model for this analysis consisted of a flat plate, with a sharp leading edge, at zero angle of attack, to which a 10° wedge was joined by a curved surface of constant radius of curvature. Computations were carried out for $M_\infty = 6.0$ and $Re = 60,000$ per inch for each of three similar models of different lengths so that the trailing edges had different effects on the interaction and post-reattachment regions. The analytical results compare very favorably with experiment; however, the base pressure must be known to perform the calculations.

7.2. Base and Wake Flow Fields

Complete flow-field descriptions for the base and near-wake regions of a blunt body are not readily available from analytical procedures. Predicting the flow in the base region requires that the analysis of the flow in the separation region be coupled to the analysis of the flow in the reattachment region. Even when the Reynolds number in the free stream is very high, both viscous and inviscid flow in the base region must be considered simultaneously; otherwise, either the description of the flow field is indeterminate (two-dimensional flow) or the results do not yield a closed recirculation region (axisymmetrical flow).

7.2.1. Base Flow Fields

The flow in the base region of a blunt body can be understood by use of the model shown in Fig. 22. This model, first introduced by Chapman et al. ([41]), starts where the boundary layer separates from the body and forms a mixing layer that induces a region of low-speed reverse flow (recirculation). Chapman assumed that the mixing layer's thickness at the

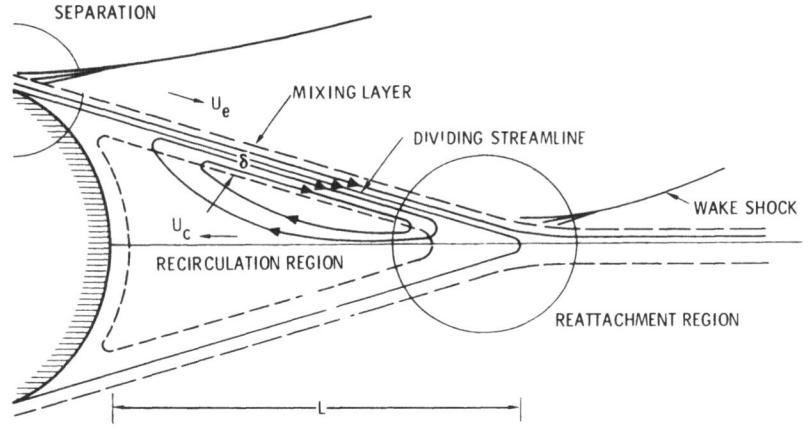

Fig. 22. Flow in the base region of a blunt body at high Reynolds number.

separation point is zero. However, in later models of the flow in the base region, the mixing layer is assumed to have a finite, rather than zero, thickness at the separation point [e.g., see [50]].

In order to make valid assumptions in the analysis of Chapman's model, it is important to note the order of magnitude of the parameters in the base region. If the mixing layer is assumed to behave like a conventional laminar boundary layer, the thickness of the mixing layer δ, when compared with the length of the mixing layer L, will be of the order

$$\delta/L = O(\text{Re}^{-1/2}) \tag{53}$$

where Re is $U_e L/\nu_e$. Thus if L is similar in magnitude to the body dimension, as the Reynolds number becomes large, δ will become small (a thin mixing region).

The mass flow entrained in the mixing layer is of the order $\varrho_e U_e \delta$ and, consistent with the requirement for the conservation of mass, this must be of the same order as the mass flow in the reverse flow region, $\varrho_c U_c L$. Therefore

$$\varrho_c U_c L/\varrho_e U_e \delta = O(1) \tag{54}$$

Combining the two foregoing expressions gives

$$\varrho_c U_c/\varrho_e U_e = O(\text{Re}^{-1/2}) \tag{55}$$

For large Reynolds numbers (with $\varrho_c \sim \varrho_e$) $U_c \ll U_e$, and the recirculation region is made up essentially of dead air.

A measure of the viscous forces in this region is $\mu_c U_c/L$, which, when divided by the dynamic pressure term, $\varrho_c U_c^2$, and combined with Eqs. (54) and (55), shows that

$$\frac{\mu_c U_c/L}{\varrho_c U_c^2} = \frac{\mu_e}{\varrho_c U_c L} \frac{\mu_c}{\mu_e} = O\left(\frac{\mu_c}{\mu_e} \frac{1}{\text{Re}} \frac{L}{\delta}\right) = O\left(\frac{\mu_c}{\mu_e} \text{Re}^{-1/2}\right) \tag{56}$$

Thus for reasonably large Reynolds numbers, the viscous forces in the recirculation region are small compared with the inertial forces, and the recirculation flow is assumed to be inviscid. On this basis, the pressure variation in the reverse-flow region, Δp_c, is of the same order as the dynamic pressure term, $\varrho_c U_c^2$, which, when combined with Eq. (55), shows that

$$\Delta p_c/\varrho_e U_e^2 = O(\text{Re}^{-1}) \tag{57}$$

This result confirms that the recirculation region can be assumed to be a constant-pressure, dead-air region.

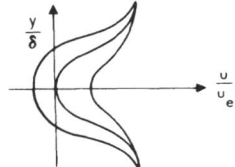

Fig. 23. Typical Falkner–Skan velocity profiles used in analysis of the flow behind a
blunt body.

A solution for the case of a constant-pressure mixing layer between
high-speed flow and dead air has been obtained by Chapman [51] on the
basis that the initial thickness of the mixing layer is zero, and by Dennison
and Baum [50] and Kubota and Dewey [52] for an initial thickness of non-
zero.

The reattachment point (Fig. 22) is the stagnation point of the dividing
streamline, which divides the recirculating flow from the flow streaming
down into the wake. Chapman assumed that compression in the reattachment region is so rapid that viscous effects are negligible. The pressure at
this stagnation point is some ratio of the pressure behind the reattachment
shock wave. Once the value of this ratio is assumed, the base pressure is
uniquely determined. It is pointed out by Dennison and Baum that with
this model for the base flow, the base pressure is independent of Reynolds
number for a given body shape, even when the initial mixing-layer thickness
is not zero.

7.2.2. Near-Wake Flow

Flow in the region behind a blunt body has been analyzed by Reeves
and Lees [53], Webb et al. [54], and Grange et al. [55]. The approach in
these studies was to assume that boundary layer equations were valid
throughout the region, even though the base region is thick compared with
the usual boundary layer thickness,* and to apply the integral-momentum
method for a viscous boundary layer interacting with external supersonic
flow. The resultant solutions were based on the use of a family of Falkner–
Skan velocity profiles, with the velocity gradient, rather than velocity itself,
vanishing at the symmetry axis (Fig. 23). Since this method is incapable
of yielding a solution for the flow near the body, the solution obtained

---

* The only significant contribution made to the flow analysis by the recirculation region
 is that it balances the mass flow. This balance is included in even the boundary-layer
 approximation by use of the exact continuity equation. As was shown earlier, the
 recirculation region makes very little dynamic contribution to the flow.

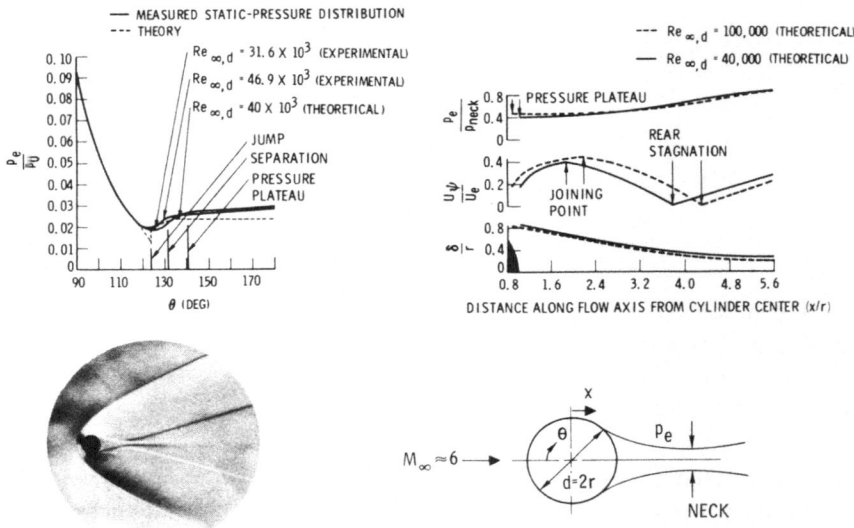

Fig. 24. Comparison of measured and predicted pressure distributions for the base and near-wake regions of a cylinder normal to the flow.

must be joined to the solution for the flow near the separation point; then a constant-pressure, Chapman-type solution is obtained for some station between the body and the reattachment point. The total solution is achieved by a patching process in which the mass flow, the velocity of the dividing streamline, the pressure, and the thickness of the viscous region are all matched at a common point—and the downstream boundary conditions are satisfied. This approach is an attempt to improve the Chapman theory by replacing Chapman's assumption of inviscid recompression in the reattachment region. The pressure distribution predicted by this method for a circular cylinder normal to the flow is compared with experimental data ([56,57]) in Fig. 24.

7.2.3. Comparison of Base and Wake Flow

It should be stressed that the methods derived here apply only to blunt bodies; different approaches and different orders of magnitude apply to slender bodies. For example, Table II shows a comparison of the relative magnitude of flow parameters in the separated and base-flow regions for blunt-nose and slender bodies (Fig. 25) in hypersonic flow. The Mach number of the inviscid flow surrounding the base flow is relatively low (3–4) for blunt bodies, but very high for slender bodies in hypersonic flow.

TABLE II

Comparison of the Base and Wake Flow Characteristics of Blunt and Slender Bodies

Parameter	Body shape	
	Blunt	Slender
Local Mach No.	Low	Very high
Local Reynolds No.	High	Relatively low
Viscous layer	Thin	Thick
$\partial p/\partial y$ Effect	Negligible	Important
Configuration	25a	25b

The local pressure, hence the Reynolds number based on the local flow properties, is high for blunt bodies and relatively low for slender bodies. Furthermore, the viscous layer in the base region of a slender body becomes comparable to the base dimension of the body, and the boundary layer concept may then have to be modified; i.e., the transverse pressure gradient becomes important in the separation zone of a slender body, and shock waves may appear in the viscous layer.

8. WAKE FLOWS

Like the boundary layer problem, wakes can be divided into three general classifications: laminar flow, transition, and turbulent flow. In general, changes in properties in the direction of the stream are small compared with changes across the wake, and the lateral pressure gradient is negligible. Thus the flow in the wake can be treated much like the flow in the boundary layer.

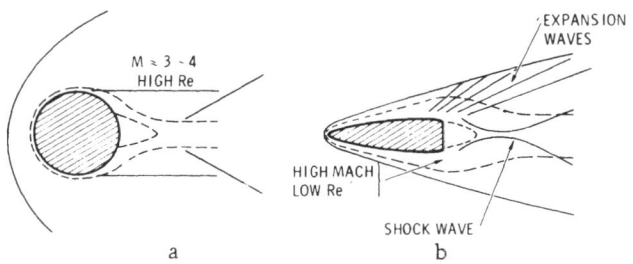

Fig. 25. (a) Blunt and (b) slender body configurations.

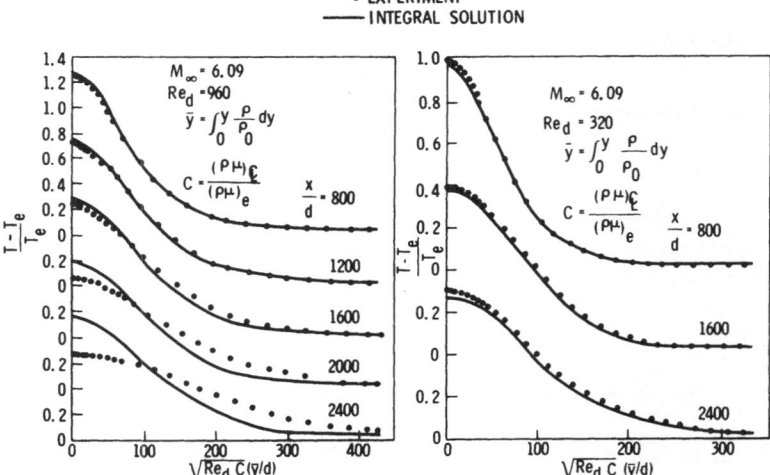

Fig. 26. Comparison of theoretical and experimental data for the temperature profiles in the two-dimensional wake behind a right-circular cylinder.

As with the boundary layer problem, laminar flow in the wake is the most amenable to analysis. The boundary-layer equations can be solved with initial conditions obtained from the solution of the base-flow region, and the boundary conditions determined from the inviscid-flow solution. In the analysis of the turbulent wake, experimental results are used to estimate the "eddy" viscosity for the equations, yielding results that are acceptable for engineering purposes. Since wake flows are very unstable compared with boundary layers, the transition from laminar to turbulent flow in the wake can be estimated from the result of laminar-flow stability theory.

McCarthy and Kubota [57] have compared theory and experiment (Fig. 26) for the temperature distribution in the wake of the two-dimensional circular cylinder shown in Fig. 24. The theory [first presented by Kubota [58]] is based on the boundary layer equations, which are converted to incompressible form by a Howarth-type transformation,

$$\bar{y}(x, y) = \frac{u_e}{u_\infty} (\mathrm{Re}_d)^{1/2} \int_0^y \frac{\varrho}{\varrho_\infty} \frac{dy}{d} \qquad (58)$$

$$\bar{x}(x) = \int_0^x \frac{\varrho_e \mu_e u_e}{\varrho_\infty \mu_\infty u_\infty} \frac{dx}{d} \qquad (59)$$

and linearized by the Oseen approximation:

$$U = u_e + u, \qquad u <\!<\!< u_e \qquad (60)$$

The measured temperature profiles of Fig. 26 compare well with linearized theory so long as the wake is laminar, i.e., when Reynolds number and distance along the wake centerline are below critical values. For $Re_d = 320$, the measured profile starts to deviate from the laminar approximation at $x/d \approx 2000$, indicating the beginning of a transition region; for $Re_d = 960$, good agreement between experiment and theory exists for $x/d \approx 1500$, again indicating the beginning of a transition region. It should be noted that, for clarity, the origin in Fig. 26 has been shifted downward for each profile at a different x/d; thus each profile on the abscissa approaches zero at high values without interfering with the other profiles.

The beginning of the transition region in the wake of a cylinder in a hypersonic stream is shown in Fig. 27 for the inner and outer wakes ([59]). The inner wake is created by the viscous boundary layer on the cylinder, and the outer wake is created by the bow shock wave. This very strong, curved bow shock produces high entropy changes and high temperatures. The solid curves were obtained by use of the theory of hydrodynamic stability of small disturbances in laminar wakes, as described by Behrens ([59,60]). The data shown here correspond to a hundredfold amplification of the initial disturbances. Although not shown in this illustration, transition in the wake of a blunt axisymmetrical body behaves in a similar manner ([61,62]).

The turbulent inner wake grows rapidly in the downstream direction

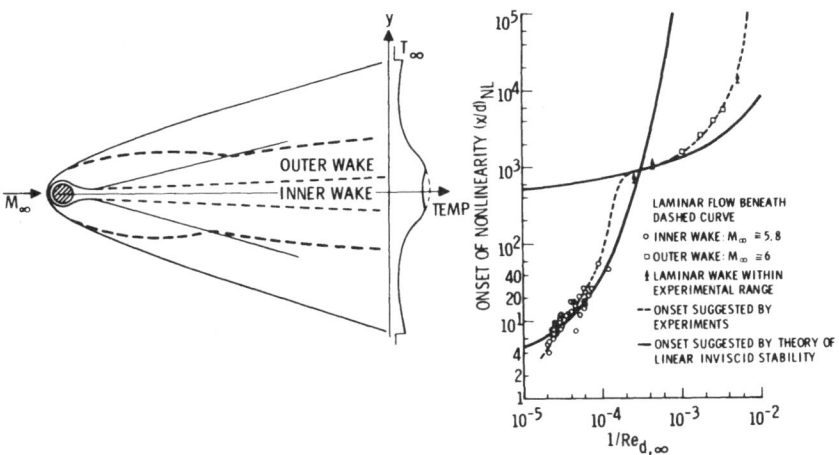

Fig. 27. Prediction of the onset of transition in the far wake and comparison with experiment ([59]). (Reprinted by permission of the American Institute of Aeronautics and Astronautics, New York.)

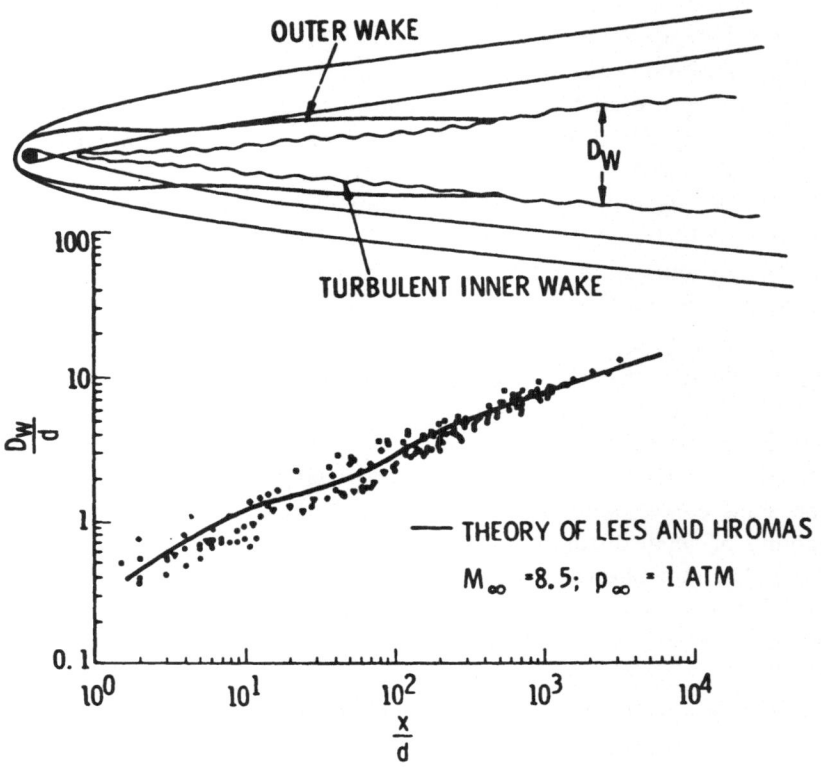

Fig. 28. Predicted width of turbulent-wake flow fields and comparison with experiment ([34]). (Reprinted by permission of the American Institute of Aeronautics and Astronautics, New York.)

and it soon "swallows" the laminar outer wake. The rate of growth of the inner wake depends on this swallowing effect, because the turbulent viscosity is a function of the total momentum defect in the wake. Lees and Hromas [63]) analyzed this interaction of the turbulent inner wake and the laminar outer wake by using the momentum-integral method.

In Fig. 28, theoretical predictions of wake widths for spheres are compared with experimental data ([64]). Experimental results were obtained in free-flight ranges, and the wake widths were measured from shadowgraphs of the wake. General features of the turbulent wake (such as wake thickness and velocity defect) are predicted by the theory with good accuracy. The microfeatures of the turbulent wake (i.e., the levels and scales of velocity, temperature, and density fluctuations) have not been determined.

9. SUMMARY

In this chapter, emphasis has been placed on methods of analysis being used in the aerospace industry to solve practical problems associated with the flow around a blunt body traveling at hypersonic speed. At this speed, the free stream of a blunt body such as the Apollo command module is abruptly decelerated through a very strong bow shock, accelerates around the body through sonic to supersonic speed, and eventually ends up in a wake behind the body. Meanwhile, because of viscosity, the boundary layer at the surface of the body builds up and separates, and this separated viscous flow exerts a dominating influence in the base-flow and near-wake regions behind the body.

The discussion has been limited to regimes involving continuum flow, a necessary base line for solving problems that currently affect spacecraft design. Other considerations, such as dissociation, ionization, radiative effects, nonequilibrium flow, and chemical reactions, play a secondary role, at least for vehicles of current interest.

The classical methods of analyzing the hypersonic inviscid flow near the stagnation region of a blunt body have been reviewed. Each of these has unique advantages and disadvantages. In some cases, assumptions are too crude to give a sophisticated picture of the compressible flow field. In others, a transformed plane is used, and the entire flow field must be solved before any information for the physical plane is available.

The most promising technique for analyzing hypersonic blunt-body inviscid flow is that involving time-dependent flow. A finite-difference approximation is assumed, and convergence takes place as a function of time. Convergence from a given set of initial conditions to the steady-flow solution is automatic, and internal shocks are self-generating. Although the machine time required is high and integration instabilities must be dealt with, the method has been used very successfully and is highly accurate.

Once the problem in the vicinity of the stagnation point is solved, standard techniques involving the method of characteristics are adequate to solve the inviscid flow field in the supersonic region. Again, the machine time necessary for this solution can be expensive, but the techniques are straightforward and well understood.

In the analysis of the attached boundary layer, techniques are available for estimating its nature with some accuracy. These techniques are restricted in application, however, and usually require fundamental approximations— such as concerning the initial velocity and enthalpy profiles. For turbulent flow, additional assumptions, involving the nature of the turbulent viscosity, must be made. Encouraging results have been obtained recently through the application of three viscosity laws (law of the wall, overlap layer, and velocity-defect layer) for different regimes in the boundary layer.

In practice, all the tools described here are used extensively to predict the flow field and parameters in the separation region. Usually, no tool is sufficient by itself. The reason for this is that "universal" correlations are available for only very simple configurations, and analytical techniques have been developed for only planar and axisymmetrical cases; moreover, wind-tunnel tests often do not duplicate actual flight conditions. In addition, viscous flow is internally coupled with external flow, and the inviscid–viscous interaction must be considered.

The base-flow and near-wake regions of a blunt body have been described by using Chapman's model (or modifications of it), which has been shown to involve a mixing layer and a recirculation region of essentially dead inviscid flow. Such models have been used to predict the base pressure behind a hypersonic vehicle, and these results from the analysis of the base-flow region behind the body are required as the initial condition in solving the wake flow. The latter flow can be treated much like the boundary layer, because changes in the parameters in the direction of the stream are dominant. Furthermore, like boundary layer flow, the wake has been described as having laminar flow, transition, or turbulent flow. Techniques are available for analyzing each of these types, although the detailed state properties and microfeatures of the urbulent flow have not been accurately predicted.

NOTATION

$\bar{a}$	velocity of sound
a	summation index
a	arbitrary coefficient
C	Chapman–Rubesin constant $= (\mu/\mu_e)(T_e/T)$
C_D	drag
C_f	skin friction coefficient
C_L	lift coefficient
C_M	moment coefficient
C_p	specific heat at constant pressure
CFL	Courant–Friedrichs–Lewy term
D	drag
d	diameter

E	total energy
g	general functional relationship
H	total enthalpy
h	altitude
h	static enthalpy
i, I	incremental position
j, J	incremental position
j	index defining number of dimensions to be analyzed
K	constant
K	Knudsen number
k	conductivity
k	constant
L	lift
M	Mach number
Δn	stream-tube width
p	pressure
Pr	Prandtl number
$\dot{q}$	heat flux
R	radius
r	radius
Re	Reynolds number
Re_d	Reynolds number based on diameter
Re_x	Reynolds number based on distance x
$\mathrm{Re}_{\delta*}$	Reynolds number based on δ^*
Re_θ	Reynolds number based on momentum thickness
s	summation index
s	distance
T	temperature
t	time
U	velocity
U_τ	friction velocity $= \tau_w/\varrho$
u	x component of local velocity
v	y component of velocity
$\overline{W}$	interpolation factor defined by Eq. (25)
w	z component of velocity
x, r, θ	curvilinear coordinates
x, y, z	Cartesian coordinates

Greek Letters

α	angle of attack
α, β, γ	arbitrary constants
β	wedge angle
γ	ratio of specific heats
Δ	shock-wave standoff distance
Δ	increment
δ	boundary layer thickness
δ^*	boundary layer displacement thickness
ε	density ratio (ϱ_1/ϱ_∞)
θ	angle
θ	boundary layer momentum thickness
λ	mean free path
μ	arcsin(1/M)
μ	kinematic viscosity
ν	dynamic viscosity
ϱ	density
σ	shock angle relative to free-stream velocity
τ	shear stress
ϕ	stream function for three-dimensional flow field
ϕ	velocity-potential function
ψ	flow-field stream function

Superscript

coordinate axes rotated through an angle

Subscripts

b	at the body wall
c	in the recirculation region
e	edge of boundary layer (local inviscid condition)
eff	effective value
geo	geometrical relationship
i	condition of incipient separation
i, j	arbitrary elements of a set
mol	molecular value
sep	separation point

w	value at wall
0	condition at stagnation point
1	conditions aft of shock wave
2	conditions downstream from bow shock wave
∞	free-steam condition
θ	based on momentum thickness
ψ	values measured along the dividing stream line

REFERENCES

1. S. A. Schaff, "Mechanics of Rarefied Gases," NAVORD Report 1488, Vol. 5 (February 1959).
2. R. F. Probstein and N. H. Kemp, "Viscous Aerodynamic Characteristics in Hypersonic Rarefied Gas Flow," Avco, Report RR48 (December 1959).
3. W. D. Hayes, and R. F. Probstein, *Hypersonic Flow Theory: Vol. 1, Inviscid Flows*, Academic Press, New York (1966).
4. M. D. Van Dyke, "The Supersonic Blunt-Body Problem—Review and Extension," *J. Aerospace Sci.* **25** (8), 485–496 (August 1958).
5. H. Lomax, and M. Inouye, "Numerical Analysis of Flow Properties about Blunt Bodies Moving at Supersonic Speeds in an Equilibrium Gas," NASA TR R-204 (July 1964).
6. P. Garabedian and H. Lieberstein, "On the Numerical Calculation of Detached Bow Shock Waves in Hypersonic Flow," *J. Aerospace Sci.* **25** (8), 109–118 (1968).
7. H. G. Webb, Jr., H. S. Dresser, B. K. Adler, and S. A. Waiter, "Inverse Solution of Blunt-Body Flow Fields at Large Angle of Attack," *AIAA J.* **5** (6), 1079–1085 (June 1967).
8. W. W. Joss, "Application of the Inverse Technique to the Flow over a Blunt Body at Angle of Attack," NASA CR-445 (April 1966).
9. R. Swigart, "Hypersonic Blunt-Body Flow Fields at Angle of Attack," *AIAA J.* **2** (1), 115–117 (1964).
10. G. E. Kaatari, "Shock Envelopes of Blunt Bodies at Large Angle of Attack," NASA TN D-1980 (December 1963).
11. W. D. Hayes, "Rotational Stagnation-Point Flow," *J. Fluid Mech.*, **19**, Part 3, 366–374 (1964).
12. O. M. Belotserkovskii, "Flow with a Detached Shock Wave about a Symmetrical Profile," *Prikl. Matem. i Mekhan.* **22** (2), 206–219 (1958); translated in *Appl. Math. Mech.* **1958**, pp. 279–296.
13. G. Waldman, "Integral Approach to the Yawed Blunt-Body Problem," AIAA Paper No. 65-28 presented at AIAA 2nd Aerospace Sciences Meeting, New York, 1965.
14. F. G. Gravalos, I. H. Edelfelt, and H. W. Emmons, "The Supersonic Flow about a Blunt Body of Revolution for Gases at Chemical Equilibrium," in: *Proceedings of the Ninth International Astronomical Congress*, Vol. 1, Springer-Verlag, Berlin (1959).
15. S. H. Maslen, and W. E. Moeckel, "Inviscid Hypersonic Flow Past Blunt Bodies," *J. Aeron. Sci.*, **24** (9), 683–693 (1957).

16. H. G. Webb, Jr., and D. M. Schrello, "Final Report: Flow-Field Prediction and Analysis for Project RAM—Phase II," Space Division, North American Rockwell Corporation, SID 64-548 (July 1964).

17. I. O. Bohachevsky, and R. E. Mates, "A Direct Method for Calculation of the Flow about an Axisymmetric Blunt Body at Angle of Attack," *AIAA J.*, **4** (5), 776–782 (1966).

18. L. Fox, "Numerical Solutions of Ordinary and Partial Differential Equations," Pergamon Press, New York (1962).

19. G. Moretti, and M. Abett, "A Time-Dependent Computational Method for Blunt-Body Flows," *AIAA J.* **4** (2), 2136–2141 (December 1966).

20. H. G. Webb, Jr. and H. S. Dresser, "Unsteady Flow over Axisymmetric Blunt Bodies at Zero Angle of Attack," Space Division, North American Rockwell Corporation, SD 67-881 (September 1967).

21. P. D. Lax, "Weak Solutions of Nonlinear Hyperbolic Equations and Their Numerical Computations," *Commun. Pure Appl. Math.*, **7**, 159–193 (1954).

22. P. D. Lax, and B. Wendroff, "Systems of Conservation Laws," *Commun. Pure Appl. Math.* **13**, 217–237 (1960).

23. R. D. Richtmyer, *Difference Methods for Initial-Value Problems*, Interscience, New York (1957).

24. G. Moretti, and G. Bleich, "Three-Dimensional Flow around Blunt Bodies," AIAA Paper No. 67-222 presented at AIAA 5th Aerospace Sciences Meeting, New York, January 23–26, 1966.

25. A. D. Wood, J. F. Springfield, and A. J. Pallone, "Chemical and Vibrational Relaxation of an Inviscid Hypersonic Flow," *AIAA J.* **2** (10), 1697–1705 (October 1964).

26. B. T. Chu, "Wave Propagation and the Method of Characteristics in Reacting Gas Mixtures with Application to Hypersonic Flow," Wright Air Development Center, TN-57-213, ASTIA Doc. AD 118350 (May 1957).

27. C. W. Chu, A. F. Niemann, Jr., and S. A. Powers, "An Inviscid Analysis of the Plume Created by Multiple Rocket Engines and a Comparison with Available Schlieren Data—Part I: Calculation of Multiple Rocket Engine Exhaust Plumes by the Method of Characteristics," AIAA Paper No. 66-651 presented at AIAA Second Propulsion Joint Specialist Conference, Colorado Springs, Colorado, June 13–17, 1966.

28. J. V. Rakich, "Three-Dimensional Flow Calculation by the Method of Characteristics," *AIAA J.* **5** (10), 1906–1908 (October 1967).

29. E. C. Knox, and C. H. Lewis, "A Comparison of Experimental and Theoretically Predicted Pressure Distributions and Force and Stability Coefficients for a Spherically Blunted Cone at $M_\infty \approx 18$ and Angles of Attack," ARO Inc., AEDC-TR-65-234 (February 1966).

30. S. V. Patankar, "Heat and Mass Transfer in Turbulent Boundary Layers," PhD thesis, Imperial College of Science and Technology, University of London (June 1967).

31. J. Valensi, R. Michel, and D. Guffroy, "Résultats Expérimentaux et Théoriques sur le Transfert de Chaleur au Bord d'Attaque des Ailes à Forte Flèche en Hypersonique," AGARD Report No. 97 (May 1965).

32. H. Schlichting, *Boundary-Layer Theory*, 6th ed., McGraw-Hill Book Co., New York (1968).

33. C. R. Cohen, and E. Reshotko, "Similar Solutions for the Compressible Laminar Boundary Layer with Heat Transfer and Pressure Gradient," NACA TR 1293 (1956).

34. G. L. Mellor, "Incompressible Turbulent Boundary-Layer Equations with Arbitrary Pressure Gradients and Divergent or Convergent Cross Flows," *AIAA J.* **5** (9), 1570–1579 (September 1967).

35. S. A. Waiter, and L. P. LeBlanc, "Solution of the Boundary-Layer Equation by an Implicit Variable-Step-Size Finite-Difference Technique," Space Division, North American Rockwell Corporation, SD 68-635 (1968).

36. S. A. Waiter, and R. B. Anderson, "Solution of the Laminar Boundary-Layer Equations by an Implicit Finite-Difference Technique," Space Division, North American Rockwell Corporation, SD 67-587 (September 1967).

37. A. Roshko, and G. J. Thomke, "Correlation for Incipient-Separation Pressure," McDonnell Douglas Corporation, Report DAC-59800 (May 1966).

38. Z. Popinski, and C. F. Ehrlich, "Development Design Methods for Predicting Hypersonic Aerodynamic Control Characteristics," Lockheed California Company, Technical Report AFFDL-TR-66-85 (1966).

39. W. L. Hankey, "Prediction of Incipient Separation in Shock-Boundary-Layer Separation," *AIAA J.* **5** (2), 355–356 (February 1967).

40. D. A. Needham, "A Note of Hypersonic Incipient Separation," *AIAA J.* **5** (12), 2284–2285 (December 1967).

41. D. R. Chapman, D. M. Kuehn, and H. K. Larson, "Investigation of Separated Flows in Supersonic and Subsonic Streams with Emphasis on the Effect of Transition," NACA Report 1356 (1958).

42. G. E. Gadd, D. W. Holder, and J. D. Regan, "An Experimental Investigation of the Interaction between Shock Waves and Boundary Layers," *Proc. Royal Soc. (London) Ser. A* **226**, 227–253 (1954).

43. S. M. Bogdonoff, and C. E. Kepler, "Separation of a Supersonic Turbulent Boundary Layer," Department of Aeronautical Engineering, Princeton University, Report 249 (January 1954).

44. L. P. LeBlanc and H. G. Webb, Jr., "Boundary-Layer Separation in a Supersonic Stream," Space Division, North American Rockwell Corporation, SID 61-72 (1961).

45. J. G. Lewis, T. Kubota, and L. Lees, "Experimental Investigation of Supersonic Laminar, Two-Dimensional Boundary-Layer Separation in a Compressible Corner with and without Cooling," *AIAA J.* **6** (1), 7–14 (January 1968).

46. N. Curle, "The Effects of Heat Transfer on Laminar Boundary-Layer Separation in Supersonic Flow," *Aero. Quart.* **12** Part 4, 309–336 (1961).

47. E. E. Zukowski, "Turbulent Boundary-Layer Separation in front of a Forward-Facing Step," *AIAA J.* **5** (10), 1746–1753 (October 1967).

48. L. Lees and B. L. Reeves, "Supersonic Separated and Reattaching Laminar Flows: I—General Theory and Application to Adiabatic Boundary Layer Shock–Wave Interactions, *AIAA J.* **2** (11), 1907–1920 (1964).

49. D. R-S. Ko and T. Kubota, "Supersonic Laminar Boundary Layer along a Two-Dimensional Adiabatic Curved Ramp," AIAA Paper 68-105 presented at 6th Aerospace Sciences Meeting, New York, 22–24 January 1968.

50. M. R. Dennison and E. Baum, "Compressible Free Shear Layer with Finite Initial Thickness," *AIAA J.* **1** (2), 342–349 (February 1963).

51. D. R. Chapman, "Laminar Mixing of a Compressible Fluid," NACA Report 958 (1950).

52. T. Kubota and C. F. Dewey, Jr., "Momentum Integral Methods for the Laminar Free Shear Layer," *AIAA J.* **2** (4), 625–629 (April 1964).

53. B. L. Reeves and L. Lees, "Theory of the Laminar Near Wake of Blunt Bodies in Hypersonic Flow," *AIAA J.* **3** (11), 2061–2074 (1965).

54. W. H. Webb, R. J. Golik, F. W. Vogenitz, and L. Lees, "A Multimoment Integral Theory for the Laminar Supersonic Near Wake," *Proceedings of the 1965 Heat Transfer and Fluid Mechanics Institute*, Stanford University Press, Stanford, California (1965).

55. J. Grange, J. M. Klineberg, and L. Lees, "Laminar Boundary-Layer Separation and Near-Wake Flow for a Smooth Blunt Body at Supersonic and Hypersonic Speeds," *AIAA J.* **5** (6), 1089–1097 (June 1967).

56. J. F. McCarthy, Jr., "Hypersonic Wakes," California Institute of Technology, GALCIT Hypersonic Research Project, Memo 67 (1962).

57. J. F. McCarthy, Jr. and T. Kubota, "A Study of Wakes behind a Circular Cylinder at $M = 5.7$," *AIAA J.* **2** (4), 629–636 (April 1964).

58. T. Kubota, "Laminar Wake with Streamwise Pressure Gradient," California Institute of Technology, GALCIT Hypersonic Research Project, Memo No. 9 (May 1962).

59. W. Behrens, "The Far Wake behind Cylinders at Hypersonic Speeds: I, Flowfield," *AIAA J.* **5** (12), 2135–2141 (1967).

60. W. Behrens, "The Far Wake behind Cylinders at Hypersonic Speeds: II, Stability," *AIAA J.*, **6** (2), 225–232 (February 1968).

61. W. G. Clay, M. Labitt, and R. E. Slattery, "Measured Transition from Laminar to Turbulent Flow and Subsequent Growth of Turbulent Wakes," *AIAA J.* **3** (5), 837–841 (May 1965).

67. L. N. Wilson, "Far-Wake Behavior of Hypersonic Spheres," *AIAA J.* **5** (7), 1238–1244 (July 1967).

63. L. Lees, and L. Hromas, "Turbulent Diffusion in the Wake of a Blunt-Nosed Body at Hyprsonic Speeds," *J. Aerospace Sci.*, **29**, 976 (1962).

64. R. Knystautas, "Growth of the Turbulent Inner Wake behind 3-in. Diameter Spheres," *AIAA J.* **2** (8), 1485–1486 (1964).

Chapter 6

RAREFIED GAS DYNAMICS

S. A. Schaaf

Professor of Engineering Science
University of California, Berkeley

INTRODUCTION

Rarefied gas dynamics is concerned with flows at such low density that the molecular mean free path is not negligible. Under these conditions, the gas no longer behaves as a continuum. Important modifications in aerodynamic and heat transfer characteristics occur which are ascribable to the basic molecular structure of the gas.

This branch of gas dynamics has been the subject of many investigations since the time of Maxwell. Most of the early studies were related to very low speed flows, and usually to "internal" geometries—pipes, ducts, orifices, and the like—in connection with vacuum problems. The results of this classical research are to be found in the standard texts on kinetic theory. Since World War II, however, a revival of interest in the field has occurred due to applications to very high-altitude, high-speed flight.

It is convenient to subdivide rarefied gas dynamics into four different flow regimes. These are called "free-molecular flow," "near-free-molecular flow," "transition flow," and "slip flow," corresponding, respectively, to extremely rarefied, highly rarefied, moderately rarefied, and only slightly rarefied gas flows. This subdivision is desirable because the four flow regimes exhibit quite different phenomena and the basic theoretical approaches are entirely different. Since "rarefied" is a relative term, the demarcation of these four subdivisions is not characterized by absolute pressure or gas density levels, but rather in terms of the ratio of the mean free path λ to some dimension L characteristic of the flow field. The ratio λ/L is called the Knudsen number, Kn. Free-molecular flow corresponds to very large Knudsen number, $\text{Kn} > 10$; slip flow corresponds to Knudsen numbers in the range $0.01 < \text{Kn} < 0.1$; while the transitional regimes lies in between, with $0.10 < \text{Kn} < 10$. These values are, of course, arbitrary.

Phenomena do not change abruptly. However, they seem to correspond pretty well with present experimental evidence.

The mean free path, the sound speed a, and the kinematic viscosity ν are related by $\nu \sim \lambda a$. Hence the Knudsen number is expressible in terms of the two basic parameters of ordinary continuum gas dynamics, namely the Mach number $M = V/a$ and the Reynolds number $Re = VL/\nu$, where V is the reference gas velocity. This relation is $Kn \sim M/Re$, which is basic for a discussion of rarefied gas dynamics.

The literature on rarefied gas dynamics has become much too extensive to cover in a chapter in this book. Detailed summaries are available in ([1,2]) and in the proceedings of the first five symposia on rarefied gas dynamics ([3]).

1. ELEMENTS OF KINETIC THEORY

The basic assumption of kinetic theory is that a gas can be idealized as consisting of simple molecules moving freely with respect to each other except during collisions. Most of the more advanced results are also confined to the assumption that the molecules possess no internal degrees of freedom. The basic concept of kinetic theory is the molecular distribution function $f(\mathbf{x}, \mathbf{c}, t)$, which is the number density in phase space of molecules with position $\mathbf{x}$, velocity $\mathbf{c}$, at the time t. The distribution function obeys a conservation equation, known as the Maxwell–Boltzmann equation,

$$\frac{\partial f}{\partial t} + \sum_{i=1}^{3} c_i \frac{\partial f}{\partial x_i} = \int\!\!\!\int\!\!\!\int_{-\infty}^{\infty} d\mathbf{c}' \int\!\!\int d\omega \{|\,\mathbf{c}' - \mathbf{c}\,|\, I(\mathbf{c}' - \mathbf{c}, \theta)(f_1 f_1' - ff')\} \quad (1)$$

where $I(\mathbf{c}' - \mathbf{c}, \theta)$ is the differential cross section for the scattering of a molecule with velocity $\mathbf{c}'$ through an angle θ within the solid angle $d\omega$, and $\mathbf{c}_1$ and $\mathbf{c}_1'$ are the velocities after collision of two molecules whose velocities before collision were $\mathbf{c}$ and $\mathbf{c}'$. It has also been assumed that no external forces are acting.

The Maxwell–Boltzmann equation is a nonlinear integro-partial differential equation. The left side represents the rate of change of f due to convection. The right-hand side, known as the collision integral, represents the rate of change of f due to molecular collisions. The validity of the equation requires that the gas density be not so great that multiple collisions or the volume occupied by the molecules become important, nor so small that ensemble averages fluctuate too much. Both of these requirements are generally satisfied for gas flows of interest to the mechanics of rarefied gases.

The equilibrium solution of Eq. (1), i.e., the solution for which $f = f(\mathbf{c})$, only, is

$$f = f^0 = \frac{n}{(2\pi RT)^{3/2}} \exp\left[-\frac{(c_1 - u_1)^2 + (c_2 - u_2)^2 + (c_3 - u_3)^2}{2RT} \right]$$

where n is the number density, R the gas constant per unit molecular weight, T the absolute temperature, and u the macroscopic gas velocity. The function f^0 is known as the Maxwellian or equilibrium distribution.

The nonequilibrium case has been the subject of many investigations, and is still by no means completely resolved. For all molecular models (monatomic), and regardless of the magnitude of the departure from equilibrium, the five basic conservation equations of continuum gas dynamics, apply, namely,

$$\frac{\partial \varrho}{\partial t} + \sum_{i=1}^{3} \frac{\partial (\varrho u_i)}{\partial x_i} = 0$$

$$\frac{\partial u_i}{\partial t} + \sum_{j=1}^{3} \left(u_j \frac{\partial u_i}{\partial x_j} + \frac{1}{\varrho} \frac{\partial p_{ij}}{\partial x_j} \right) = 0; \qquad i = 1, 2, 3 \qquad (2)$$

$$\frac{\partial p}{\partial t} + \sum_{i=1}^{3} \left(\frac{\partial (pu_i)}{\partial x_i} + \frac{2}{3} \frac{\partial q_i}{\partial x_i} + \frac{2}{3} \sum_{j=1}^{3} p_{ij} \frac{\partial u_i}{\partial x_j} \right) = 0$$

where p_{ij} and q_i are moments of the distribution function, identifiable as stresses or heat flux components,

$$p_{ij} = m \int\!\!\int\!\!\int_{-\infty}^{\infty} (c_i - u_i)(c_j - u_j) f \, d\mathbf{c} \qquad (3)$$

$$q_i = \tfrac{1}{2} m \int\!\!\int\!\!\int_{-\infty}^{\infty} (c_i - u_i)(\mathbf{c} - \mathbf{u})^2 f \, d\mathbf{c} \qquad (4)$$

The conservation equations are obtained from the Maxwell–Boltzmann equation by utilizing the five collisional invariants. Equations (2) do not form a determined set, however, since they have more dependent unknowns than five. One of the tasks of kinetic theory is to provide the necessary supplementary relations.

For an unbounded region and for infinitesimal gradients in the macroscopic quantities n, T, and u, considered as functions of space and time, Eq. (1) may be linearized. For the case of "Maxwellian molecules," i.e., central force fields which repel each other with the inverse fifth power of their distance of separation, the eigenfunctions of the linearized collision integral operator are known, so that rigorous solutions may be obtained.

Then

$$f = f^0\left[1 - \sum_{i=1}^{3} \sum_{j=1}^{3} \frac{p\delta_{ij} - p_{ij}}{2pRT} (c_i - u_i)(c_j - u_j)\right.$$

$$\left. - \sum_{i=1}^{3} \frac{q_i}{pRT} (c_i - u_i)\left(1 - \frac{(\mathbf{c} - \mathbf{u})^2}{5RT}\right)\right] \tag{5}$$

and the Navier–Stokes and Fourier heat conduction formulas (for the case of no bulk viscosity), namely,

$$p_{ij} = \delta_{ij}\left(p + \frac{2}{3}\mu \sum_{\varkappa=1}^{3} \frac{\partial u_\varkappa}{\partial x_\varkappa}\right) - \mu\left(\frac{\partial u_i}{\partial x_j} + \frac{\partial u_j}{\partial x_i}\right) \tag{6}$$

$$q_i = -\varkappa \, \partial T/\partial x_i \tag{7}$$

are obtained. The quantities μ and $\varkappa$ are given by

$$\mu = 0.243(2m/A)^{1/2}mRT, \qquad \varkappa = 15R\mu/4 \tag{8}$$

and are identified as the coefficients of viscosity and thermal conductivity.

For other molecular models, these results are only approximately valid; in particular, other expressions for the transport properties, i.e., the coefficients of viscosity and thermal conductivity, are obtained. Attempts to extend these calculations to the nonlinear case of finite departure from equilibrium have been made by many investigators. Included in these formulations are the Burnett equations, which seek to find correction terms to the viscous stress tensor and heat flux vector by expanding the Maxwell–Boltzmann equation in a perturbation series in λ, and the thirteen-moment equations, which increase the number of macroscopic variables to include the viscous stresses and heat flux as dependent variables with the necessary additional partial differential equations.

The foregoing remarks pertain to the state of the gas sufficiently far from a solid boundary so as to be unaffected by it in any direct way. The essential phenomena of rarefied gases, however, are found mostly in the region relatively "near," i.e., within a few mean free paths, of the solid boundaries. A knowledge of the physics of the interaction of gas molecules and solid surfaces is thus of primary importance. Mathematical developments have somewhat outstripped the basic physics in this area—as will be seen presently. What is really required is the probability law for the reflection of a molecule incident on a surface back into the gas; i.e., a joint distribution function

$$g(\mathbf{c}_r, \mathbf{c}_i) \tag{9}$$

which would indicate the number density of molecules reflected from a surface with velocity c_r, given that they were incident on the surface with velocity c_i. Present knowledge of the mechanisms of surface interaction is very far from being able to furnish such information. The following paragraphs will summarize the present state of knowledge in this area, and indicate the various approximations which are used in lieu of more complete information.

The study of surface interaction has been quite extensive. The phenomena of physical and chemical adsorption, of surface catalysis, and of sputtering have all been very widely investigated. Typical results are described in the textbooks. Little of this very extensive research appears to be of direct assistance for the problem at hand, however, beyond indicating quite clearly that surface interaction is influenced by the composition and temperature of the gas, by the chemical and physical structure and temperature of the solid, and perhaps most importantly, by the nature of the film of adsorbed vapor which usually exists on the actual surface. Only a few of these parameters have been systematically investigated in terms of use for the precise formulation of boundary conditions.

In particular, present data are confined to incident energies which are either in the thermal range of a few hundredths of a volt, up to a volt or less, or to energies of 100 V or more. The range of interest for applications to aerodynamics of satellites, 5–25 V, is thus almost completely untouched.

Direct investigations of surface interaction have been carried out using molecular-beam techniques. The results indicate only the overall mass flux in the various directions. To date, it has not been possible to measure the energy distribution of the reemitted molecules. From inspection of the molecular-beam data, it is apparent that there is a tendency for the reemission to follow the cosine law of diffuse reflection for typical technical surfaces. Such reemission is not always the case, however. The reemission from a freshly-cleaned LiF crystal, for example, departs very greatly from the cosine law. It should also be noted that even for technical surfaces, large departures from energy accommodation at the surface are regularly observed. The results just presented are typical of a growing body of empirical information. However, they do not yet permit prediction of interaction phenomena in terms of the reemission probability function of Eq. (9).

Lacking detailed knowledge of surface interaction, recourse must still be had to partial formulations in terms of overall macroscopic averages. These formulations were originated by Maxwell, but have been somewhat extended recently. For gasdynamic purposes, it is necessary to determine —as a minimum—the overall transfer of energy and of normal and tangential momentum as a gas flows past a solid surface. This is done in terms of three accommodation coefficients defined by

$$\alpha = (e_i - e_r)/(e_i - e_w) \qquad (10)$$

$$\sigma = (\tau_i - \tau_r)/\tau_i \qquad (11)$$

$$\sigma' = (p_i - p_r)/(p_i - p_w) \qquad (12)$$

where e_i, τ_i, and p_i are fluxes of energy and of tangential and normal momentum, respectively, incident on the surface; e_r, τ_r, and p_r are the fluxes of these quantities reemitted from the surface; and e_w and $p_w(\tau_w = 0)$ are the fluxes which would be reemitted by a gas in complete Maxwellian equilibrium with the surface. The quantities α, σ, and σ' are functions of the various parameters which affect surface interaction, including the velocity (magnitude and direction) of the gas flow past the surface element. However, the concept of describing the overall interaction in terms of these quantities is useful only if the coefficients are reasonably constant for a given gas and surface combination. This appears to be the case for technical surfaces, at least for the quantities α and σ. The molecular-beam data, however, suggest that σ' might well be sensitive to the inclination of the gas velocity vector to the surface.

Many measurements of α and somewhat fewer of σ have been made. Tabulations of representative values are listed in Tables I and II. The approximate nature of such results, in view of the foregoing discussion, should be kept in mind. All determinations have been made by means of macroscopic measurements, often with indifferent attention paid to the state of the surface. Some of these measurements were made in the slip-flow regime, while others were made in the free-molecular flow regime. For the former, an additional uncertainty exists as to the relation between the measured quantities and the supposed accommodation coefficient parameters, since no rigorous solution of the Maxwell–Boltzmann equation in the vicinity of a solid surface exists.

The foregoing accommodation-coefficient formulation of the phenomena of surface interaction is a more or less natural outgrowth of the treatment originarlly suggested by Maxwell. He postulated an interaction model in which incident molecules were reflected either "specularly," i.e., with reversal of the normal component of velocity and no change in the tangential components, or "diffusely," i.e., at random. A single parameter f was introduced to indicate the fraction of those molecules reflected diffusely, the remaining fraction, $1 - f$, being assumed to reflect specularly. Since both energy and momentum fluxes are of interest, it soon developed that

TABLE I

Thermal Accomodation Coefficients α*

Gas	Surface	Adsorbed gas	Temp. (°C)	α
H_2	Pt, bright	—	—	0.32
H_2	Pt, black			0.74
O_2	Pt, bright			0.81
O_2	Pt, black			0.93
N_2	Pt	—	—	0.50
N_2	W			0.35
Air	Flat lacquer on bronze	—	—	0.88–0.89
Air	Polished bronze			0.91–0.94
Air	Machined bronze			0.89–0.93
Air	Etched bronze			0.93–0.95
Air	Polished cast iron			0.87–0.93
Air	Machined cast iron			0.87–0.88
Air	Etched cast iron			0.87–0.96
Air	Polished aluminum			0.87–0.95
Air	Machined aluminum			0.95–0.97
Air	Etched aluminum			0.89–0.97
He	W	—	—	0.025–0.057
He	Ni, not flashed			0.20
He	Ni, flashed			0.085
He	W, flashed	—	—	0.17
He	W, flashed	—	—	0.12
He	W, not flashed	—	—	0.53
A	W, flashed			0.82
A	W, flashed			0.46
A	W, not flashed			1.00
He	H_2 adsorbed on W	—	—	0.041
He	N_2 adsorbed on W			0.064
He	W clean system			0.02
He	W	None	30	0.0164
He	W	None	−120	0.0130
He	W	None	−196	0.0109
Ne	W	None	30	0.0412
Ne	W	None	−120	0.0426
Ne	W	None	−196	0.0495
A	W	None	30	0.271
A	W	None	−120	0.300
A	W	None	−196	0.549
Xe	W	None	30	0.773
Xe	W	None	−120	0.878

* Compiled by F. C. Hurlbut.

TABLE I (*continued*)

Gas	Surface	Adsorbed gas	Temp. (°C)	α
Xe	W	None	-196	0.942
He	K	None	25	0.083
A	K	None	25	0.444
He	K	None	-196	0.042
A	K	None	-196	0.41
He	W	O (composite film)	32	0.107
Ne	W	O (composite film)	32	0.204
H_2	W	O	32	0.201
H_2	W	None	32	0.105
He	W	N (monolayer)	32	0.040
Ne	W	N	32	0.117
H_2	W	N	32	0.642
CO_2	W	CO_2	32	0.990
He	Glass	None	29.3	0.31

TABLE II

Tangential Momentum Accommodation Coefficients σ*

Gas	Surface	α
Air	Machined brass	1.00
CO_2	Machined brass	1.00
Air	Old shellac	1.00
CO_2	Old shellac	1.00
Air	Hg	1.00
Air	Oil	0.90
CO_2	Oil	0.92
H_2	Oil	0.93
Air	Glass	0.89
He	Oil	0.87
Air	Fresh shellac	0.79
Air	Ag_2O	0.98
He	Ag_2O	1.00
H_2	Ag_2O	1.00
O_2	Ag_2O	0.99
Air	Oil on machined Al	0.90
N_2	Adsorbed gas on oil film on Al	0.8–0.93
O_2	Adsorbed gas on oil film on Al	0.8–0.95
A	Adsorbed gas on oil film on Al	0.8–0.95
Air	Adsorbed gas on oil film on Al	0.60–0.93
N_2	Glass	0.93–0.97

* Compiled by F. C. Hurlbut.

one parameter was not adequate, and the three accommodation coefficients were introduced. The case of "completely diffuse" reflection corresponds to, but is of course not identical with, $\alpha = \sigma = \sigma' = 1$, while the case of "completely specular" reflection corresponds to $\alpha = \sigma = \sigma' = 0$.

Several other paramatrized models of surface interaction have been introduced in connection with the problem of predicting aerodynamic forces on artificial satellites. Some of these introduce empirical constants to account for real-gas effects such as dissociation during the interaction. Others are based upon molecular-beam data. At the present time, however, the accommodation-coefficient treatment is the most readily available method for dealing with the phenomena of surface interaction.

The formulation is adequate for free-molecular-flow calculations providing that the configurations have convex geometry. For concave bodies, it must be extended.

For slip flow, a detailed hypothetical surface-interaction model must be used in conjunction with the Maxwell–Boltzmann equation. Near a solid boundary, infinitesimal gradients of the macroscopic parameters permit linearization of the Maxwell–Boltzmann equation. Rigorous, closed-form solutions—even for Maxwell molecules—have not yet been obtained, but iteration-type computations lead to boundary conditions for the macroscopic variables of the "slip-velocity" and "temperature-jump" type. These will be discussed more completely below.

A number of attempts have been made to obtain explicit solutions of the Maxwell–Boltzmann equation for special problems for arbitrary values of the gas density. These methods proceed along three general lines: (1) the Maxwell–Boltzmann equation is linearized and solutions are sought in terms of expansions, in terms of the eigenfunction of the linearized collision operator, or in terms of "half-range" functions which explicitly distinguish between molecules incident on and reflected from the solid surfaces; (2) the Maxwell–Boltzmann equation is replaced by a set of moment or transport equations obtained from an assumed form for the distribution function; (3) the Maxwell–Boltzmann equation is replaced by a slightly more tractable equation which, presumably, retains the fundamental properties of the correct equation, namely

$$\frac{\partial f}{\partial t} + \sum_{i=1}^{3} c_i \frac{\partial f}{\partial x_i} = \frac{f - f^0}{\tau} \tag{13}$$

where τ is a relaxation time which can be adjusted to give a "best" approximation to the exact Maxwell–Boltzmann equation. This is known as the "model" equation.

The first two of these methods turn out to be nearly equivalent so far as the actual calculations are concerned. Both have two fundamental difficulties: (1) the equations for the expansion coefficients, or for the moments, do not separate, so that the complete Maxwell–Boltzmann equation is replaced by an infinite set of equations for an infinite number of unknowns, and (2) the choice of boundary conditions to associate with any finite approximation to this infinite set appears to be quite arbitrary.

The foregoing discussion has been confined to a gas consisting of molecules possessing no internal degrees of freedom. Since most of the interest in the mechanics of rarefied gases is in connection with the flow of atmospheric air, it is necessary to consider the problems which arise because of internal degrees of freedom, such as the rotational motions of the N_2 and O_2 molecules. The collision integral in the Maxwell–Boltzmann equation must be modified to include transfer from one rotational state to another by the colliding molecules. At normal temperatures, only the rotational states about the two principal axes of greatest moment of inertia are involved. Other modes of internal motion, such as rotation about the third axis, or vibration, are all below the quantum-mechanical threshold of excitation. Even the rotational modes must be treated quantum-mechanically.

The indicated calculations have so far only been sketched, to obtain the form of the results, with only rough estimates of the coefficients involved. The case of infinitesimal departure from equilibrium involves an additional quantity τ', the relaxation time for the equalization of internal energy between rotational and translational degrees of freedom. If τ' is very small, one obtains the familiar Navier–Stokes equations, except with a bulk viscosity term, i.e., the normal stress in the fluid is given by

$$p = R\varrho T - \mu' \sum_{i=1}^{3} \partial u_i / \partial x_i \qquad (14)$$

This appears to be the appropriate formulation for air at normal temperatures. For other gases, or perhaps for air at extreme temperatures, there may be more than one relevant relaxation time, some or all of which are long compared to the translational relaxation time μ/p. A number of methods for treating these cases have been proposed, but will not be discussed here, since they lie outside the scope of this chapter.

In connection with the mention of internal degrees of freedom, attention should be called to the nature of the energy accommodation coefficient α for a gas possessing such internal freedom. There would appear to be no reason to suppose that all such degrees of freedom are accommodated to the same extent, so that several different quantities such as α(translational), α(rotational), and α(vibrational) might be introduced. Experiments, with air at least, have not been sufficiently accurate to warrant such distinction, however, so that all possible modes of energy storage are lumped together for the values of α quoted in Table I.

2. FREE-MOLECULAR FLOW

Free-molecular flow is the subdivision of rarefied gas dynamics corresponding to the lowest densities. The basic assumption for theoretical calculations is that intermolecular collisions can be neglected. The fluxes of mass, momentum, and energy incident on and reemitted from a surface element can be treated separately and do not interfere with each other. The incident flux is entirely unaffected, even in supersonic flow, by the presence of the surface.

For a configuration with convex geometry, the analysis is particularly simple, because all incident molecules originate in the undisturbed gas flow. This case will be considered first. The geometric parameters are defined in Fig. 1. It is assumed the flow past the surface element is in Maxwellian equilibrium, with density ϱ, temperature T, and macroscopic velocity V, inclined at an angle of attack θ to the surface element. The distribution function is then

$$f = \frac{\varrho}{m(2\pi RT)^{3/2}} \exp\left[-\frac{(c_1 - V\sin\theta)^2 + (c_2 + V\cos\theta)^2 + c_3{}^2}{2RT}\right] \quad (15)$$

The number flux dN_i incident on the surface element dS is given by

$$dN_i = \left(\int_{-\infty}^{\infty}\int_{-\infty}^{\infty}\int_{0}^{\infty} fc_1\, dc_1\, dc_2\, dc_3\right) dS \quad (16)$$

$$= f_m\left(\frac{RT}{2\pi}\right)^{1/2}\{\exp[-(S\sin\theta)^2] + \sqrt{\pi}\,(S\sin\theta)[1+\mathrm{erf}(S\sin\theta)]\}\, dS \quad (17)$$

where S denotes the "molecular speed ratio"

$$S = V/(2RT)^{1/2} \quad (18)$$

and $\mathrm{erf}(x)$ is the error function.

The quantity S is proportional to the Mach number M through the relation

$$S = \mathrm{M}(\gamma/2)^{1/2} \quad (19)$$

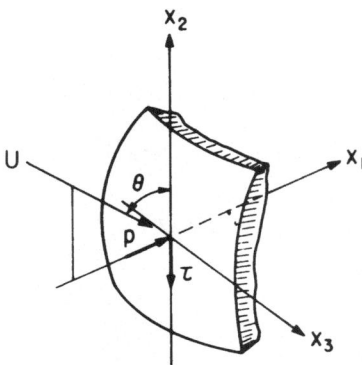

Fig. 1. Coordinate system for free-molecular flow.

where γ is the specific-heat ratio C_p/C_v. For all of the following equations, it will be assumed that equilibrium conditions have been obtained at the surface, with no "ingassing" or "outgassing." Hence the number flux dN_r reemitted from the surface element is equal to the number flux incident, dN_i,

$$dN_r = dN_i \tag{20}$$

This is a generally applicable assumption for aerodynamic problems, where the convective flux is typically large compared to any outgassing or ingassing effect. These latter persist typically for hours, however, and can be of considerable importance for problems of enclosed geometry, such as the flow of gas through a duct.

The fluxes of normal momentum, p_i (pressure), and tangential momentum, τ_i (shear stress), incident on the surface element are given by

$$p_i = \int_{-\infty}^{\infty} \int_{-\infty}^{\infty} \int_{0}^{\infty} mc_1{}^2 f \, dc_1 \, dc_2 \, dc_3 \tag{21}$$

$$= (\varrho V^2/2\sqrt{\pi}\, S^2)\{(S \sin \theta) \exp \left[-(S \sin \theta)^2\right]$$

$$+ \sqrt{\pi}\, [\tfrac{1}{2} + (S \sin \theta)^2][1 + \mathrm{erf}(S \sin \theta)]\} \tag{22}$$

$$\tau_i = -\int_{-\infty}^{\infty} \int_{-\infty}^{\infty} \int_{0}^{\infty} mc_1 c_2 f \, dc_1 \, dc_2 \, dc_3 \tag{23}$$

$$= \frac{\varrho V^2 \cos \theta}{2\sqrt{\pi}\, S} \{\exp[-(S \sin \theta)^2] + \sqrt{\pi}\, (S \sin \theta)[1 + \mathrm{erf}(S \sin \theta)]\} \tag{24}$$

The flux of incident energy E_i for a gas possessing j internal degrees of freedom (for air, at normal temperatures, $j \approx 2$) must be considered in two parts. These are $E_{i,\text{tr}}$, the incident flux of translational energy, and $E_{i,\text{int}}$, the incident flux of energy associated with the internal degrees of freedom.

Thus

$$E_i = E_{i,\text{tr}} + E_{i,\text{int}} \tag{25}$$

The first of these is given by

$$E_{i,\text{tr}} = \int_{-\infty}^{\infty} \int_{-\infty}^{\infty} \int_{0}^{\infty} \tfrac{1}{2}mc^2 c_1 \, dc_1 \, dc_2 \, dc_3 \tag{26}$$

$$= \varrho RT \left(\frac{RT}{2\pi}\right)^{1/2} \{(S^2 + 2) \exp[-(S \sin \theta)^2]$$

$$+ \sqrt{\pi}\,(S^2 + \tfrac{5}{2})(S \sin \theta)[1 + \text{erf}(S \sin \theta)]\} \tag{27}$$

The second is calculated by using the fact that in equilibrium, each internal mode possesses, on the average, $\tfrac{1}{2}RT$ units of energy per molecule. Hence

$$E_{i,\text{int}} = \frac{j}{2}mRT \, dN_i/dS \tag{28}$$

$$E_{i,\text{int}} = \frac{\varrho(5 - 3\gamma)}{\sqrt{\pi}\,(\gamma - 1)} \left(\frac{RT}{2}\right)^{3/2}$$

$$\times \{\exp[-(S \sin \theta)^2] + \sqrt{\pi}\,(S \sin \theta)[1 + \text{erf}(S \sin \theta)]\} \tag{29}$$

by use of Eq. (17) and the relation $j = (5 - 3\gamma)/(\gamma - 1)$. The net pressure p, shear stress τ, and heat flux Q delivered to the surface must also include that due to the reemitted molecules. Denoting these fluxes by p_r, τ_r, and E_r, one writes

$$p = p_i + p_r \tag{30}$$

$$\tau = \tau_i - \tau_r \tag{31}$$

$$Q = E_i - E_r \tag{32}$$

As indicated previously, the present state of knowledge as to the physics of surface interaction does not permit the direct calculation of the quantities p_r, τ_r, and E_r. Rather, recourse must be had to the empirical accommodation-coefficient formulation, according to which [see Eqs. (24)–(26)], these quantities are given by

$$p_r = (1 - \sigma')p_i + \sigma'p_w \tag{33}$$

$$\tau_r = (1 - \sigma)\tau_i \tag{34}$$

$$E_r = (1 - \alpha)E_i + \alpha E_w \tag{35}$$

The quantities p_w and E_w denote the fluxes of normal momentum and of energy which would be carried by a reemitted stream of molecules in complete Maxwellian equilibrium with the surface. They are thus given by

$$p_w = \tfrac{1}{2}m(2\pi RT_w)^{1/2} \, dN_i/dS \tag{36}$$

$$E_w = (4 + j) \frac{mRT_w}{2} \frac{dN_i}{dS} = \frac{\gamma + 1}{2(\gamma - 1)} mRT_w \frac{dN_i}{dS} \tag{37}$$

The final expressions are thus

$$p = \frac{\varrho V^2}{2S^2} \left\{ \left[\frac{2 - \sigma'}{\sqrt{\pi}} S \sin\theta + \frac{\sigma'}{2}\left(\frac{T_w}{T}\right)^{1/2} \right] \exp[-(S\sin\theta)^2] \right.$$

$$+ \left[(2 - \sigma')\left(S^2\sin^2\theta + \frac{1}{2}\right) + \frac{\sigma'}{2}\left(\frac{\pi T_w}{T}\right)^{1/2}(S\sin\theta) \right]$$

$$\left. \times \, [1 + \mathrm{erf}(S\sin\theta)] \right\} \tag{38}$$

$$\tau = \frac{\sigma\varrho V^2 \cos\theta}{2S\sqrt{\pi}} \left\{ \exp[-(S\sin\theta)^2] + \sqrt{\pi}\,(S\sin\theta)[1 + \mathrm{erf}(S\sin\theta)] \right\} \tag{39}$$

$$Q = \alpha\varrho RT\left(\frac{RT}{2\pi}\right)^{1/2}\left(\left[S^2 + \frac{\gamma}{\gamma - 1} - \frac{\gamma + 1}{2(\gamma - 1)}\frac{T_w}{T}\right]\{\exp[-(S\sin\theta)^2]\right.$$

$$\left. + \sqrt{\pi}\,S\sin\theta[1 + \mathrm{erf}(S\sin\theta)]\} - \frac{1}{2}\exp[-(S\sin\theta)^2] \right) \tag{40}$$

It will be observed that the pressure, but not the shear stress, depends on the surface temperature. Aerodynamic forces in free-molecule flow are thus generally functions of the surface temperature, as well as the gas density and velocity and the configuration geometry. It is interesting to compare the expression for pressure and shear stress given above, in the limit of infinite speed ratio, with those of Newtonian flow, an approximation often used for hypersonic continuum flow. For Newtonian flow, it is assumed that all incident normal momentum is communicated to the surface, while no tangential momentum is. Thus (with M, $S \rightarrow \infty$)

$$p_{\text{Newtonian}} = \varrho V^2 \sin^2 \theta \tag{41}$$

$$\tau_{\text{Newtonian}} = 0 \tag{42}$$

Limiting expressions for free-molecule flow depend, as indicated, on the surface temperature. Two extremes are of interest, an "ambient-temperature" surface, for which $T_w = T$, and an "equilibrium-temperature" surface, for which $T_w = T_{\text{eq}}$, the temperature at which $Q = 0$. The first corresponds approximately to free-flight conditions at very high speed, for which radiation cooling of the surface is dominant, while the latter corresponds approximately to unheated wind-tunnel conditions. The corresponding expressions are

$$p_{\text{ambient}} \rightarrow \varrho V^2 (2 - \sigma') \sin^2 \theta \tag{43}$$

$$p_{\text{equilibrium}} \rightarrow \varrho V^2 \left\{ (2 - \sigma') \sin^2 \theta + \sigma' \sin \theta \left[\frac{\pi(\gamma - 1)}{2(\gamma + 1)} \right]^{1/2} \right\} \tag{44}$$

$$\tau \rightarrow \sigma \varrho V^2 \sin \theta \cos \theta \tag{45}$$

It will be observed that the free-molecular flow results correspond to those of Newtonian flow only insofar as the pressure for the "ambient" temperature condition, and for completely diffuse reflection ($\sigma = \sigma' = 1$), is concerned. The shear stress and the pressure for other surface temperatures or other types of surface interaction are different.

The foregoing expressions for the forces and heat transfer acting on an element of surface area have been applied by many authors to the calculation of gross characteristics of most of the elementary geometric configurations. It has usually been necessary to assume T_w, σ, σ', and α all to be constant over the entire surface. This is generally valid only in the sense that the corresponding values of these four quantities which appear in the final results are actually appropriately weighted averages. For a surface which is actually completely diffusely reflecting (e.g., a highly porous surface) and for which the internal heat conduction is sufficiently rapid, the assumption that $\alpha = \sigma = \sigma' = 1$ and $T_w = \text{const}$ may become quite accurate. No actual surfaces are known which are specularly reflecting, but, in principle, characteristics for a configuration possessing such an idealized surface would be accurately predicted by the results with $\alpha = \sigma = \sigma' = 0$. In view of these limitations, it does not seem worthwhile to reproduce the detailed calculations which have been made.

3. SLIP FLOW

Slip flow is the part of rarefied gas dynamics corresponding to only slight rarefaction and for which the departure from continuum gas behavior is slight. It takes its name from the phenomenon of slip, i.e., the gas immediately adjacent to a solid surface may possess a finite velocity with respect to it. There is also a temperature of the gas next to a surface which will in general differ from the surface temperature. These are boundary condition effects.

Elementary derivations of these relations are to be found in the texts on kinetic theory. The slip velocity condition, e.g., arising from flow past the surface can be treated in the following manner. Far from the surface, i.e., a few mean free paths, the gas is supposed in motion with a constant velocity gradient du/dy arising from a constant shear stress related by $\tau = \mu \times du/dy$ (see Fig. 2). It is the extrapolation of this macroscopic gas velocity to $y = 0$, i.e., u_0, which is known as the slip velocity. The distribution function far from the surface is given by a special case of Eq. (5), namely,

$$f = \frac{n}{(2\pi RT)^{3/2}} \exp\left[-\frac{(c_x - u)^2 + c_y^2 + c_z^2}{2RT}\right]\left[1 - \frac{1}{2pRT}\mu\frac{du}{dy}(c_x - u)c_y\right] \tag{46}$$

Since the gas flow is steady, τ is constant, and, in particular, the value of τ far from the surface is equal to the value of τ on the surface. Neglecting the effect of molecular collisions in the so-called Knudsen layer within a few mean free paths of the wall, Eq. (46) is used to calculate the flux of momentum from the gas onto the surface, i.e.,

$$\tau = \int_{-\infty}^{\infty}\int_{-\infty}^{0}\int_{-\infty}^{\infty} mc_x c_y f \, dc_x \, dc_y \, dc_z \tag{47}$$

$$= \tfrac{1}{2}\mu(du/dy) + \tfrac{1}{4}\varrho u_0 \bar{V}, \qquad \bar{V} = 2\mu/\varrho\lambda \tag{48}$$

Of this flux, a fraction σ is transmitted. One then has

$$\tau = \mu\frac{du}{dy} = \sigma\left[\frac{1}{2}\mu\frac{du}{dy} + \frac{1}{2}\mu\frac{u_0}{\lambda}\right] \tag{49}$$

or

$$u_0 = \frac{2-\sigma}{\sigma}\lambda\frac{du}{dy} \tag{50}$$

Similar analyses, including the effect of a temperature gradient in the direction of flow and for the transfer of energy, lead to the usual elementary

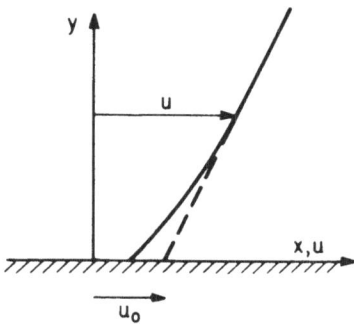

Fig. 2. Coordinate system for slip velocity.

boundary conditions

$$u_0 = \frac{2-\sigma}{\sigma} \lambda \frac{du}{dy} + 3\left(\frac{RT}{8\pi}\right)^{1/2} \lambda \frac{1}{T} \frac{dT}{dx} \tag{51}$$

and

$$T_0 - T_w = \frac{2-\alpha}{\alpha} \frac{2(\gamma-1)}{\gamma+1} \frac{\varkappa}{R\mu} \lambda \frac{dT}{dy} \tag{52}$$

The second term in the slip-velocity condition is known as the thermal creep. It is of considerable importance for many vacuum instrumentation problems, since it can induce a pressure variation along a tube even when there is no net flow. The currents induced by this term also give rise to forces which are known as the "slip-flow radiometer forces."

The limitations and deficiencies in the derivation of Eqs. (51) and (52), as sketched above, are quite apparent. The analysis is clearly limited to infinitesimal departure from equilibrium, i.e., the velocity and temperature gradients, and likewise the ratio of the mean free path to a distance characteristic of a macroscopic change in the velocity and temperature, are all infinitesimal. The principal deficiency in the analysis is the calculation of the flux incident on the surface by using the molecular distribution function characteristic of the gas far from the surface. In the actual case, of course, intermolecular collisions between molecules originating in the gas and those reemitted from the surface will alter the distribution function in the vicinity of the surface. Attempts to meet this difficulty by solving the linearized Maxwell–Boltzmann equation in this region have been formulated by several authors; the "model" equation has also been utilized. These investigations have necessarily been based upon idealized models of the surface interaction, in fact, on the one-parameter "specular or completely diffuse" model. It has also been impossible to obtain complete solutions of the problem, even for Maxwellian molecules; rather, recourse must be had to successive approximations in terms of increasing numbers of eigenfunctions, or half-range functions. The results agree with Eqs. (51) and (52) with $\alpha = \sigma$, in the "first approximation." For higher approximations, the form of these equations is unchanged, except that the quantities $(2 - \sigma)/\sigma$ or $(2 - \alpha)/\alpha$ are replaced by increasingly complex expressions (rational fractions of σ or α). Since only the gross coefficient can be measured, the question cannot be resolved by experiment. It should perhaps be specifically noted, however, that the parameters α and σ obtained from slip-flow measurements and the use of Eqs. (51) and (52) are really slightly different quantities from those defined for free-molecular flow. Experiments have not been sufficiently extensive nor precise to shed much light on this, although there is perhaps a suggestion that the coefficient approximately given by $(2 - \sigma)/\sigma$ in slip flow is somewhat larger than for the same gas and surface when the determination is in free-molecular flow. This is in line with the theoretical results.

In summary, then, the modifications to be made in the boundary conditions in slip flow are thus at least moderately well established. They are true noncontinuum effects which appear as small corrections to the continuum no-slip or temperature-jump boundary conditions. They reduce to the continuum results smoothly as λ approaches zero. Some uncertainty exists as to the exact numerical values of the coefficients, but these quantities are in any event still of an essentially empirical nature.

The remaining basic problem in slip flow is that of specifying the correct differential equations to associate with these boundary conditions. It was originally expected that some modification in the Navier–Stokes equations of continuum mechanics would be appropriate, e.g., the Burnett or thirteen-moment systems. This now appears probably not to be the case. The reasons for preferring the Navier–Stokes equations for slip-flow gas dynamics are, briefly: (1) the results obtained from the various approximate solutions to the linearized Maxwell–Boltzmann equation for special problems at arbitrary density agree, for small λ, with the predictions of the Navier–Stokes equations and the slip and temperature-jump boundary conditions of Eqs. (51) and (52); (2) the experimental evidence obtained from shock-wave structure, from sound dispersion and absorption, and from flow in pipes and between rotating cylinders agree with this formulation; (3) the existence of a bulk viscosity stress term associated with the internal degrees of freedom of the molecules of the gas (N_2 or O_2 for air) is an established first-order effect which is included in the Navier–Stokes formulations for these gases, but which has not been accounted for in the more elaborate investigations.

The slip-flow regime is characterized by the fact that rarefaction effects (probably due primarily to the changed boundary conditions only) are small corrections to continuum flow behavior. It is important to note that the continuum flow referred to is not the idealized inviscid flow of hydrodynamics, but rather a highly viscous flow. From the basic relation λ/L

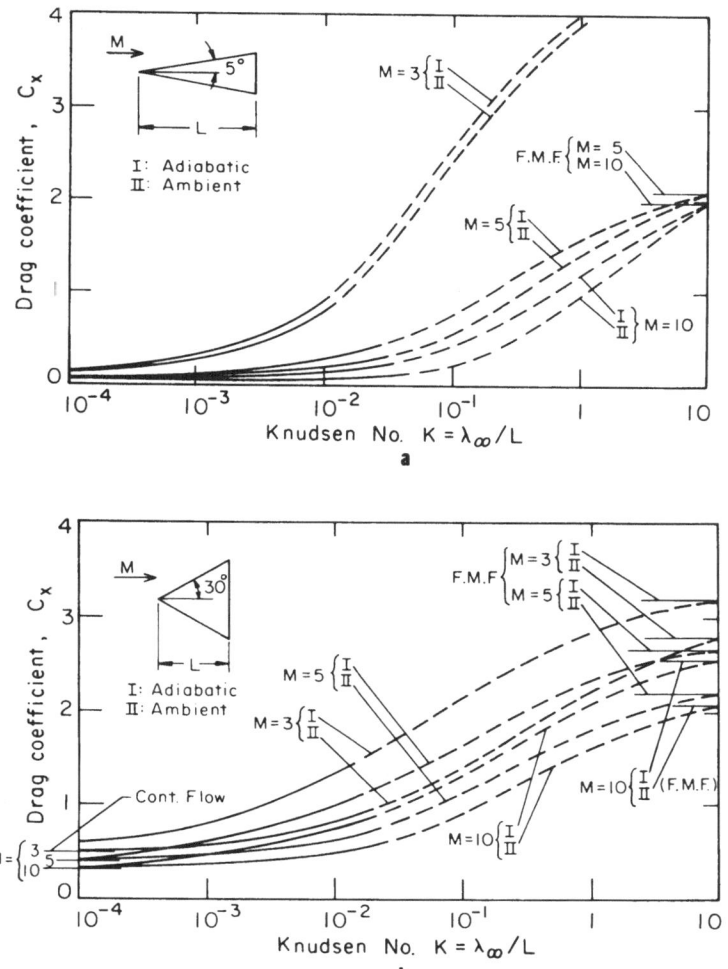

Fig. 3. Dependence of drag on temperature, Mach number, and bluntness. (a) Drag coefficient for a slender wedge. (b) Drag coefficient for a blunt wedge.

$\approx M/Re$, it is clear that λ/L is of non-negligible importance only when the Reynolds number Re is very small compared to normal values (the Mach number M being restricted to moderate values in the present chapter). Depending on M, the values of Re to ensure slip flow conditions will be in the Stokes–Oseen range of very small Re, the laminar-boundary-layer range of moderately large Re, or somewhat intermediate values. For all these cases, and increasingly as the Reynolds number decreases with the corresponding further penetration into the realm of rarefied gas dynamics, viscous effects are of great importance. Skin friction becomes comparable to surface pressure. Heat transfer coefficients increase. Surface pressure is disturbed from its inviscid value by the displacement effect of the viscous layer adhering to the surface. These are the dominant fluid-mechanical effects in the slip-flow regime, in contrast with the relatively small corrections introduced by the altered boundary conditions. It should also be noted that these primarily continuum viscous effects are sensitive to both heat transfer and geometry effects. Highly cooled boundary layers are much thinner than adiabatic or heated boundary layers, with consequent large differences in the skin-friction and displacement pressure effects for different thermal conditions. Blunt objects are also much less sensitive to viscous effects than are sharp slender objects. It was the practice formerly to depict the various regimes of rarefied gas dynamics in terms of a M–Re diagram with boundary curves in terms of characteristic values of the various Knudsen numbers. For example, the demarcation between slip flow and continuum flow has frequently been taken as $M/\sqrt{Re} = 0.01$. This sort of presentation has been abandoned in the present chapter because of the much greater complexity which now appears to exist. For example, the diagram should actually be at least three-dimensional, with some sort of thermal parameter, say T/T_w, as a third axis (even this would be entirely inadequate if real-gas effects are to be included). Also, as remarked above, blunt or slender objects are characterized by quite different variations in the relative importance of both continuum viscous and noncontinuum effects (see Fig. 3).

REFERENCES

1. S. A. Schaaf, "Mechanics of Rarefied Gases," in: *Handbuch der Physik*, part 2, **8** (1963), pp. 591–624.
2. M. H. Kogan, *Dynamics of Rarefied Gases and Kinetic Theory*, Moscow (1967).
3. *Rarefied Gas Dynamics*, Proceedings of the (first five) Symposia (nine vols.), Academic Press, New York.

Chapter 7

FUNDAMENTALS OF RADIATION GAS DYNAMICS

S. I. Pai

Research Professor
Institute for Fluid Dynamics and Applied Mathematics
University of Maryland

INTRODUCTION *

In gas dynamics, we study fluid mechanics simultaneously with heat transfer problems. In general, there are three basic modes of heat transfer: heat convection, heat conduction, and thermal radiation. If the temperature of a gas is not too high and the density is not too low, the transfer of heat by thermal radiation is usually negligibly small in comparison with that by conduction and convection. Hence in ordinary gas dynamics, the thermal radiation effects are always neglected. In the present space age, there are many technological developments of interest—hypersonic flight, gas-cooled nuclear reactors, power plants for space exploration needs, fission and fusion reactions—in which the temperature is very high and the density is rather low. As a result, thermal radiation becomes an important mode of heat transfer. A complete analysis of such a high-temperature flow field should be based upon a study of the gasdynamic field and the thermal radiation field simultaneously. We use the term "radiation gas dynamics" ([1]) for such a new branch of fluid dynamics.

There are three different thermal radiation effects on the flow field of a high-temperature gas: (1) radiation stresses, (2) radiation energy density E_R, and (3) heat flux of radiation q_R. The exact expressions for these radiation terms are very complicated and will be derived later. In order to show the relative importance of heat transfer by radiation and that by convection, we consider the radiation terms based on optically-thick approximations in comparison with the corresponding terms in ordinary

---

* Symbols used in this chapter are defined on pp. 306–309.

gas dynamics. We shall show later that the radiation energy density and the radiation pressure are usually of the same order of magnitude. If we consider the radiation pressure, we should also consider the radiation energy density. The radiation energy density per unit mass is

$$E_R/\varrho = a_R T^4/\varrho \tag{1}$$

where a_R is known as the Stefan–Boltzmann constant, which is 7.67×10^{-15} erg cm^{-3} °K^{-4}, T is the absolute temperature in °K, and ϱ is the density of the gas. This radiation energy density should be compared with the internal energy $U_m = c_v T$ of the fluid, where c_v is the specific heat at constant volume of the fluid. We take $\varrho = 1.23 \times 10^{-6}$ g/cm^3, which is 1/1000 of the value of air at standard sea level, and $c_v = 7 \times 10^6$ erg-g-°K, which is the value for a diatomic gas such as air. If $T = 10^5$ °K, the radiation energy density and the internal energy are of the same order of magnitude. Since radiation energy density is proportional to T^4 while the internal energy is proportional to T, if we change the temperature by one order of magnitude, say $T = 10^4$, the radiation energy density will be three orders of magnitude smaller than that of internal energy. Hence at $T = 10^4$ °K, in our example, the radiation energy density and radiation stresses are negligible in comparison with internal energy and the gas pressure.

For the radiation equilibrium condition, the radiation flux is

$$q_R = \sigma T^4/\varrho \tag{2}$$

where $\sigma = c a_R/4 = 5.75 \times 10^{-5}$ erg cm^{-2} sec^{-1} °K^{-4} is known as Stefan–Boltzmann constant for radiative transfer and c is the velocity of light, $c = 3 \times 10^{10}$ cm/sec in free space. The radiation flux should be compared with the heat flux by convection:

$$q_v = U c_v T \tag{3}$$

where U is a typical flow velocity. If we take $U = 6.5 \times 10^4$ m/sec, which is an average speed of a satellite, we find that at $T = 10^4$ °K, q_R and q_v are of the same order of magnitude. Again, since q_R is proportional to T^4 and q_v is proportional to T, at low temperatures such as $T = 10^3$ °K, q_R will be negligibly small in comparison with q_v.

From the above rough estimate, our conclusion is that if the density of the gas is of the order of 1/1000 of the density of air at sea level, at $T = 10^4$ °K or higher, we should consider the radiation flux in our study of gas dynamics, and at $T = 10^5$ °K or higher, we should consider both the radiation energy density and radiation flux. When the radiation terms are of the same order of magnitude as the corresponding terms of ordinary gas dynamics, we should use more accurate formulas to calculate these radiation terms than those of Eqs. (1) and (2).

1. FUNDAMENTALS OF RADIATIVE TRANSFER

We may consider the thermal radiation as either a stream of photons or as electromagnetic waves. When we consider the thermal radiation as a stream of photons, we use the relativistic Boltzmann equation to study the motion of these particles [1]. When we consider the thermal radiation as electromagnetic waves, we may use geometrical optics to study the behavior of thermal radiation from the macroscopic point of view. The results of these two approaches are the same [see [1]]. In this chapter, we consider only the case of continuum theory and we shall use the treatment of geometrical optics. The thermal radiation may be expressed in terms of a specific intensity I_ν which is defined as follows:

$$I_\nu = \lim_{d\sigma_0, d\nu, dt, d\omega \to 0} \left(\frac{dE_\nu}{d\sigma_0 \cos \theta \, d\nu \, dt \, d\omega} \right) \qquad (4)$$

where I_ν is a function of time t, spatial coordinates, direction θ, which is the angle between the direction of the ray and the normal of area $d\sigma_0$, and the frequency of the wave ν. The amount of radiative energy flowing through the area $d\sigma_0$ in the frequency range ν to $\nu + d\nu$ in the direction of the ray L which makes an angle θ with the normal of $d\sigma_0$ within a solid angle $d\omega$ in the time interval dt is dE_ν. The total amount of energy radiated over the whole spectrum is

$$dE = \int_0^\infty (dE_\nu/d\nu) \, d\nu = I \cos \theta \, d\sigma_0 \, d\omega \, dt \qquad (5)$$

where

$$I = \int_0^\infty I_\nu \, d\nu \qquad (6)$$

is the integrated thermal radiation intensity.

If the specific intensity I_ν is known, we may easily calculate the effects of thermal radiation on the flow, which are:

1. The flux of heat energy by thermal radiation is $q_R{}^i$, i.e.,

$$q_R{}^i = \int_{4\pi} In^i \, d\omega \qquad (7)$$

where n^i is the directional cosine of the radiation ray with respect to the ith axis. We should add the divergence of this radiative heat flux in the energy equation of gas dynamics, i.e.,

$$Q_R = \nabla \cdot \mathbf{q}_R \tag{8}$$

where ∇ is the gradient operator, $\nabla = \mathbf{i}(\partial/\partial x) + \mathbf{j}(\partial/\partial y) + \mathbf{k}(\partial/\partial z)$, and $\mathbf{i}$, $\mathbf{j}$, and $\mathbf{k}$ are the unit vectors in the x, y, and z directions, respectively. In general, Q_R is a differentio-integral expression and the energy equation in radiation gas dynamics is an integrodifferential equation.

2. The radiation energy density E_R. The radiation energy density within the frequency range ν to $\nu + d\nu$ is

$$U_\nu = (1/c) \int_{4\pi} I_\nu \, d\omega \tag{9}$$

where c is the velocity of light. The radiation energy density for the whole spectrum is then

$$E_R = \int_0^\infty U_\nu \, d\nu = (1/c) \int_{4\pi} I \, d\omega \tag{10}$$

This radiative energy should be added to the total energy of a gas, and it behaves in a manner similar to the internal energy of the gas. Under special conditions, Eq. (10) may be reduced to Eq. (1).

3. Radiation stress tensor. The ijth component of the radiation stress tensor is

$$\tau_R^{ij} = -(1/c) \int_{4\pi} I n^i n^j \, d\omega \tag{11}$$

We may define a radiation pressure p_R such that

$$p_R = -\tfrac{1}{3}(\tau_R^{11} + \tau_R^{22} + \tau_R^{33}) = (1/3c) \int_{4\pi} I \, d\omega = \tfrac{1}{3} E_R \tag{12}$$

It is evident that the radiation pressure and the radiation energy density are always of the same order of magnitude, as we mentioned in the introduction.

Our next problem is how to find the specific intensity I_ν for a given problem. For the macroscopic treatment, the specific intensity I_ν may be expressed in terms of two overall coefficients of the medium: one is the absorption coefficient of radiation k_ν, and the other is the emission coefficient of radiation j_ν. They are defined as follows:

Absorption Coefficient k_ν of Radiation

The loss of specific intensity along the ray of radiation over a distance ds in a medium of density ϱ is given by the relation

$$dI_\nu = -\varrho k_\nu I_\nu \, ds \tag{13}$$

The integration of Eq. (13) gives

$$I_\nu(s) = I_\nu(s_0) \exp\left(-\int_{s_0}^{s} \varrho k_\nu \, ds\right) = I_\nu(s_0) \exp(-\tau_\nu) \tag{14}$$

[where τ_ν is known as the optical radiation thickness of the layer $(s - s_0)$ and s_0 is a reference point where the specific intensity is $I_\nu(s_0)$], and

$$L_{R\nu} = 1/\varrho k_\nu \tag{15}$$

is the radiation mean free path. Hence the optical thickness is a dimensionless distance which indicates the effective length in the absorption of radiation. For a given physical length $(s - s_0)$, if τ_ν is large, the medium is said to be optically thick, while if τ_ν is small, the medium is said to be optically thin.

The absorption coefficient k_ν is a function of the temperature and density of the medium as well as of the frequency ν. In general, k_ν consists of two parts: one is the true absorption and the other is that due to scattering. In the macroscopic theory of radiation gas dynamics, we assume that k_ν is a given function of temperature and density of the medium and of the frequency ν. The determination of k_ν can be made by microscopic theory or experiment. The absorption coefficient in radiation gas dynamics has a position similar to other transport coefficients, such as these of viscosity, thermal conductivity, etc., in ordinary gas dynamics. It may be considered as a new transport coefficient.

Emission Coefficient j_ν of Radiation

The radiation energy emitted from a mass dm is

$$dE_e = j_\nu \, dm \, d\omega \, d\nu \, dt \tag{16}$$

As above, the emission coefficient j_ν consists of two parts: one is the true emission and the other is that due to scattering.

If we have both the absorption coefficient and the emission coefficient of a medium, the conservation of radiative energy gives the radiative transfer equation which governs the specific intensity I_ν, i.e.,

$$dE_o - dE_i = dE_e + dE_a - dE_t \tag{17}$$

The difference between the outgoing energy dE_o and the incoming radiative energy dE_i must be equal to the sum of the energy emitted dE_e and the energy absorbed dE_a minus the net change of radiative energy in the volume with time. Eq. (17) in terms of k_ν and j_ν is

$$\frac{1}{c}\frac{\partial I_\nu}{\partial t} + n^i\frac{\partial I_\nu}{\partial x^i} = \varrho k_\nu(J_\nu - I_\nu) \tag{18}$$

where

$$J_\nu = j_\nu/k_\nu \tag{19}$$

is the radiation source function.

One of the most difficult problems in radiation gas dynamics is to determine the radiation source function. At the present stage of investigation, we usually use the assumption of local thermodynamic equilibrium, which may be considered as a first approximation of the actual case. After we know more about the results under the local-thermodynamic-equilibrium condition, we study the source function in the nonequilibrium condition.

Under the complete-thermodynamic-equilibrium condition, the specific intensity of thermal radiation is given by the Planck radiation function B_ν which is also known as the blackbody radiation function,

$$B_\nu(\nu, T) = \frac{2h\nu^3}{c^2}\frac{1}{\exp(h\nu/kT) - 1} \tag{20}$$

where h is the Planck constant, $h = 6.62 \times 10^{-27}$ erg sec, and k is the Boltzmann constant, $k = 1.379\ 10^{-16}$ erg/°K. The properties of B_ν had been known before Planck found the correct expression (20). For instance, Kirchhoff knew that under thermodynamic equilibrium,

$$I_\nu = j_\nu/k_\nu = B_\nu(T) \tag{21}$$

The function $B_\nu(T)$ must be a function of temperature and independent of the material. Wien found the displacement law

$$U_\lambda/T^5 = G(c/\lambda T) \tag{22}$$

where $U_\lambda\,d\lambda = U_\nu\,d\nu$ and λ is the wavelength and $c = \lambda\nu$. At low frequencies,

$$B_\nu(\nu, T) = (1/c^2)8\pi\nu^2 kT \tag{23}$$

(Rayleigh–Jeans law). Of course, Eq. (23) is not valid for high frequencies, which causes the ultraviolet breakdown of the Rayleigh–Jeans law.

If we assume that the gas is in local thermodynamic equilibrium, i.e., the emission is determined by the local temperature [see ([1])], Eq. (18) becomes

$$\frac{1}{c}\frac{\partial I_\nu}{\partial t} + \frac{\partial I_\nu}{\partial s} = \varrho k_\nu'(B_\nu - I_\nu) \tag{24}$$

where

$$k_\nu' = k_\nu[1 - \exp(-h\nu/kT)] \tag{25}$$

and s is the distance along a radiation ray.

In general, we have to solve the radiative transfer equation (18) or (24) with other fundamental equations in radiation gas dynamics with the radiation terms in integral forms. These differentio-integral equations are very difficult to solve. In order to get some essential features of the effects of thermal radiation, some approximations have been used. They are discussed in some detail in ([1]). Here, we shall list a few of the most common ones:

a. Optically-Thick Medium.
When the radiation mean free path $L_{R\nu}$ is very small in comparison with a typical dimension of the flow field, the solution of Eq. (24) may be expressed as follows:

$$I_\nu = B_\nu - L_{R\nu}(n^i\, \partial B_\nu/\partial x^i) + O(L_{R\nu}^2) \tag{26}$$

If we neglect terms of higher order than $L_{R\nu}^2$, the radiation terms in gas-dynamic equations can be easily evaluated, i.e.,

$$E_R = a_R T^4 = 3p_R \tag{27}$$

and all the shearing stresses of radiation vanish. The expression (27) is the formula we used before to estimate the value of radiation energy density. The radiative heat flux in the present optically-thick medium is

$$\mathbf{q}_R = D_R \nabla E_R \tag{28}$$

where

$$D_R = c/(3\varrho K_R) \tag{29}$$

is the Rosseland diffusion coefficient of radiation and

$$K_R = \int_0^\infty \frac{\partial B_\nu}{\partial T}\, d\nu \bigg/ \int_0^\infty \frac{1}{k_\nu'}\frac{\partial B_\nu}{\partial T}\, d\nu \tag{30}$$

is the Rosseland mean absorption coefficient (^{2}). For the optically-thick medium, the differentio-integral equations are reduced to differential equations. These fundamental equations are essentially the same as the ordinary Navier–Stokes equations of a compressible fluid with a few more terms.

b. One-Dimensional Approximation.

If the radiation mean free path is not small, we have to use more terms in the series expansion of Eq. (26). The number of terms we should use is open to question, and we do not know about the convergence of this series. Hence it is better to solve the integrodifferential equations or to use some other approximations. In many practical problems, such as boundary layer flows, the gradient of temperature in one direction is much larger than those in the other directions. Hence we may assume that the specific intensity I_ν is essentially a function of this predominant coordinate, say y, i.e., $I_\nu(y, \theta)$, and the state variables and the absorption coefficient k_ν' are functions of y only and independent of θ or ϕ. If we define the optical thickness τ_ν in terms of y, i.e.,

$$\tau_\nu = \int_0^y \varrho k_\nu' \, dy \tag{31}$$

we may carry out the integration in angular coordinates θ and ϕ for the solution of I_ν. Hence the integral expressions of radiative terms will be simplified. We shall discuss the actual expression later.

c. Gray-Gas Approximation.

The absorption coefficient k_ν' is in general a function of frequency ν. As a result, the integral expressions of radiation terms may not be integrated analytically. A gray-gas approximation has been used to assume that the absorption coefficient, and then the optical thickness, is independent of frequency. Actually, we use some mean value of the absorption coefficient for the overall flow phenomena. The Rosseland mean absorption coefficient (30) is independent of frequency, and may be used for an optically-thick medium. The Planck absorption coefficient K_P is the one used for a finite radiation mean free path, and is defined as (^{3})

$$K_p = \int_0^\infty k_\nu' B_\nu \, d\nu \bigg/ \int_0^\infty B_\nu \, d\nu = \left(\int_0^\infty k_\nu' B_\nu \, d\nu \right) \bigg/ B \tag{32}$$

By the gray-gas approximation, the radiation terms may be integrated with respect to frequency ν.

2. FUNDAMENTAL EQUATIONS OF RADIATION GAS DYNAMICS

In radiation gas dynamics, we have seven unknowns: the temperature T, pressure p, and density ϱ of the gas, the velocity vector $\mathbf{q}$ with three components u^i, and the specific radiation intensity I_ν. We use the following seven fundamental equations for these unknowns:

a. Equation of State. The ideal-gas law may be used as the equation of state,

$$p = \varrho RT \tag{33}$$

where R is the gas constant.

b. Equation of Continuity. The conservation of mass gives the equation of continuity,

$$\frac{\partial \varrho}{\partial t} + \frac{\partial \varrho u^i}{\partial x^i} = 0 \tag{34}$$

The summation convention is used for the repeated tensorial indices i.

c. Equations of Motion. The conservation of momentum gives the equations of motion; in vector form this is given by

$$\varrho \, D\mathbf{q}/Dt = -\nabla p + \nabla \cdot \tau_S + \nabla \cdot \tau_R + \mathbf{F} \tag{35}$$

where $D/Dt = (\partial/\partial t) + \mathbf{q} \cdot \nabla$ is the total derivative with respect to time t. The viscous stress tensor τ_s has as ijth component

$$\tau_s^{ij} = \mu\left(\frac{\partial u^i}{\partial x^j} + \frac{\partial u^j}{\partial x^i}\right) - \frac{2}{3}\,\mu\,\frac{\partial u^k}{\partial x^k}\,\delta^{ij} \tag{36}$$

where $\delta^{ij} = 0$ if $i \neq j$, $\delta^{ij} = 1$ if $i = j$, and μ is the coefficient of viscosity of the gas, which is a function of temperature.

The radiation stress tensor τ_R has its ijth component given by Eq. (11). Hence Eq. (35) is an integrodifferential equation.

In (35) $\mathbf{F}$ is the body force, which may consist of the gravitational force $\mathbf{F}_g = \varrho \mathbf{g}$, where $\mathbf{g}$ is the gravitational acceleration, and the electromagnetic force $\mathbf{F}_e$, which may be written as (^{4})

$$\mathbf{F}_e = \varrho_e \mathbf{E} + \mathbf{J} \times \mathbf{B} \tag{37}$$

where ϱ_e is the excess electric charge, $\mathbf{E}$ is the strength of the electric field, $\mathbf{J}$

is the electric current density, and **B** is the magnetic induction. If we include the electromagnetic forces in our study, we have to add the corresponding equations for the electromagnetic variables. For simplicity, in the present chapter we shall neglect the body force **F**.

d. Energy Equation. The conservation of energy gives the energy equation as follows:

$$\varrho \, D\bar{e}_m/Dt = \nabla \cdot (p\mathbf{q}) + \nabla \cdot (\mathbf{q} \cdot \tau_S) + \nabla \cdot (\mathbf{q} \cdot \tau_R) + \nabla \cdot (\varkappa \nabla T)$$
$$+ \nabla \cdot \mathbf{q}_R + Q \tag{38}$$

where $\bar{e}_m = U_m + \frac{1}{2}q^2 + \phi + (E_R/\varrho)$ is the total energy of the gas per unit mass, with U_m the internal energy of the gas per unit mass, $\frac{1}{2}q^2$ is the kinetic energy of the gas per unit mass, ϕ is the potential energy of the gas per unit mass, and E_R is the radiation energy density per unit volume, which is given by Eq. (10).

The term on the left-hand side of Eq. (38) is the rate of change of the total energy of the gas per unit volume. The first term on the right-hand side of Eq. (38) is the work done by the pressure of the gas; the second term is the energy dissipated by the viscous stresses; the third term is the energy dissipated by the radiation stresses; the fourth term is the energy transfer by heat conduction, with $\varkappa$ as the coefficient of heat conductivity; the fifth term is the heat transfer by thermal radiation, which is given by Eq. (7); and the last term is the energy input by other heat sources, such as those due to chemical reaction, an electromagnetic field, etc. In this chapter, we shall neglect Q for simplicity.

e. Radiative Transfer Equation. The specific radiation intensity I_ν is governed by the radiative transfer equation (18) or (24). For most of the flow problems, we do not consider very-high-frequency phenomena, so that the unsteady term $(1/c)(\partial I_\nu/\partial t)$ is negligibly small in comparison with the spatial variation terms because the velocity of light c is a very large quantity. Hence we use the following equation for the specific intensity I_ν:

$$l \frac{\partial I_\nu}{\partial x} + m \frac{\partial I_\nu}{\partial y} + n \frac{\partial I_\nu}{\partial z} = \frac{\partial I_\nu}{\partial s} = \varrho k_\nu{}'(B_\nu - I_\nu) \tag{39}$$

where l m, and n are the direction cosines of the radiation ray s with respect to the x, y, and z axes, respectively, and s is the distance along the radiation ray. The reduced absorption coefficient $k_\nu{}'$ is considered as a given function of the state variables.

Equation (39) may be integrated with respect to the variable s, and we have

$$I_\nu(s) = I_\nu(s_0) \exp[-\tau_\nu(s, s_0)] + \int_{s_0}^{s} B_\nu(s_1)\{\exp[-\tau_\nu(s, s_1)]\}\varrho k_\nu' \, ds_1 \quad (40)$$

where

$$\tau_\nu(s, s_1) = \int_{s_1}^{s_0} \varrho k_\nu' \, ds' \quad (41)$$

and s_0 is an initial point on the ray of radiation whose specific intensity $I_\nu(s_0)$ is known. The determination of $I_\nu(s_0)$ depends on the boundary conditions of our problem, which will be discussed later. With the help of Eq. (40), the radiation terms $\mathbf{q}_R$, etc., may be expressed as integrals of the state variables T and ϱ and our fundamental equations of radiation gas dynamics are the system of integrodifferential equations (33)–(35) and (38). Since such a system of integrodifferential equations is too complicated to solve for practical problems, we have to make reasonable approximations so that the fundamental equations may be simplified into forms which can be analyzed. The following are some of these approximations:

1. All the well-known approximations of gas dynamics may be used. For instance, we may consider (a) the inviscid and nonheat-conducting flow. Outside the boundary layer or other transition regions, the viscosity and heat conductivity may be neglected when the Reynolds number of the flow is high. In our case, we should add the radiation terms to the equations of an inviscid and nonheat-conducting fluid; and (b) boundary layer flow. In the boundary layer region, the well-known Prandtl boundary layer approximations may be applied.

2. Some approximations on the radiation terms may be used: (a) When the ratio of the radiation pressure to the gas pressure R_p, which may be called radiation pressure number, is small, the radiation energy density and the radiation stresses may be neglected. We need to consider the radiative heat flux only. Many of the current aerospace problems satisfy such a condition, as we will discuss later. (b) We may simplify Eq. (40) and then the radiation terms according to the value of radiation mean free path or other approximations. If the radiation mean free path is small, we may use Eqs. (27) and (28) for the radiation terms. The fundamental equations of radiation gas dynamics for an optically-thick medium are then

$$p = \varrho R T \quad (42a)$$

$$\partial \varrho / \partial t + \nabla \cdot (\varrho \mathbf{q}) = 0 \quad (42b)$$

$$\varrho \, Dq/Dt = -\nabla(p + p_R) + \nabla \cdot \tau_S + \mathbf{F} \tag{42c}$$

$$\varrho \, D\bar{e}_m/Dt = -\nabla \cdot [\mathbf{q}(p+p_R)]+\nabla \cdot (\mathbf{q} \cdot \tau_S)+\nabla \cdot [(\varkappa + \varkappa_R)\nabla T]+Q \tag{42d}$$

where

$$\varkappa_R = 4D_R a_R T^3 \tag{43}$$

is the coefficient of heat conductivity by thermal radiation and p_R, E_R, and K_R are given by Eqs. (27) and (30).

For a large radiation mean free path, we may use the one-dimensional approximation and the gray-gas approximation, and the integrals of specific intensity and of the radiation terms with respect to the angular variable θ and the frequency ν may be carried out. For instance, the radiative heat flux in the predominant direction y is given by the following formula:

$$q_R = 2\pi \int_\tau^{\tau_2} B(t)\varepsilon_2(t - \tau) \, dt - 2\pi \int_0^\tau B(t)\varepsilon_2(\tau - t) \, dt$$

$$+2q_R(\tau_2)\varepsilon_3(\tau_2 - \tau) - 2q_R(0)\varepsilon_3(\tau) \tag{44}$$

where (1)

$$\tau = \int_0^y \varrho k_\nu' \, dy$$

and the boundaries for y are $y = 0$, $\tau = 0$ and $y = L$, $\tau = \tau_2$; (2) the exponential integral ε_n is defined by the following formula:

$$\varepsilon_n(t) = \int_1^\infty m^{-n}e^{-mt} \, dm = \int_0^1 z^{n-2}e^{-t/z} \, dz \tag{45}$$

and (3)

$$B(t) = \int_0^\infty B_\nu(\nu, T) \, d\nu = (2\pi^4 k^4/15c^2h^3)T^4 = (\sigma/\pi)T^4 \tag{46}$$

Equation (44) gives

$$\frac{\partial q_R}{\partial y} = \frac{\partial \tau}{\partial y} \frac{\partial q_R}{\partial \tau} = \frac{c a_R}{L_{Rp}} \left[\frac{1}{2} \int_0^{\tau_2} T^4(t)\varepsilon_1(|\, t - \tau \,|) \, dt - T^4(\tau) \right]$$

$$+ \frac{1}{L_{Rp}} \, [2q_R(\tau_2)\varepsilon_2(\tau_2 - \tau) + 2q_R(0)\varepsilon_2(\tau)] \tag{47}$$

where $q_R(0)$ and $q_R(\tau_2)$ are the values of q_R at $\tau = 0$ and $\tau = \tau_2$, respectively. The integral expression (47) should be used in the energy equation.

If we replace the exponential integral by an exponential such that

$$\varepsilon_2(t) = \tfrac{1}{3}m^2 e^{-mt} \tag{48}$$

and apply the condition that the limiting values of an optically-thick medium and an optically-thin medium are satisfied, Eq. (47) gives the following relation for the radiative heat flux q_R:

$$\frac{d^2 q_R}{d\tau^2} - 3q_R + 16\sigma T^3 \frac{dT}{d\tau} = 0 \tag{49}$$

We should solve Eq. (49) with the gasdynamic equations simultaneously by considering q_R as a dependent variable.

For the three-dimensional case of a gray gas, Eq. (49) may be replaced by the following formula:

$$\frac{1}{\varrho K} \frac{\partial}{\partial x_i} \left(\frac{1}{\varrho K} \frac{\partial q_{R_j}}{\partial x_j} \right) - 3q_{R_i} + 16\sigma T^3 \frac{\partial T}{\partial x_i} \frac{1}{\varrho K} = 0 \tag{50}$$

where K is the mean absorption coefficient of the medium.

3. INITIAL AND BOUNDARY CONDITIONS OF RADIATION GAS DYNAMICS (¹)

For every particular problem of radiation gas dynamics, we have certain initial and boundary conditions. Our problem is to find solutions of the fundamental equations of radiation gas dynamics which satisfy these initial and boundary conditions. By initial conditions, we mean the velocity distribution and the state of the gas in the flow field as well as the specific intensity of radiation at a certain initial time $t = 0$ for the whole space. Customarily, in radiation gas dynamics, we do not give the spatial distribution of these initial conditions, but we only require that the initial values be consistent with the boundary condition at $t = 0$ and the fundamental equations. Hence we need to examine the boundary conditions only. By boundary conditions, we mean the velocity distribution and the state of the gas as well as the specific radiation intensity at the boundary of the domain considered at the time $t \geqq 0$.

In radiation gas dynamics, we have to consider the boundary conditions of both the gasdynamic field and the radiation field.

The boundary conditions of the gasdynamic field depend on the mean free path of the gas particles. When the mean free path is negligibly small in comparison with a typical length of the flow field, the gas may be considered as a continuum. The no-slip boundary condition is a good approximation. Under this condition, we have the situation that across a surface separating a body and a fluid or two fluids, the velocity components, the stresses, the temperature, and the heat flux are all continuous. In all problems discussed in this chapter, we consider the case for which the gas may be considered as a continuum, and the no-slip boundary conditions will be used.

When the mean free path of the gas is not small, even though the gas may still be considered as a continuum, there will be slip on the boundary. The velocity and the temperature of the gas at the wall may be different from those of the wall itself. This is known in rarefied gas dynamics as slip flow.

When the mean free path of the gas is larger than the typical length of the flow field, the gas cannot be considered as a continuum. We now have free-molecular flow. The boundary conditions depend on the smoothness of the wall. When the free molecules strike the wall, it may reflect diffusely or specularly. In general, some of the molecules will reflect diffusely and some will reflect specularly.

Even though the no-slip boundary conditions are the correct boundary conditions for the gasdynamic field when the mean free path of the gas is very small, such no-slip conditions may be relaxed under certain conditions, particularly in the flow field far away from the boundary or transition region at very high Reynolds numbers. Under these conditions, where the viscous effect is small, we may assume that the gas is inviscid. In the inviscid flow field, a surface of discontinuity is allowable. The no-slip condition should be relaxed. Two types of discontinuity surface may occur in the inviscid flow field: the shock wave and the vortex surface. Across a shock wave, there is a discontinuity in the velocity component normal to the shock front, but a continuity in the tangential velocity. On the other hand, across a vortex surface, there is a discontinuity in the tangential velocity component, but a continuity in the normal velocity component.

The boundary conditions of the radiation field depend on the radiation mean free path. When the radiation mean free path is very small, i.e., the optically-thick medium, all the radiation terms can be expressed in terms of the temperature by the Rosseland approximation if we assume that local thermodynamic equilibrium is attained. Under this condition, we do not need explicitly the boundary condition for the specific radiation intensity and we need only the boundary condition for the temperature. Hence the consideration of gasdynamic boundary conditions is sufficient for this case of radiation gas dynamics.

When the radiation mean free path is small but not negligibly small, we find that the fundamental equations of radiation gas dynamics for the optically-thick medium, Eqs. (42), are sufficient to describe the flow field away from the boundary, but the no-slip condition of temperature will not be satisfied. Hence, in analogy to the slip flow of rarefied gas dynamics, we have a temperature jump in this case. This temperature jump depends on the boundary condition of the specific radiation intensity.

For a finite radiation mean free path, we have to consider the boundary conditions of the radiation field by studying the interaction of radiation at the interface of two media. When radiation strikes a surface, part of the radiative energy may penetrate into the second medium and part of the radiative energy may reflect back into the first medium. Hence an incident ray may result in a transmitted ray and a reflected ray. However, radiation gas dynamics involves macroscopic analysis. We do not study each ray individually, but a large number of rays statistically. We have to use a statistical average to consider the actual surface conditions in engineering problems. The properties of a surface may be expressed in terms of the following three overall coefficients:

1. The absorption coefficient of the surface a_ν, which is equal to the emissivity coefficient of the surface e_ν. In general, these coefficients are functions of the frequency of the radiation rays.

2. The reflection coefficient of the surface r_ν.

3. The transparency coefficient of the surface tr_ν.

These three coefficients are connected by the following equation:

$$a_\nu + r_\nu + \mathrm{tr}_\nu = 1 \tag{51}$$

The values of each of these coefficients depend on the condition of the surface, and should be determined experimentally.

There are a few cases which are of interest in the general discussion of radiation gas dynamics:

a. Rough Opaque Surface. For such a surface, the coefficient of transparency tr_ν is always zero, and we have

$$a_\nu + r_\nu = 1 \tag{52}$$

b. Gray Surface. For such a surface, all three coefficients a_ν, r_ν, and tr_ν are independent of the frequency ν.

c. Black Surface. For such a surface, both the coefficients of reflection and of transparency are zero, and then we have

$$a_\nu = e_\nu = 1 \tag{53}$$

The specific intensity of this kind of surface under thermodynamic equilibrium is then

$$I_\nu(T_w) = n^2 B_\nu(T_w) \tag{54}$$

where T_w is the temperature of the surface or wall. Since for most engineering problems, the index of refraction n is very close to unity, we may take $n = 1$, and the specific intensity of a black wall is then

$$I_\nu(T_w) = B_\nu(T_w) \tag{55}$$

For a nonblack but opaque surface, the specific intensity on the wall under the thermodynamic-equilibrium condition with $n = 1$ is

$$I_\nu(T_w) = e_\nu B_\nu(T_w) = (1 - r_\nu) B_\nu(T_w) \tag{56}$$

Equations (55) and (56) may be used to determine the boundary conditions of the specific radiation intensity.

4. SIMILARITY PARAMETERS OF RADIATION GAS DYNAMICS [1]

In general, the fundamental equations of radiation gas dynamics form a system of nonlinear integrodifferential equations. It is extremely difficult to solve these equations for given boundary conditions. Even in the simple case of an optically-thick medium, where the integrodifferential equations reduce to a set of differential equations (42), the fundamental equations are still much more complicated than the Navier–Stokes equations of a compressible fluid without radiation effects. There is no general method for finding the solution to these nonlinear equations. In order to bring out the essential features of the flow problems of radiation gas dynamics, it is desirable to find the important parameters which characterize the flow problems.

These parameters in dimensionless form are known as similarity parameters because they show the relative effects of various forces and transfers in the flow field. From these parameters, one may have some guide as to how the flow with radiation effects differs from those without radiation effects. Thus we may divide the radiation flow field into various regions, and proper approximations may be applied to these regions so that practically important flow problems can be solved.

These parameters are also useful in correlating experimental results. In experimental investigations, it frequently happens that the test model is of a different size than the actual body, that the test fluid is in a different thermodynamic state than the actual fluid, or that the test fluid is under conditions different than those actually encountered, and we have to know the relation between the test condition and the actual situation. This relation depends mainly on the important parameters of the problem. If we know these parameters, it will be easy to correlate the experimental data.

There are two methods of finding these important parameters: one is known as inspection analysis, and the other is known as dimensional analysis. We shall not discuss these methods, but only give the relevant results, i.e., the parameters of importance in radiation gas dynamics. The reader may refer to [1,2] for these methods.

The dimensionless parameters of radiation gas dynamics may be divided into two groups: one group consists of all the parameters of ordinary gas dynamics, and the other group consists of special parameters which are due to the radiation effects. We are going to study these parameters according to such a division as follows:

4.1. Dimensionless Parameters of Ordinary Gas Dynamics

In the following, we shall discuss only those parameters of ordinary gas dynamics which we shall use in Chapter 9:

a. The Time Parameter R_t

$$R_t = \frac{t_0}{L/U} \qquad (57)$$

This parameter characterizes the time scale of the flow problem with respect to the flow velocity and a dimension of the flow field. The quantity L/U may be considered as a characteristic time of the flow field. In the steady flow problem, R_t is infinite because the typical time t_0 is infinite, so that the unsteady term $\partial Q/\partial t = 0$. However, for the convection terms, the value of R_t should be taken as unity. In general, we may assume that the value of R_t in ordinary gas dynamics is of the order of unity. Only in the case of high-frequency phenomena such as flutter problems of high-frequency wave motion may the value of R_t be very small.

b. The Ratio of Specific Heats γ. This is the ratio of specific heat at constant pressure to that at constant volume. It is a measure of the relative complexity of the molecules of the gas, because the specific heat of a gas is closely related to its molecular structure. The exact expressions for specific heats in terms of state variables of a gas should be derived on the basis of statistical mechanics. For simplicity, we shall take it as a constant.

c. Mach Number M

$$\text{M} = \frac{U}{a} = \frac{\text{Flow velocity}}{\text{Sound speed}}, \tag{58}$$

where sound speed $a = (\gamma RT)^{1/2}$. The Mach number is a measure of the compressibility of the gas due to a high velocity. It is easy to show that the ratio of the density variation of a gas to the velocity variation is, to a first approximation, proportional to the square of the Mach number of the flow. Hence for very small Mach number, the variation of the density of a gas is negligible. For large Mach number, the effect of compressibility, i.e., the variation of density of the gas, is important. When $\text{M} < 1$, the flow is said to be subsonic, and when $\text{M} > 1$, the flow is supersonic. The flow field of a subsonic flow differs greatly from that of a supersonic flow. For instance, a shock wave occurs only in the supersonic flow field for the steady case, but never in the corresponding subsonic case. When Mach number M is of the order of unity, the flow is considered to be transonic; when the Mach number M is much larger than unity, the flow is hypersonic. Many new features of the flow field in the regions of transonic flow and hypersonic flow differ from those of ordinary subsonic flow and supersonic flow. Hence the Mach number is one of the most important parameters in ordinary gas dynamics, as well as in radiation gas dynamics. Because of the variation of local Mach number in the flow field, the flow may be subsonic in one part of the flow field and supersonic in another part, with a transonic flow field between them.

The sound speed of a gas will be increased by the effect of thermal radiation. With radiation effects, we have an effective sound speed and a corresponding effective Mach number, which is the ratio of the flow velocity to the effective sound speed with radiation effect included.

d. Reynolds Number Re

$$\text{Re} = \frac{UL\varrho}{\mu} = \frac{UL}{v_g} = \frac{\text{Inertial force}}{\text{Viscous force}} \tag{59}$$

This is the most important parameter for the fluid dynamics of a viscous fluid. When the Reynolds number is small, the viscous force is predominant and the effect of viscosity is important in the whole flow field. When the Reynolds number is large, the inertial force is predominant and the effect of viscosity is important only in a narrow region near a solid boundary or some other restricted region, which is known as a boundary layer region or transition region. Outside these transition or boundary layer regions, the flow may be considered as inviscid. If the Reynolds number is enormously large, the flow becomes turbulent.

e. Prandtl Number Pr

$$Pr = \frac{c_p \mu}{\varkappa} = \frac{\nu_g}{(\varkappa/c_p \varrho)} = \frac{\text{Kinematic viscosity}}{\text{Thermal diffusivity}} \qquad (60)$$

The thickness of the boundary layer of the velocity field of a laminar flow is proportional to the square root of the kinematic viscosity ν_g, while the thickness of the thermal boundary layer is proportional to the square root of the thermal diffusivity $(\varkappa/c_p \varrho)^{1/2}$. Hence the Prandtl number Pr shows the relative importances of the heat conductivity and viscosity of the fluid. As a first approximation, the Prandtl number gives the square of the ratio of the thickness of the velocity boundary layer to that of the thermal boundary layer.

In radiation gas dynamics, the effective thermal conductivity of a flow field depends on both the heat conduction and the thermal radiation. In general, the thermal radiation increases the effective thermal conductivity and decreases the effective Prandtl number.

4.2. Dimensionless Parameters of Radiation Gas Dynamics

There are a few parameters which are due essentially to the effects of thermal radiation; these are given below:

f. Relativistic Parameter R_r

$$R_r = \frac{U}{c} = \frac{\text{Flow velocity}}{\text{Velocity of light}} \qquad (61)$$

For ordinary gas particles, the relativistic parameter is usually very small. Hence the relativistic effect on the flow of gas particles is negligible. However, the thermal radiation effect is due to the motion of photons, which move with the speed of light c. As a result, the relativistic parameter plays an important role in heat transfer by thermal radiation.

g. Knudsen Radiation Number Kn_r

$$Kn_r = L_R^* = \frac{L_R}{L} = \frac{\text{Radiation mean free path}}{\text{Characteristic flow length}} \tag{62}$$

This number characterizes the distance traveled by photons before they are absorbed by the molecules in the gas. When the Knudsen radiation number is very small, we say that the medium is optically thick. When the Knudsen radiation number is very large, we say that the medium is optically thin.

h. Radiation Pressure Number R_p

$$R_p = \frac{a_R T^4}{3p} = \frac{\text{Radiation pressure}}{\text{Gas pressure}} \tag{63}$$

In ordinary gas dynamics, this number is usually very small. Hence in ordinary gas dynamics, the effects of radiation are negligible. However, for the condition of very high temperature and low pressure, this number will be large. When R_p is not negligibly small, we have to consider the effects of radiation pressure and radiation energy density. Even when R_p is small, the radiation heat flux may not be negligible, as we shall see later. Figure 1 shows the values of R_p at various pressures and temperatures. Ordinarily, R_p is very small except at very high temperatures (over 40,000 °K) and very low pressures.

We may obtain other parameters from various combinations of the above eight parameters. These new parameters may be more appropriate for special problems. Some of these new parameters are given below:

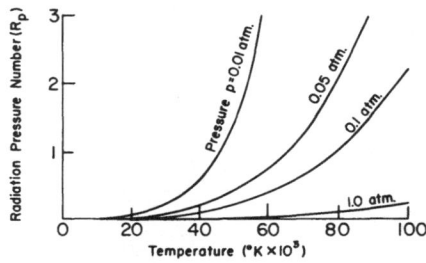

Fig. 1. Radiation pressure number as a function of temperature and pressure.

a. Peclet Number Pe

$$\text{Pe} = \frac{UL}{(\varkappa/c_p\varrho)} = (\text{Pr})(\text{Re}) \tag{64}$$

The Peclet number plays the same role with respect to the thermal boundary layer as the Reynolds number plays with respect to the velocity boundary layer.

b. Knudsen Number Kn_f

$$\text{Kn}_f = \frac{\text{Mean free path of a gas}}{\text{Characteristic length}} = \frac{L_f}{L} = 1.255\sqrt{\gamma}\,\frac{M}{\text{Re}} \tag{65}$$

where the mean free path of a gas L_f may be taken as follows:

$$L_f = 1.255\sqrt{\gamma}\,(v_q/a) \tag{66}$$

In most of the flow problems in this chapter, we consider that the Knudsen number is very small, so that the gas may be considered as a continuum.

c. Radiative Flux Number R_f

$$R_F = \frac{\text{Radiative heat flux}}{\text{Heat conduction flux}} \tag{67}$$

Since the exact expression for radiative heat flux depends on the Knudsen radiation number, the expression for the radiative flux number also depends on the Knudsen radiation number. For the optically thick case, i.e., small value of Knudsen radiation number Kn_r, the radiative heat flux is given by Eq. (28). The corresponding radiative flux number is

$$R_{F1} = \frac{\varkappa_R}{\varkappa} = 4\frac{\gamma-1}{\gamma}\,(\text{Pr})(\text{Re})L_R{}^*\,\frac{R_p}{R_r} \tag{68}$$

For the case of an optically-thin gas, the radiative heat flux is given by Eq. (44). The proper radiative flux number is

$$R_{F2} = \frac{ca_R T^4 L^2}{L_R \varkappa T} = 3\frac{\gamma-1}{\gamma}\,(\text{Pr})(\text{Re})\,\frac{R_p}{L_R{}^*R_r} \tag{69}$$

The main difference between R_{F1} and R_{F2} lies in the parameter, Knudsen radiation number $L_R{}^*$. For a very small radiation mean free path, $L_R{}^* \ll 1$, the radiative flux number R_{F1} is proportional to the Knudsen radiation number; while for large Knudsen radiation number, $L_R{}^* \gg 1$, the radiative flux number is inversely proportional to $L_R{}^*$. Hence for both very large and very small $L_R{}^*$, the radiative heat flux tends to be zero.

We may also define the radiative flux number as the ratio of radiative heat flux to the convective heat flux, i.e.,

$$R_F{}' = \frac{\text{Radiative heat flux}}{\text{Convective heat flux}} = R_F \frac{T}{c_p \varrho UTL} = \frac{R_F}{(\text{Pr})(\text{Re})} \qquad (70)$$

It is evident that we may have two different forms for $R_F{}'$ depending on whether we use R_{F1} or R_{F2}.

It is interesting to notice that all the radiative flux numbers are directly proportional to the radiation pressure number R_p and inversely proportional to the relativistic parameter R_r. Usually, both R_p and R_r are small in many practical problems. Hence we may neglect the effect of radiation energy density and radiation pressure, but not the radiative heat flux. In the following sections, we shall discuss several cases which satisfy this condition.

5. RADIATION MEAN FREE PATH

When we have radiative heat flux, one of the most important physical properties of the gas is the radiation mean free path $L_{R\nu}$. This property represents a new diffusive phenomenon in the medium in addition to the momentum diffusivity by viscosity and the thermal diffusivity by thermal conductivity. Many new and interesting phenomena may be introduced by this radiative diffusivity. The value of the radiation mean free path should be determined either by experiments or from some microscopic theory.

In general, the radiation mean free path $L_{R\nu}$ is a function of the frequency ν of the radiation and the state variables of the medium. If we consider the variation of the radiation mean free path with frequency, the gas is said to be nongray and the radiative heat flux should be expressed in integral form. In engineering problems, it is advisable to use some simple expression of the average value of the radiation mean free path over the whole frequency spectrum. For an optically-thick medium, we may use the Rosseland approximation, i.e., $L_R = \varrho K_R$, where K_R is the Rosseland mean absorption coefficient given by Eq. (30). For an optically-thin medium or case where the radiation mean free path is finite, we may use the Planck mean value $L_P = \varrho K_P$, where K_P is the Planck mean absorption coefficient given by Eq. (32).

It should be noticed that the value of the radiation mean free path itself does not determine whether the medium is optically thick or thin.

We have to compare the radiation mean free path to the representative length L which characterizes the flow field investigated in order to determine whether the medium may be regarded as optically thick or thin. In other words, the optical thickness defined by Eq. (14) or a similar expression determines the optical properties of the medium. In fact, for a given state of a medium, the Planck mean free path of radiation L_p is usually smaller tahn the Rosseland mean free path of radiation L_R. For high-temperature air (up to a temperature of 20,000 °K), we may take $L_R = 8.3 L_p$ as a first approximation.

After we have taken the average of the radiation mean free path over the whole spectrum, we should find some simple formula for the variation of the average radiation mean free path with temperature and density or pressure. Since the absorption coefficient increases with the density, the radiation mean free path decreases with an increase of density. The variation of radiation mean free path with temperature is very complicated. At low temperature, say, $T < 10,000$ °K, the radiation mean free path decreases with increase of temperature, while at very high temperature, ($T > 100,000$ °K), it increases with the temperature, and there is a minimum in the intermediate temperature range depending on the density and the composition of the medium. If the temperature range in the flow field is not too large, a power law for the average values of the radiation mean free path may be used, i.e.,

$$\frac{L_R}{L_{R,0}} = \frac{L_p}{L_{p,0}} = \left(\frac{T_0}{T}\right)^{m_1}\left(\frac{\varrho_0}{\varrho}\right)^{m_2} \tag{71}$$

where the subscript 0 refers to the values at some reference condition. The powers m_1 and m_2 should be so chosen that formula (71) gives the best fit with the opacity data over the temperature range considered. For instance, in the temperature range of 7000–12,000 °K, the following values may be used for air: $m_1 = 4.4$, $m_2 = 1$, $\varrho_0 = 1.23$ g/cm³, $T_0 = 10,000$ °K, and $L_{p,0} = 0.5$ m. For higher temperatures, in the neighborhood of 20,000 °K, we may take $m_1 = 2.5$ and $m_2 = 1.$ At very high temperatures, as in some astrophysical problems, $m_1 = -\frac{7}{2}$ and $m_2 = 2$ have been used.

If we consider a large range of temperature including the minimum value of the mean radiation, free path the following formula may be used:

$$\frac{L_{R,0}}{L_R} = \frac{L_{p,0}}{L_p} = \frac{L_{p,1,0}}{L_{p,1}} + \frac{L_{p,2,0}}{L_{p,2}} \tag{72}$$

where $L_{p,1}$ is the radiation mean free path given by Eq. (71), which is good fcr the low-temperature range, i.e., m_1 is positive; while L_{p2} is the corresponding value for the high-temperature range, i.e., m_1 is negative. Other fcrmulas have also been used (see [1]).

6. WAVE MOTION IN A RADIATING GAS [1]

We are now going to study the wave motion in a viscous, heat-conducting, radiating gas. The wave motion will bring out many new characteristic features of radiation gas dynamics which differ considerably from those of ordinary gas dynamics. The properties of a wave in a gas depend on the amplitude of the wave. The simplest type of wave is the wave of infinitesimal amplitude. One of the main features of these infinitesimal-amplitude waves is that the superposition principle is applicable to them. Mathematically speaking, we may linearize the equations which govern these waves. Since the resultant equations are linear, the sum of two solutions of them is also a solution. Thus we may study any typical solution of these wave equations to find the general features of wave propagation.

For waves of finite amplitude, the shape of the wave will distort as the wave propagates, while the infinitesimal-amplitude wave will maintain its shape when it propagates. When the distortion of the wave is large, ordinary waves will develop into shock waves, in which a large change of physical variables occurs in a very thin region. We shall discuss the shock wave in a radiating gas in the next section.

Now we consider small-amplitude waves in an optically-thick medium in which the Rosseland approximations may be used [5]. We assume that originally the gas is at rest with a pressure p_0, a temperature T_0, and a density ϱ_0. The gas is perturbed by a small disturbance, so that in the resultart disturbed motion, we have

$$u = u(x, t), \qquad v = v(x, t), \qquad w = w(x, t)$$
$$p = p_0 + p'(x, t), \qquad T = T_0 + T'(x, t), \qquad \varrho = \varrho_0 + \varrho'(x, t) \tag{73}$$

where u, v, and w are, respectively, the perturbed x, y, and z components of velocity, and the prime refers to the perturbed quantities of the state of the gas. For simplicity, we assume that all the perturbed quantities are functions of only one spatial coordinate x and of time t. Thus we consider the wave propagation in the x direction. Substituting Eq. (73) into the fundamental equations of radiation gas dynamics (42) and neglecting the higher-order terms in the perturbed quantities, we have the following linear equations for the wave motion in radiation gas dynamics:

$$\frac{p'}{p_0} = \frac{T'}{T_0} + \frac{\varrho'}{\varrho_0} \tag{74a}$$

$$\frac{\partial \varrho'}{\partial t} + \varrho_0 \frac{\partial u}{\partial x} = 0 \tag{74b}$$

$$\varrho_0 \frac{\partial u}{\partial t} = -\frac{\partial p'}{\partial x} + 4RR_p\varrho_0 \frac{\partial T'}{\partial x} + \frac{4}{3}\mu \frac{\partial^2 u}{\partial x^2} \tag{74c}$$

$$\varrho_0 \frac{\partial v}{\partial t} = \mu \frac{\partial^2 v}{\partial x^2} \tag{74d}$$

$$\varrho \frac{\partial w}{\partial t} = \mu \frac{\partial^2 w}{\partial x^2} \tag{74e}$$

$$c_p^*\varrho_0 \frac{\partial T'}{\partial t} - 4RR_pT_0 \frac{\partial \varrho'}{\partial t} = \frac{\partial p'}{\partial t} + \varkappa^* \frac{\partial^2 T'}{\partial x^2} \tag{74f}$$

where $R_p = a_R T_0^3/(3R\varrho_0)$ is the radiation pressure number of undisturbed flow,

$$c_p^* = c_p + 12RR_p \tag{75}$$

is the effective specific heat at constant pressure, including the radiation effect, and

$$\varkappa^* = \varkappa + 12RR_p\varrho_0 D_R = \varkappa + \varkappa_R \tag{76}$$

is the effective coefficient of heat conductivity, including the radiation effect.

Examining the linearized equations (74), we see that the perturbed quantities may be divided into two groups:

1. *Transverse waves.* In our problem, there is no distinction between the y and z directions. Hence v and w are governed by similar equations, which show the wave propagation of variables in the direction perpendicular to the x axis.

2. *Longitudinal waves.* The four variables u, p', T', and ϱ' interact, and we have to solve the four linearized equations simultaneously.

We look for a periodic solution in which all the perturbed quantities are proportional to

$$\exp[i(\omega t - \lambda x) = \exp[-i\lambda_R(x - Vt)] \exp(\lambda_i x) \tag{77}$$

where ω is a given real angular frequency, $\lambda = \lambda_R + i\lambda_i$ is the complex wave number, $i = \sqrt{-1}$, and

$$V = \omega/\lambda_R \tag{78}$$

is the speed of wave propagation. Substituting the perturbed quantities in the form of Eq. (77) into Eq. (74), we obtain the dispersion relation $\lambda(\omega)$ for both the transverse and longitudinal waves.

a. Transverse Waves. The dispersion relation for the transverse waves v and w is

$$v_g \omega^2 + i\lambda = 0 \tag{79}$$

where $v_g = \mu/\varrho_0$ is the coefficient of kinematic viscosity of the gas. This is the well-known damped wave in a viscous fluid [1]. This wave is independent of the compressibility effect of the medium.

b. Longitudinal Waves. The dispersion relation of the longitudinal waves obtained from Eqs. (74a)–(74c) and (74f) is

$$\varkappa^* \left(\frac{1}{\varrho_0} + \frac{4}{3} i\omega \frac{v_g}{\varrho_0} \right) \lambda^4 - \left\{ \frac{\omega^2 \varkappa^*}{p_0} + \frac{4}{3} \frac{v_g \omega^2}{T_0(\gamma - 1)} [1 + 12(\gamma - 1)R_p] \right.$$
$$\left. - i\omega(c_p + 20RR_p + 16RR_p^2) \right\} \lambda^2 - \frac{i\omega^3}{T_0(\gamma - 1)} [1 + 12(\gamma - 1)R_p] = 0 \tag{80}$$

Equation (80) is a quadratic equation in λ^2, and there are two roots of λ^2, which represent two modes of the longitudinal waves: one is the sound wave in a viscous, heat-conducting, radiating gas, and the other is the heat wave in this medium. Both the sound wave and the heat wave are affected by the thermal radiation effects. There are two types of thermal radiation effect: one is due to radiation pressure and radiation energy density, and is characterized by the radiation pressure number R_p, and the other is due to the radiative heat flux, and is characterized by the radiation heat conductivity $\varkappa_R$.

Let us consider the effects of R_p first. For an inviscid, non heat-conducting fluid without radiative heat flux, i.e., $v_g = 0$ and $\varkappa^* = 0$, Eq. (80) gives for the sound wave in a radiating gas with the speed of propagation C_R by

$$C_R^2 = \frac{\omega^2}{\lambda^2} = a_0^2 \frac{1 + 20[(\gamma - 1)/\gamma]R_p + 16[(\gamma - 1)/\gamma]R_p^2}{1 + 12(\gamma - 1)R_p} \tag{81}$$

where $a_0 = (\gamma RT)^{1/2}$ is the ordinary sound speed without the radiation effects of either the radiation pressure or the radiation energy density. For a given temperature T_0, the radiation sound speed increases with the radiation pressure number R_p. At very high values of R_p, C_R increases with the square root of R_p. This is the mode that corresponds to the ordinary sound speed.

The effects of $\varkappa^*$ are twofold. First, it introduces some damping in the sound wave discussed above. Second, it introduces another mode of the longitudinal wave which corresponds to the heat wave of an ordinary gas. In an ordinary gas, if the heat conductivity $\varkappa$ is different from zero, we have a heat wave besides the sound wave. Since the thermal radiation increases the effective value of heat conductivity, we will have this heat wave even if the ordinary heat conductivity is negligible. When we consider the case of an inviscid and nonheat-conducting gas with a small amount of radiative heat flux so that $\varkappa^*$ is a small quantity, Eq. (80) gives the following two roots of λ as a first approximation:

1. *Radiation sound wave*:

$$\lambda_1 = \pm \frac{\omega}{C_R} [1 - i\omega f(R_p)D_R] \tag{82}$$

2. *Radiation heat wave*:

$$\lambda_2 = \pm \frac{\omega}{C_R} \left[\frac{g(R_p)}{\omega D_R} \right]^{1/2} (-1 + i) \tag{83}$$

where $f(R_p)$ and $g(R_p)$ are functions of R_p. Equation (82) shows that the first-order effect of radiative heat flux is the introduction of damping in the sound wave without changing its speed of propagation. Equation (83) shows that the speed of propagation of the heat wave is proportional to C_R and the square root of D_R, and the damping factor of the heat wave is inversely proportional to $\sqrt{D_R}$. Hence when D_R tends to zero, the heat wave is damped, and we have one undamped radiation sound wave.

For an optically-thin medium, or a medium with finite radiation mean free path, the waves are more complicated than for the case of an optically-thick medium [see (¹)]. Because in general the coefficient of absorption of radiation is a function of the frequency of the radiation wave, the speed of propagation and the damping of various modes in a radiating gas depend not only on R_p and D_R, but also on the frequency of the wave (⁶). For instance, if we consider the case of very small R_p, i.e., $R_p \to 0$, with small but finite $(1/L_p)$, we still have two modes even if $R_p \approx 0$. The speed of a sound wave is independent of R_p, but its value varies with the frequency of the wave. When the frequency is very small, the speed of the sound wave in a radiating gas is equal to that of the adiabatic sound wave of ordinary gas dynamics, i.e., $a_0 = (\gamma RT)^{1/2}$. As the frequency increases, the sound speed decreases until a value a little higher than the isothermal sound speed $(RT)^{1/2}$ is reached. After the minimum of the sound speed is reached, further increase of frequency will again cause the sound speed to increase. At very high frequency, the speed of the sound wave again becomes the adiabatic sound speed. The speed of the heat wave increases with the frequency.

7. SHOCK WAVES

It is well known that waves of finite amplitude may develop into shock waves, across which large changes of the velocity and state variables occur. Since shock waves are important in high-speed flow, where thermal radiation is also important, it is interesting to see what the effects of thermal radiation are on the shock waves. We consider first a normal shock wave in an optically-thick gas. We choose the coordinate system such that the shock wave is stationary. In our system, the gas flow is parallel to the x axis and has the x component of velocity u only. Both the velocity and the state variables in front of the shock are uniform; there is a sharp transition region in which there is a large variation in velocity and state variables; and, finally, far behind the shock, the velocity and state variables become uniform, but at different values from those in front of the shock. The fundamental equations which govern the flow field with this normal shock are as follows:

$$\rho u = \text{const} = m \tag{84a}$$

$$mu + p_t - \tfrac{4}{3}\mu \, du/dx = \text{const} = mC_1 \tag{84b}$$

$$mh_R + up_t - \tfrac{4}{3}\mu u(du/dx) - \varkappa^*(dT/dx) = \text{const} = mC_2 \tag{84c}$$

where $p_t = p + p_R$, $h_R = \tfrac{1}{2}u^2 + c_v T + (E_R/\rho)$, and $\varkappa^* = 4D_R a_R T^3 + \varkappa$.

In the analysis of the shock wave, we need to know two things: (1) the Rankine–Hugoniot relations across the shock, and (2) the shock transition regions.

7.1. Rankine–Hugoniot Relations in Radiation Gas Dynamics

The Rankine–Hugoniot relations connect the values of the two uniform states, that in front of the shock (subscript 1), and that behind the shock (subscript 2). In these uniform states, we have

$$du/dx = dT/dx = 0 \tag{85}$$

Now we introduce the following dimensionless variables:

$$\xi = u/u_1, \qquad T^* = RT/u_1^2, \qquad P = T_1^* = 1/\gamma M_1^2, \qquad M_1 = u_1/a_1$$
$$a_1 = (\gamma RT_1)^{1/2}, \qquad Q = a_R u_1^6/(R\varrho_1^4), \qquad R_p = \tfrac{1}{3}Q\xi T^{*3} = p_R/p \tag{86}$$

Substituting Eq. (86) into Eqs. (84), we have the following relations for ξ and T^*:

$$\xi^2 + (1 + R_p) - [1 + P(1 + R_{p,1})]\xi = 0 \tag{87a}$$

$$-\frac{1}{2}\,\xi^2 + \left(\frac{1}{\gamma - 1} + 3R_p\right)T^* + [1 + P(1 + R_{p,1})]\xi$$
$$-\left[\frac{1}{2} + \frac{\gamma P}{\gamma - 1} + 4R_{p,1}P\right] = 0 \tag{87b}$$

If we eliminate T^* from Eqs. (87a,b), we have

$$(\xi - 1)\left[\xi - \frac{8R_p + r^2 + 1}{7R_p + r^2}\,(1 + R_{p,1})fP + (1 + R_{p,1})\right]$$
$$= (\xi - 1)(\xi - \xi_2) = 0 \tag{88}$$

where $r^2 = (\gamma+1)/\gamma - 1), f = [\xi - g(R_p)]/(\xi - 1)$, and $g(R_p) = (R_p+1)$ $\times (8R_{p,1} + r^2 + 1)/(R_{p,1} + 1)(8R_p + r^2 + 1)$. There are two roots of Eq. (88). The root unity represents the velocity of the original flow, i.e., no shock. The other root, ξ_2, represents the velocity of the gas behind a normal shock wave. The formal expression for ξ_2 is

$$\xi_2 = \frac{\gamma_e - 1}{\gamma_e + 1} + \frac{2\gamma_e P_t}{\gamma_e + 1} \tag{89}$$

where

$$\gamma_e = \frac{4(\gamma - 1)R_{p,2} + \gamma}{3(\gamma - 1)R_{p,2} + 1} \tag{90}$$

is the effective ratio of specific heats in radiation gas dynamics, and

$$P_t = (1 + R_{p,1})f(R_{p,2})p \tag{91}$$

is the effective value of P in radiation gas dynamics. When $R_{p,2} = 0, \gamma_e = \gamma$, and $P_t = P$, we have the Rankine–Hugoniot relation across a normal shock without the radiation effect. When $R_{p,2}$ is very large, $\gamma_e = \frac{4}{3}$ for all values of γ.

Since both γ_e and P_t are functions of $R_{p,2}$ and $R_{p,2}$ depends on ξ_2, we have to find the value of ξ_2 for a given set of initial conditions P and $R_{p,1}$ by the method of successive approximations.

It is of interest to find the value of ξ_2 for a few limiting cases:

a. Low-Temperature Case.
If the temperatures both in front of the shock and behind it are not too high, we have $R_{p,1} \approx R_{p,2} \approx 0$. Hence $\gamma_e \approx \gamma$ and $P_t \approx P$. Equation (89) becomes the normal shock relation of ordinary gas dynamics.

b. Weak Shock in a High-Temperature Gas.
If the temperature of the gas is initially very high, $R_{p,1}$ is then not negligible. If, in addition, the shock-wave strength is weak, $R_{p,2}$ will be approximately equal to $R_{p,1}$. Hence in Eq. (89), we may write $\gamma_e = \gamma_{e,1}$ and $P_t = P_{t,1}$. The effects of radiation on the uniform state behind a weak shock in this case are: (1) the value of γ is replaced by the effective value $\gamma_{e,1}$, i.e.,

$$\gamma_{e,1} = \frac{4(\gamma - 1)R_{p,1} + \gamma}{3(\gamma - 1)R_{p,1} + 1} \tag{92}$$

and (2) the value of P changes into the effective value $P_{e,1}$, i.e., the gas pressure is replaced by the total pressure, which is the sum of the gas pressure and the radiation pressure. When the shock strength is infinitesimally small, we have

$$u_1^2 = \gamma_{e,1} \frac{p_1 + p_{R,1}}{\varrho_1} = C_R^2 \tag{93}$$

This formula (93) is another way to define a radiation sound speed C_R, which is practically identical to that given by Eq. (81).

c. Very Strong Shock in a Cold Gas.
For this case, $R_{p,1} \ll 1$, but $R_{p,2} \gg 1$. We then have

$$\xi_2 = \frac{1}{7} + \frac{27}{70 M_1^2} \tag{94}$$

where we take $\gamma = \frac{5}{3}$. In the limit of $M_1 = \infty$, $\xi_2 = \frac{1}{7}$ and $\gamma_{e,2} = \frac{4}{3}$.

d Shock Wave in a Very Hot Plasma.
For this case, $R_{p,1} \gg 1$ and we have

$$\xi_2 = \frac{1}{7} + \frac{1}{6 M_{e,1}} \tag{95}$$

where $M_e = u/C_R$. It is of interest here to find the temperature jump across the shock. In the limiting case of very large $M_{e,1}$, we have

$$T_2/T_1 = 1.033\sqrt{M_{e,1}} \tag{96}$$

Without the radiation effect, it is well known that at very high shock Mach number, the temperature jump across a normal shock is proportional to the square of the Mach number. However, now with the radiation effect, the temperature jump is proportional to the square root of the Mach number. Hence there is a large difference between the two cases.

7.2. Shock-Wave Structure [8]

In the previous section, we discussed the uniform states in front of and behind a shock wave. Actually, there is a transition region in which the flow variables change gradually from the value in front of the shock to that behind it. For an optically-thick medium, we should solve the complete set of equations (84a–c). We have to consider viscosity, heat-conduction, and radiative heat transfer. The main feature of the present problem is the introduction of a new diffusive transport property, the radiation mean free path L_{R_ν}, in addition to the viscosity and heat conductivity occurring in the shock-structure problem of ordinary gas dynamics, where the mean free path of the gas L_f plays an important role. Since the relative values of L_R and L_f may vary greatly, new phenomena can occur, particularly when L_R is much larger than L_f. The problem is very similar to that of the shock structure in a chemically-reacting medium, in which the relaxation phenomena are important and in which a relaxation length L_r may be introduced to represent the distance to attain the chemical equilibrium condition in the shock transition region. If L_r is much larger than L_f, a considerable portion of the shock transition region, which is usually referred to as an inviscid, nonheat-conducting tail, is determined by the new diffusive property due to the chemical reaction only. If we completely neglect the viscosity and heat conductivity in a chemically-reacting gas, we first obtain a shock discontinuity in temperature and velocity as if it were a nonchemically-reacting gas, and the kinetic temperature immediately behind this shock discontinuity is higher than the final equilibrium temperature. We then have a long transition region, the tail, in which the temperature decreases from the overshot value to its final equilibrium value as if the gas were an inviscid, nonheat-conducting medium. Such a shock is referred to as partly-dispersed shock. If the shock is sufficiently weak, we may obtain a continuous solution of the shock structure from the Rankine–Hugoniot conditions by neglecting the viscosity and heat conductivity and including the relaxation phenomena only. This is known as a fully-dispersed shock.

A similar situation obtains for the shock structure in a radiating gas. For simplicity, let us consider the case when R_p is negligible but R_F is not. In this case, we have as a new feature the radiation mean free path L_p, where we use the Planck mean value as a representative value of our problem. If L_p is much larger than L_f, then in some portion of the shock transition region, the flow field will be determined mainly by L_p. Hence we would have phenomena similar to the partly-dispersed and fully-dispersed shocks in a chemically-reacting gas. The chemical reaction is important only behind the normal shock, while the radiative heat transfer may be important both in front of and behind the shock. Hence in a radiating gas, we may have inviscid, nonheat-conducting tails in front of the shock and behind it. For a partly-dispersed shock, the front and the rear inviscid, nonheat-conducting tails may join together by a surface of discontinuity, the ideal shock. For a fully-dispersed shock, the front and rear inviscid, nonheat-conducting tails would merge together without any discontinuity between them. In the region of the discontinuity surface, the effects of viscosity and heat conductivity must be considered, and then the discontinuity surface will be replaced by a sharp transition region in the actual case.

8. TWO-DIMENSIONAL CHANNEL FLOWS OF AN IONIZED, RADIATING GAS [9]

At very high temperatures, when the thermal radiation effects are important, the gas will be ionized. We should consider simultaneously the radiation and electromagnetic effects on the flow problem. Now we consider an ionized, radiating gas flowing in a two-dimensional channel. For simplicity, we consider only the fully-developed steady flow. All the variables, except for the pressure, are functions of the distance along the normal to the wall of the channel. The channel may be considered as two infinite parallel plates: One of the plates is at rest located at the plane $y = 0$, and the other, at the plane $y = L$, may be in uniform motion with a constant velocity U or at rest. There is a constant or zero pressure gradient in the flow direction, i.e., x direction. There is an externally-applied transverse magnetic field $H_y = H_0 = $ const in the y direction perpendicular to the plates. The lower plate is assumed to be insulated, so that the x component of the magnetic field will be zero all the time, and the magnetic field on the upper wall depends on the externally-applied electric field $E_z = E_0$ $= $ const in the direction z perpendicular to both the plates and the flow direction. The temperatures of the two plates are kept at constant values T_0 and T_1, respectively. The ionized gas is assumed to be viscous, heat conducting, thermally radiating, and electrically conducting. The flow is assumed to be steady and laminar. We assume that the flow is in local thermodynamic equilibrium and the gas is gray. Hence the radiation pressure p_R is given by the following equation:

$$p_R = \frac{2\pi}{c} B(T_0)\varepsilon_4(\tau) + \frac{2\pi}{c} \int_0^\tau B(t)\varepsilon_3(\tau - t)\, dt$$

$$+ \frac{2\pi}{c} B(T_1)\varepsilon_4(\tau_2 - \tau) + \frac{2\pi}{c} \int_\tau^{\tau_2} B(t)\varepsilon_3(t - \tau)\, dt \qquad (97)$$

where

$$\tau = \int_0^y \varrho K_R\, dy$$

is the optical thickness of the gas and K_R is a certain average value of the absorption coefficient of the gas. If τ is very large, Eq. (97) reduces to the Rosseland approximate expression. If τ is small, we obtain the following expression when we keep the terms up to the order of τ:

$$p_R = \tfrac{1}{2}[p_{Rb}(T_1)+p_{Rb}(T_0)] - \tfrac{3}{4}[p_{Rb}(T_0)\tau - p_{Rb}(T_1)(\tau_2 - \tau)]+\tfrac{1}{4}a_R \int_0^{\tau_2} T^4(t)\, dt \tag{99}$$

where p_{Rb} is the Rosseland approximate value, i.e., $p_{Rb}(T) = \tfrac{1}{3}a_R T^4$, and τ_2 is the value of τ at $y = L$.

The radiative heat flux with black plates is

$$q_R = 2\pi B(T_1)\varepsilon_3(\tau_2 - \tau) + 2\pi \int_\tau^{\tau_2} B(t)\varepsilon_2(t - \tau)\, dt - 2\pi B(T_0)\varepsilon_3(\tau)$$

$$-2\pi \int_0^\tau B(t)\varepsilon_2(\tau - t)\, dt \tag{100}$$

For an optically-thick medium, Eq. (100) reduces to the Rosseland formula, and for the optically-thin case, we have, up to the order of τ,

$$q_R \equiv \sigma(T_1^4 - T_0^4) - \tfrac{1}{2}\sigma[T_1^4(\tau_2 - \tau) - T_0^4\tau]+\tfrac{1}{2}\sigma\left[\int_\tau^{\tau_2} T^4(t)\, dt - \int_0^\tau T^4(t)\, dt\right] \tag{101}$$

Since the thermal radiation terms can be expressed in terms of the state variables T and ϱ, we have five unknowns in our problem, the x component of velocity u, the x component of magnetic field H_x, and the pressure p, density ϱ, and temperature T of the gas. The fundamental equations for these five unknowns are expressed as follows: We express all the variables in dimensionless form: the velocity u is expressed in terms of a reference velocity U, all the lengths in terms of L, and all the state variables in terms of their corresponding values at the lower wall. Hence we have:

1. *Equation of state*:

$$p^* = \varrho^* T^* \tag{102}$$

2. *The x-wise equation of motion*:

$$\mu^* \frac{du^*}{dy^*} + (\mathrm{Re})R_H H_x^* = \left(\frac{du^*}{dy^*}\right)_0 = \mathrm{const} \tag{103}$$

where $\mathrm{Re} = LU\varrho_0/\mu_0$, $R_H = \mu_e H_0/\varrho_0 U^2$ is the magnetic pressure number, and the asterisk refers to the dimensionless quantities. The magnetic field is in terms of the external field H_0.

3. *The y-wise equation of motion*:

$$p^* + R_{p,0}p_R^* + \tfrac{1}{2}\gamma \mathrm{M}^2 R_H H_x^* = 1 + R_{p,0} \tag{104}$$

where $\mathrm{M} = u/a_0$ is the Mach number and $R_{p,0}$ is the radiation pressure number.

4. *Electric current equation*:

$$dH_x^*/du^* = \mathrm{Re}_\sigma(-u^* + R_E) \tag{105}$$

where $\mathrm{Re}_\sigma = UL/\nu_H$ is the magnetic Reynolds number and $R_E = E_0/\mu_e H_0 U$ is the electric field number.

5. *Energy equation*:

$$\mathrm{M}^2 R_E(\mathrm{Re})R_H H_x^* + \mathrm{M}^2 \mu^* \frac{d}{dy^*}\left(\frac{1}{2}u^{*2}\right) + \frac{\varkappa^*}{(\gamma-1)\mathrm{Pr}}\frac{dT^*}{dy^*}$$
$$+ R_{p,0}R_F(\mathrm{Re})q_R(T^*) = \mathrm{const} = b \tag{106}$$

where R_F is the proper radiation flux number.

In general, we have to solve Eqs. (102)–(106) simultaneously. In order to bring out the essential features, we consider the case of a fluid with constant transport properties, so that $\mu^* = 1$ and $\varkappa^* = 1$. In this way, we may solve first the velocity u^* and magnetic field H_x^* from Eqs. (103) and (105), then solve the energy equation for T^*; from the values of H_x^*

and T^*, we obtain p^* from Eq. (104), and from the values of p^* and T^*, we obtain ϱ^* from Eq. (102).

It should be noted that Eq. (103) is true only when $d(p + p_R)/dx = 0$, i.e., the case of plane Couette flow. If the pressure gradient is not zero, strictly speaking, we can not assume that the state variables are functions of y only. They must be functions of both y and x, and then for a compressible fluid, all the variables including u and H_x are functions of y and x. We can only treat this problem approximately if we assume that the velocity and H_x are functions of y only, as we will see later. Hence we now consider first the case of Couette flow.

For Couette flow with a fluid of constant properties, the velocity and magnetic field distributions are well known from ordinary magnetohydrodynamics:

$$u^* = \frac{1 + R_E(\cosh R_h - 1)}{\sinh R_h} \sinh R_h y^* - R_E(\cosh R_h y^* - 1) \qquad (107)$$

and

$$H_x^* = \frac{R_E}{R_h} \sinh R_h y^* - \frac{1 + R_E(\cosh R_h - 1)}{R_h \sinh R_h} (\cosh R_h y^* - 1) \qquad (108)$$

where $R_h = (\mathrm{Re}_\sigma R_H \mathrm{Re})^{1/2}$ is the Hartmann number. The velocity and magnetic field distributions depend on both the Hartmann number R_h and the electric field number R_E. Figures 2 and 3 show, respectively, some typical velocity and magnetic field distributions in plane Couette flow according to magnetohydrodynamics.

Substituting the values of u^* and H_x^* of Eqs. (107) and (108) into the energy equation (106), we may calculate the temperature distributions

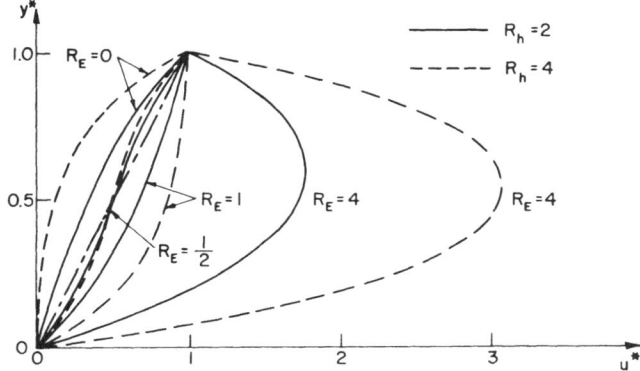

Fig. 2. Velocity distributions in plane Couette flow.

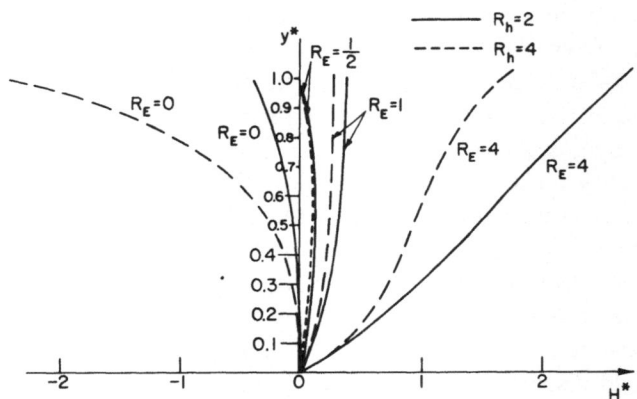

Fig. 3. Magnetic field distributions in plane Couette flow.

with a proper expression for radiative heat flux q_R. Figure 4 shows the temperature distributions for three cases: for optically-thick and for optically-thin media, and for the case without radiation effects. The most interesting result of Fig. 4 is that if we neglect completely the thermal radiation effect, the temperature will be enormously high for the case of large R_h and R_E. With radiation effects, the maximum temperature in the flow field drops a great deal. The exact temperature distribution depends on the optical properties of the gas. For an optically-thick medium, the temperature is lower than for the corresponding optically-thin medium. There is a tend-

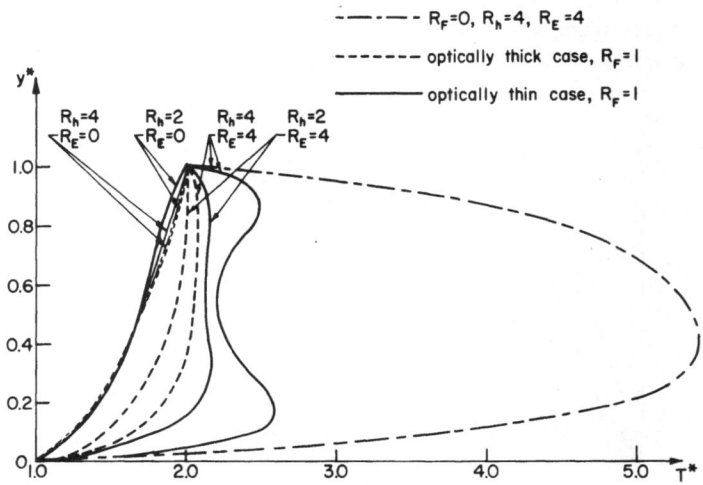

Fig. 4. Temperature distributions in the plane Couette case.

ency for the temperature to be uniform in the middle portion of the channel and to drop to the values of the plates at a rate as if there were thermal boundary layers near the plates.

For the plane Poiseuille flow, where both walls are at rest and there is a pressure gradient, all the variables should be functions of both x and y if the radiation pressure p_R is not negligible. Equation (103) should be replaced by the following equation:

$$\frac{d}{dy^*}\left(\mu^* \frac{du^*}{dy^*}\right) + \mathrm{Re}R_H \frac{dH_x^*}{dy^*} = \frac{d}{dx^*}(p^* + R_{p,0}p_R^*) \qquad (109)$$

When the pressure gradients are zero, Eq. (109) reduces to Eq. (103). In general, we do not have fully-developed flow in this problem, and all variables are functions of both x and y. Fully-developed flow may be obtained only in the case when p_R, E_R, and $\partial q_{Rx}/\partial x$ are negligibly small. In this case, $\partial T/\partial x$ is not zero because $\partial p/\partial x$ is not zero. However, we may assume that

$$\frac{\partial T}{\partial x} = \frac{T_w - T}{T_w - T_m} \frac{dT_m}{dx} \qquad (110)$$

where the mean temperature over the cross section of the channel, T_m, is defined as follows:

$$T_m = \int_0^L Tu \, dy \Big/ \int_0^L u \, dy \qquad (111)$$

We approximate the variation of temperature along the channel from the heat balance over a section of length dx as follows:

$$\varrho c_p L u_m \, dT_m = 2q_t \, dx \qquad (112)$$

where u_m is the mean velocity u across the cross section and q_t is the total heat flux at a section far away from the entrance and the exit of the channel, i.e.,

$$q_t = -\left(\varkappa \frac{\partial T}{\partial y}\right)_{y=0} - q_{Ry_{y=L}} = \left(\varkappa \frac{\partial T}{\partial y}\right)_{y=L} + q_{Ry_{y=L}} \qquad (113)$$

Substituting Eqs. (110)–(113) into the energy equation, we have

$$\frac{d}{dy}\left(\varkappa \frac{dT}{dy}\right) + \frac{dq_{Ry}}{dy} = \frac{2q_t}{L} \frac{T_w - T}{T_w - T_m} \frac{u}{u_m} - \frac{d}{dy}\left(u\mu \frac{du}{dy}\right) \qquad (114)$$

Now, Eq. (114) is a total differential equation of y only, and we have approximately the fully-developed temperature distribution $T(y)$.

9. UNSTEADY LAMINAR BOUNDARY LAYER ON AN INFINITE PLATE ([10])

One of the main problems encountered in the extremely-high-speed flight of spacecraft entering or reentering planetary atmospheres is the aerodynamic heating of the spacecraft surface. At high speeds, equal to or in excess of escape velocity, radiative heat transfer becomes a very important mode of heat transfer, and slender body shape may be preferred. Under such conditions, transpiration cooling may be desirable. In this section, we study the effects of radiative heat transfer in the transpiration cooling problem. For entry or reeentry flight, the spacecraft is decelerated, and unsteadiness of the flow field is important. In order to bring out some essential features of this complicated problem, we treat a simple model of the unsteady uniform flow $u_\infty(t)$ of a compressible and radiating gas of constant pressure p_∞ and constant temperature T_∞ over an infinitely long plate. The plate is kept at a constant temperature T_w. Fluid may be injected or sucked from the surface of the plate. In our simple model, the following assumptions are made:

1. The normal velocity on the surface of the plate is small, so that the boundary layer approximations hold.

2. The pressure in the whole flow field is constant.

3. The gas is an ideal gas, for which the ideal-gas law holds,

$$\varrho T = \varrho_\infty T_\infty \tag{115}$$

where the subscript ∞ refers to the values in the free stream.

4. The coefficient of viscosity μ of the gas is proportional to its absolute temperature T,

$$\mu/\mu_\infty = C(T/T_\infty) \tag{116}$$

where C is the well known Chapman–Rubesin constant. Equations (115) and (116) give

$$\varrho\mu = C\varrho_\infty\mu_\infty = \text{const} \tag{117}$$

5. Both the specific heat at constant temperature c_p and the Prandtl number Pr are constant.

6. The temperature T of the gas is so high that the radiative heat transfer is of the same order of magnitude as heat transfer by conduction and by convection, but the radiation pressure and radiation energy density are still negligible.

7. The gas is a gray gas, so that the reduced mass absorption coefficient k_ν' is independent of frequency, i.e., some mean value of k_ν' over the whole spectrum will be used. We further assume that the reduced mass absorption coefficient is a constant, $k_\nu' = K$. Finally, we assume that the radiation field is in local thermodynamic equilibrium.

8. The plate is a black surface, so that the emission coefficient is unity and the refelection coefficient is zero.

9. All the variables are functions of time t and the normal distance y from the plate only. This assumption is true for an infinitely long plate and is approximately true for a semiinfinite plate far away from the leading edge.

Under the above assumptions, the basic equations of our problems are:

1. *Equation of continuity*:

$$\frac{\partial \varrho}{\partial t} + \frac{\partial \varrho v}{\partial y} = 0 \tag{118}$$

2. *Equation of motion*:

$$\varrho \frac{\partial u}{\partial t} + \varrho v \frac{\partial u}{\partial y} = \varrho \frac{du_\infty}{dt} + \frac{\partial}{\partial y}\left(\mu \frac{\partial u}{\partial y}\right) \tag{119}$$

The coordinate system is fixed relative to the plate.

3. *Energy equation*:

$$\varrho c_p\left(\frac{\partial T}{\partial t} + v \frac{\partial T}{\partial y}\right) = \frac{\partial}{\partial y}\left(\varkappa \frac{\partial T}{\partial y} + q_R\right) + \mu\left(\frac{\partial u}{\partial y}\right)^2 \tag{120}$$

where q_R is the y-wise radiative heat flux given by the integral expression

$$q_R = 2\sigma\left[\int_\tau^\infty T^4(\tau^*)\varepsilon_2(\tau^* - \tau)\,d\tau^* - \int_0^\tau T^4(\tau^*)\varepsilon_2(\tau - \tau^*)\,d\tau^* - T_w^4\varepsilon_3(\tau)\right] \tag{121}$$

where the optical thickness τ is defined as follows:

$$\tau = \int_0^y \varrho K\,dy = \varrho_\infty K \int_0^y (\varrho/\varrho_\infty)\,dy = \varrho_\infty KY = Y/L_R \tag{122}$$

where the transformed normal coordinate Y is given by the formula

$$Y = \int_0^y (\varrho/\varrho_\infty)\,dy \tag{123}$$

Since the density ϱ may be expressed in terms of T by Eq. (115), we have only three unknowns u, v, and T determined by Eqs. (118)–(120) with the boundary conditions:

$$u = 0, \qquad v = v_w(t), \qquad T = T_w = \text{const} \qquad \text{for} \quad y = 0$$

$$u \rightarrow u_\infty(t), \qquad T \rightarrow T_\infty = \text{const} \qquad \text{for} \quad y \rightarrow \infty \tag{124}$$

where the subscript w refers to the value at the wall.

Equation (118) gives

$$\varrho v = \varrho_w v_w - \varrho_\infty \, \partial Y/\partial t \tag{125}$$

Now we consider a special case where a similar solution of velocity exists. We introduce the following new variables:

$$u^* = \frac{u}{u_\infty}, \qquad t^* = \frac{u_0 t}{L}, \qquad \eta = \frac{Y}{L}\left[\frac{\text{Re}_0}{C(1 - \alpha t^*)}\right]^{1/2} \tag{126}$$

where u_0 is a reference velocity, L is a characteristic length, α is a contant, and

$$\text{Re}_0 = \varrho_\infty u_0 L/\mu_\infty \tag{127}$$

is a reference Reynolds number in the free stream. For steady flow, $\alpha = 0$, and for unsteady case, we take $\alpha = -1$ without loss of generality.

The function $u_\infty(t)$ and $v_w(t)$ have the forms

$$u_\infty(t) = u_0(1 - \alpha t^*)^{-\beta} \tag{128}$$

and

$$\frac{\varrho_w v_w}{\varrho_\infty u_0}\left(\frac{R_0}{C}\right)^{1/2} = \gamma_0(1 - \alpha t^*)^{-1/2} \tag{129}$$

where β is an arbitrary constant and γ_0 is a constant dependent on the rate of fluid injection at the wall, and will be referred to as injection parameter.

For the boundary conditions (128) and (129), we have the similar solution of velocity $u^* = u^*(\eta)$ which is given by the total differential equation:

$$u^{*''} - (\tfrac{1}{2}\alpha\eta + \gamma_0)u^{*'} + \alpha\beta(1 - u^*) = 0 \tag{130}$$

where the prime refers to the differentiation with respect to η. We may integrate Eq. (130) for various values of β and γ_0. Even though the velocity

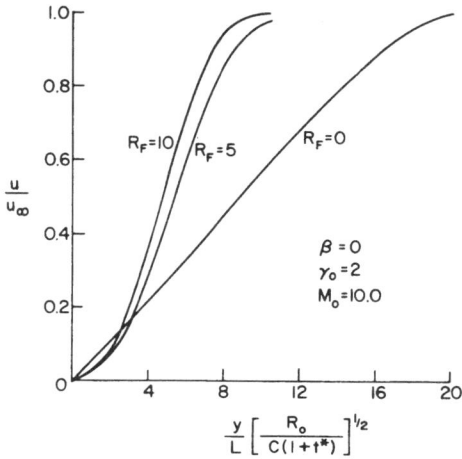

Fig. 5. Velocity distributions on an infinite
plate with injection.

distribution $u^*(\eta)$ does not depend on the temperature T explicitly, the velocity distribution in the physical plane, $u(y, t)$, depends on the temperature distribution and hence on the effect of thermal radiation. Figure 5 shows the effect of the radiative flux number on the velocity distribution in the physical plane. The velocity boundary-layer thickness decreases with increase of the radiative flux number R_F because the temperature decreases with increase of R_F, and hence the density will increase with increase of R_F. For the same amount of momentum loss, the boundary layer thickness will decrease with increase of R_F.

If we keep the temperature of the wall constant, the skin coefficient on the wall is independent of R_F, but the second derivative of velocity with respect to y varies with the thermal radiation. The velocity gradient inside the boundary layer decreases with increase of R_F, but the velocity gradient at the wall is unaffected if we keep the temperature at the wall constant. If the wall temperature is allowed to vary with the thermal radiation, the skin friction at the wall will vary with R_F. The thermal radiation tends to cause an inflection point in the boundary layer. Thus as the radiation becomes extremely large, it seems evident that an instability will develop in the gasdynamic boundary layer.

After the similar solution of velocity $u^*(\eta)$ is obtained, we may calculate the temperature distribution $T(t^*, \eta)$ from the energy equation. In general, even though the velocity is a function of η only, the temperature will be a function of both t^* and η. Only for the optically-thick medium may we have a similar solution in temperature as well as in velocity. Now we introduce the dimensionless variables

$$\zeta = c_p T/u_\infty^2, \qquad q_R^* = q_R/\varrho_\infty u_\infty^3 \tag{131}$$

the energy equation (120) in terms of t^* and η becomes

$$(1 - \alpha t^*) \frac{\partial \zeta}{\partial t^*} + (\tfrac{1}{2}\alpha\eta + \gamma_0) \frac{\partial \zeta}{\partial \eta} + 2\alpha\beta\zeta$$

$$= \frac{1}{\mathrm{Pr}} \frac{\partial^2 \zeta}{\partial \eta^2} + \left(\frac{\partial u^*}{\partial \eta}\right)^2 + \frac{u_\infty}{u_0} \left[\frac{R_0}{C}(1 - \alpha t^*)\right]^{1/2} \frac{\partial q_R^*}{\partial \eta} \tag{132}$$

In general, there will be no similar solution for temperature even if the medium is optically thick. However, if the free stream is steady, the suction is unsteady, and the medium is optically thick, we may have a similar solution and Eq. (133) becomes

$$\left(\frac{1}{\mathrm{Pr}} + R_{F1}\zeta^3\right) \frac{d^2\zeta}{d\eta^2} + \left(R_{F1} \frac{d\zeta^3}{d\eta} - \frac{1}{2}\alpha\eta - \gamma_0\right) \frac{d\zeta}{d\eta} + \left(\frac{du^*}{d\eta}\right)^2 = 0 \tag{133}$$

where

$$R_{F1} = \frac{(\gamma - 1)^3 M_0^6 16\sigma T^3 L_R}{3\mathrm{Pr} C \varkappa_\infty} \tag{134}$$

is the radiative flux number and the boundary conditions are

$$\zeta = \zeta_w \quad \text{for} \quad \eta = 0$$
$$\zeta = \zeta_\infty \quad \text{for} \quad \eta \to \infty \tag{135}$$

We have integrated Eq. (133) for various values of radiative flux number R_{F1}, Mach number M_0, and injection number γ_0. Some of the typical results are shown in Figs. 6–8. The general conclusions are as follows:

1. The peak temperature in the boundary layer due to viscous dissipation at high Mach number is suppressed if the radiation flux number is high enough.

2. This suppression of peak temperature may lead to a calculation of radiative heat flux that is too high if one neglects to consider the thermal radiation effect in solving the boundary layer equations.

3. The thermal boundary-layer thickness increases slightly, while the velocity boundary layer thickness decreases, with increase of R_F.

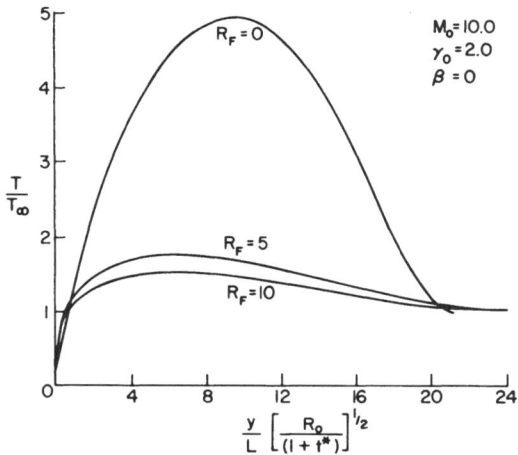

Fig. 6. Temperature distributions on an infinite plate
with injection.

4. The effectiveness of transpiration cooling decreases with increase of R_F at high Mach number.

5. The effectiveness of transpiration cooling decreases with increase of Mach number if R_F is not negligible.

6. The effectiveness of transpiration cooling is unaffected by Mach number if R_F is zero.

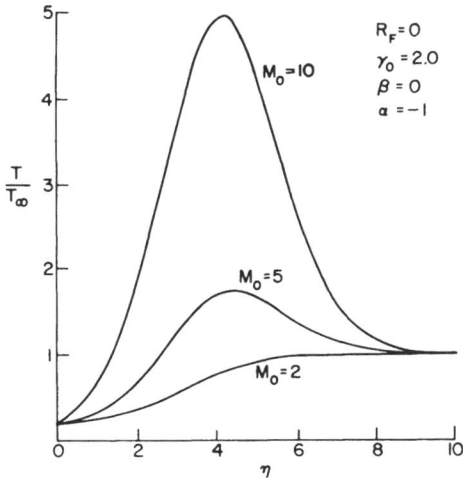

Fig. 7. Temperature distribution on an infinite
plate when radiative heat flux is negligible.

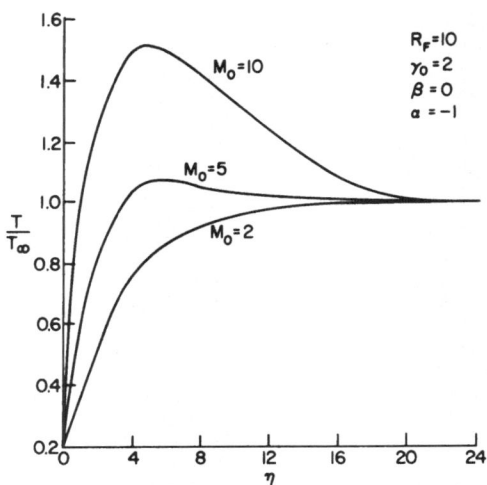

Fig. 8. Temperature distributions on an infinite
plate with radiative heat flux.

10. A UNIFORM FLOW OF A RADIATING GAS OVER A SEMIINFINITE PLATE ([11])

In the last section, we neglected the variation of variables along the flat plate. In this section, we discuss some of these x variations. We consider a steady uniform flow of velocity U and temperature T_∞ over a semiinfinite flat plate without any suction or injection at the wall. We also neglect the radiation pressure and radiation energy density, but include the radiative flux in our problem. Since the radiative heat flux may be regarded as an increase of the effective heat conductivity, the thickness of thermal boundary layer with the radiation effect included may be much larger than that of velocity boundary layer. We shall consider both the case where the thermal boundary-layer thickness is of the same order of magnitude as that of the velocity boundary layer, and the case where the thermal boundary-layer thickness is much larger than that of the velocity boundary layer. Our fundamental equations are as follows:

1. *Equation of state:*

$$p = \varrho RT \tag{136}$$

2. *Equation of continuity:*

$$\frac{\partial \varrho u}{\partial x} + \frac{\partial \varrho v}{\partial y} = 0 \tag{137}$$

3. *Equations of motion*:

$$\varrho u \frac{\partial u}{\partial x} + \varrho v \frac{\partial u}{\partial y} = \frac{\partial}{\partial y}\left(\mu \frac{\partial u}{\partial y}\right) \tag{138}$$

$$\frac{\partial p}{\partial y} = 0 \tag{139}$$

As far as the velocity field is concerned, we may use the boundary layer approximations in our problem. We also assume that the pressure is constant over the whole flow field.

4. *Energy equation*:

$$\varrho u \frac{\partial c_p T}{\partial x} + \varrho v \frac{\partial c_p T}{\partial y} = \mu\left(\frac{\partial u}{\partial y}\right)^2 + \frac{\partial}{\partial y}\left(\varkappa \frac{\partial T}{\partial y}\right) + \frac{\partial q_{Rx}}{\partial x} + \frac{\partial q_{Ry}}{\partial y} \tag{140}$$

where we apply the boundary layer approximations to the heat conduction terms but not to the thermal radiative fluxes q_{Rx} and q_{Ry}. Even though when the radiative flux number is not very large, we may use the boundary layer approximation and neglect the term $(\partial q_{Rx}/\partial x)$, when the radiative flux number is very large, we should retain the term $(\partial q_{Rx}/\partial x)$, as we shall discuss later.

First, we consider the case when the boundary layer approximation may be applied to the radiative heat flux term as well as to the velocity and heat conduction terms. Hence we may use the one-dimensional approximation for the radiative heat flux q_{Ry}. If we further assume that the gas is a gray gas and that the local thermodynamic equilibrium condition is applicable, we have

$$\frac{\partial q_{Ry}}{\partial y} = 2\sigma\varrho K_p\left[\int_0^\tau T^4 \varepsilon_1(\tau - t')\, dt' + \int_\tau^\infty T^4 \varepsilon_1(t' - \tau)\, dt - 2T^4 + T_w^4 \varepsilon_2(\tau)\right] \tag{141}$$

where K_p is the Planck mean absorption coefficient and τ is the optical thickness defined by

$$\tau = \int_0^y \varrho K_p\, dy \tag{142}$$

The mass absorption coefficient ϱK_p is a function of the pressure p and the temperature T of the gas. In the following calculations, we use the formula

$$\varrho K_p = A p^a \exp(bT - cT^2) \tag{143}$$

where A, a, b, and c are constants depending on the composition of the gas.

The boundary conditions of our problem are

$$u = v = 0, \qquad T = T_w, \qquad \varrho = \varrho_w \qquad \text{for} \quad x > 0, y = 0$$
$$u \to U, \qquad T \to T_\infty, \qquad \varrho \to \varrho_\infty \qquad \text{for} \quad x > 0, y \to \infty \tag{144}$$

where T_w is the temperature of the plate, which is a constant.

Since the thermal radiation has a larger influence on temperature distribution than on velocity distribution, we consider the similar solution of the velocity profile by using Eq. (116) for the coefficient of viscosity, and the variable Y of Eq. (123). We have the following similar solutions for the velocity profile:

$$u/U = df(\xi)/d\xi = f'(\xi) \tag{145}$$

and

$$w = -(\partial \psi/\partial x)_Y = \tfrac{1}{2}(C\mu_\infty U/x)^{1/2}(f' - f) \tag{146}$$

where

$$\xi = (Y/x)(\varrho_\infty Ux/\mu_\infty)^{1/2} \tag{147}$$

and ψ is the stream function such that

$$\partial \psi/\partial y = (\varrho/\varrho_\infty)u; \qquad \partial \psi/\partial x = -(\varrho/\varrho_\infty)v \tag{148}$$

and $f(\xi)$ is the well-known Blasius function of incompressible boundary layer flow, and has been tabulated in standard texts [e.g., [12]].

In terms of Y and x, the energy equation (140) without the x-wise radiative heat flux becomes

$$c_p u \frac{\partial T}{\partial x} + c_p \omega \frac{\partial T}{\partial Y} = \frac{C\mu_\infty}{\varrho_\infty} \left(\frac{\partial u}{\partial Y} \right)^2 + \frac{C\mu_\infty c_p}{\mathrm{Pr}\varrho_\infty} \frac{\partial^2 T}{\partial Y^2} + [T_w{}^2 \varepsilon_2(\tau) - 2T^4]2\sigma\varrho K_p$$
$$+ 2\sigma\varrho K_p \left[\int_0^\tau T^4 \varepsilon_1(\tau - t') \, dt' + \int_\tau^\infty T^4 \varepsilon_1(t' - \tau) \, dt' \right] \tag{149}$$

with the boundary conditions $(x > 0)$

$$T = T_w \qquad \text{for} \quad Y = 0$$
$$T = T_\infty \qquad \text{for} \quad Y \to \infty \tag{150}$$

Even though we have similar solutions for velocity $u(\xi)$, there will be no similarity solution for T when the absorption coefficient is finite or large. Only when the absorption coefficient is very small and the Rosseland

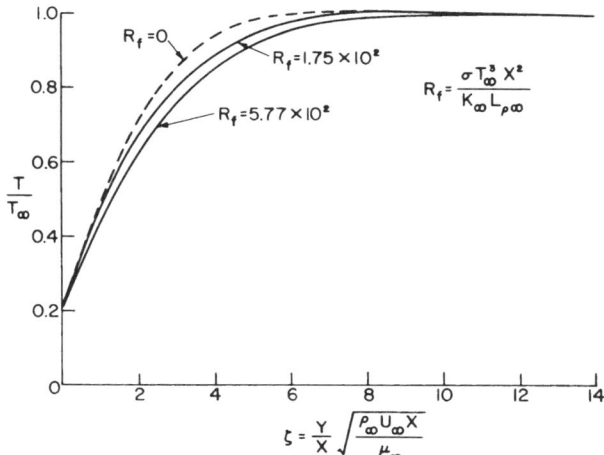

Fig. 9. Temperature distribution, $T_\infty = 10{,}000°K$, $P_\infty = 5.0$ atm, $M_\infty = 1.25$.

approximation (28) may be used, may we have similar solutions in temperature $T(\xi)$. In Eq. (149), we assume that c_p and Pr are constants.

We have numerically integrated Eq. (149) with the boundary conditions (150) using

$$\varrho K_p = 4.5 \times 10^{-7} p^{1.31} \exp(5.18 \times 10^{-4} T - 7.13 \times 10^{-7} T^2) \quad \text{cm}^{-1} \quad (151)$$

where p is atmospheres and T is in $°K$ with $Pr = 0.74$, $T_w = 2000\,°K$, and T at several temperatures. The results are shown in Figs. 9–12.

Figure 9 is the case for low Mach number $M_\infty = 1.25$. Since there are no similar solutions, the temperature profile changes as distance x increases. We use the following definition of the radiative flux number:

$$R_f = \frac{\sigma T^3 x^2}{\varkappa_\infty L_{p,\infty}} = \frac{\sigma T^3 x^2 \varrho_\infty K_{p,\infty}}{\varkappa_\infty} \quad (152)$$

When $R_f = 0$, i.e., without radiative heat flux, we have similar solutions in this case, which are shown by the dotted curves in Figs. 9–12. Figure 10 shows the corresponding case for high Mach number, $M_\infty = 15$.

In the numerical calculation of Figs. 9 and 10, we used the approximate expression

$$\varepsilon_2(z) = me^{-nz} \quad (153)$$

for the exact expression of the exponential integral and take $m = n^2/3 = 1.562$.

$R_f = 0$
$R_f = 2.5 \times 10^3$
1.21×10^4
2.25×10^4
$R_f = \frac{\sigma T_\infty^3 X^2}{K_\infty L_{p\infty}}$
$\frac{T}{T_\infty}$
$\xi = \frac{Y}{X}\sqrt{\frac{\rho_\infty U_\infty X}{\mu_\infty}}$

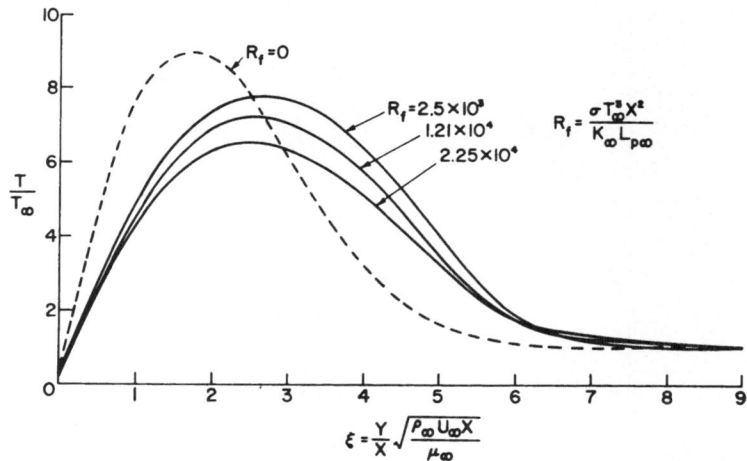

Fig. 10. Temperature distribution, $T_\infty = 20{,}000°K$, $P_\infty = 1.0$ atm, $M_\infty = 15.0$.

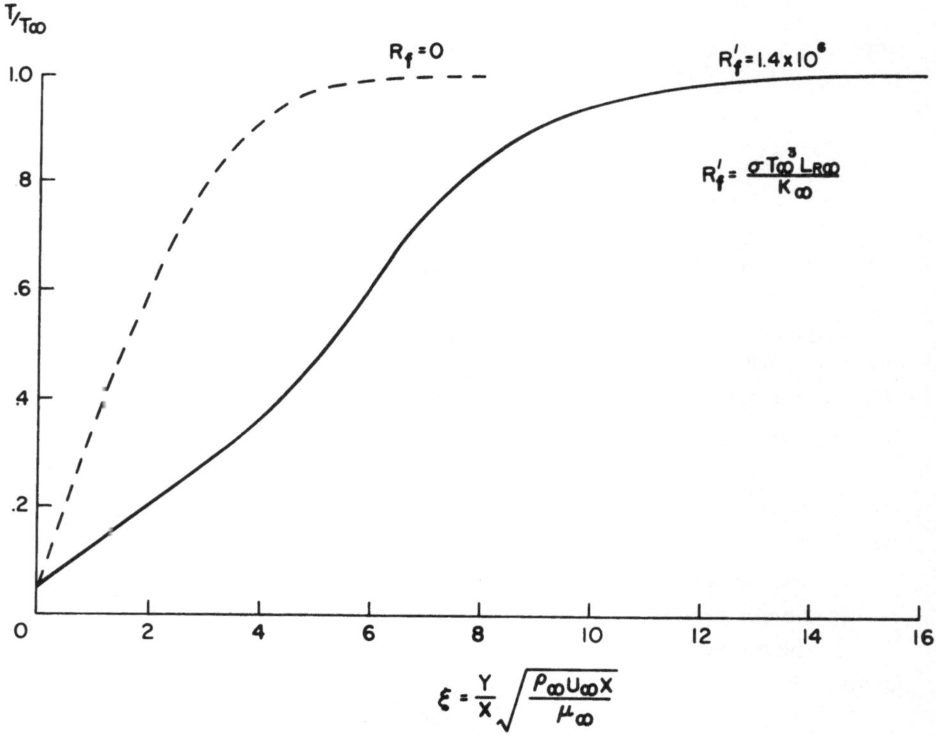

T/T_∞
$R_f = 0$
$R_f' = 1.4 \times 10^6$
$R_f' = \frac{\sigma T_\infty^3 L_{R\infty}}{K_\infty}$
$\xi = \frac{Y}{X}\sqrt{\frac{\rho_\infty U_\infty X}{\mu_\infty}}$

Fig. 11. Temperature distribution for optically-thick medium, $T_\infty = 40{,}000°K$, $P_\infty = 5.0$ atm, $M_\infty = 0.65$.

The main effects of thermal radiation are (1) to decrease the maximum temperature in the boundary layer at high Mach numbers, (2) to increase the boundary layer thickness, and (3) to decrease the slope $[d(T/T_\infty)/d\xi]_w$ at the wall.

If is interesting to compare the results of the integral expression for radiative heat flux with the approximate expression for optically-thick and optically-thin media. Figure 11 shows the case for an optically-thick gas. It was found that the general trend of the results for the optically-thick-medium approximation is the same as that for the integral expression, but quantitatively, the optically-thick-medium approximation overestimates the effect of thermal radiation.

Figure 12 shows the results with the optically-thin approximation, in which the effect of absorption is completely neglected and the radiative heat transfer term is given by

$$\partial q_{Ry}/\partial y = 4\sigma\varrho K_p(-T^4 + \tfrac{1}{2}T_w^4) \tag{154}$$

Since the wall temperature T_w is always much smaller than the local temperature T except in the neighborhood of the wall, Eq. (154) shows that

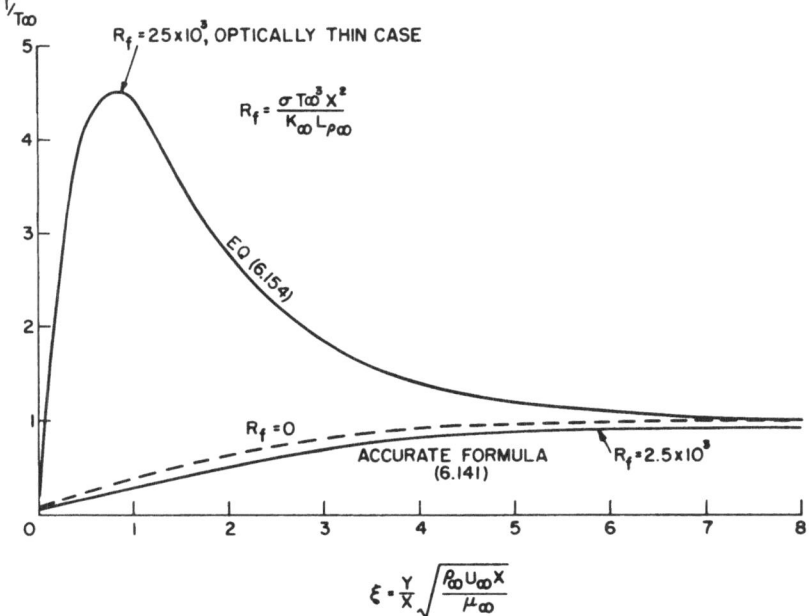

Fig. 12. Temperature distribution for optically-thin gas, $T_\infty = 20,000°K$, $P_\infty = 1.0$ atm, $M_\infty = 0.9$.

the radiative heat transfer acts as a heat source in the boundary layer. In our numerical results in Fig. 12, the optically-thin-medium approximation gives results which are entirely wrong in comparison with the results of the integral expression. Even though the optically-thin approximation is a popular one in much of the current literature, one should be very careful when using it at high temperatures.

When the temperature of the flow field is very high, the thickness of the thermal boundary layer will be so large that the x-wise radiative transfer term will not be negligible. When we include the x-wise radiative heat transfer in the energy equation, there will be an upstream influence. For instance, when the wall temperature is much lower than that of the free stream, we have a upstream temperature wake. At a given x-station upstream, the temperature is lowest at $y = 0$ and increases with y. The defect of temperature along the x axis increases from zero at infinity to the value $(T - T_w)$ at the $x = 0$.

11. STAGNATION-POINT HEAT TRANSFER IN RADIATION GAS DYNAMICS

Another interesting radiative heat transfer problem is the flow near a stagnation point of a blunt body in a hypersonic flow. The flow field in this problem may be divided into three more or less distinguishable parts as shown in Fig. 13. Region I is the shock transition region, which is very thin if the Reynolds number is large. Region II is the inviscid shock layer, in which the viscous force and heat conduction may be neglected and in which the radiative heat flux should be considered if the radiation mean

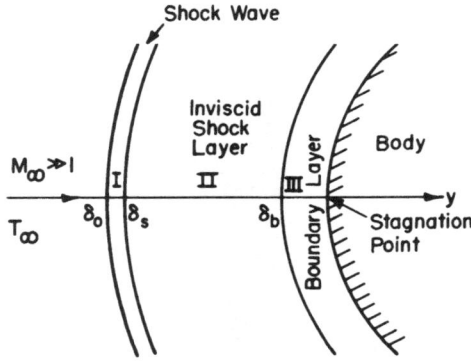

Fig. 13. Flow regions near a stagnation point of a blunt body in a hypersonic stream.

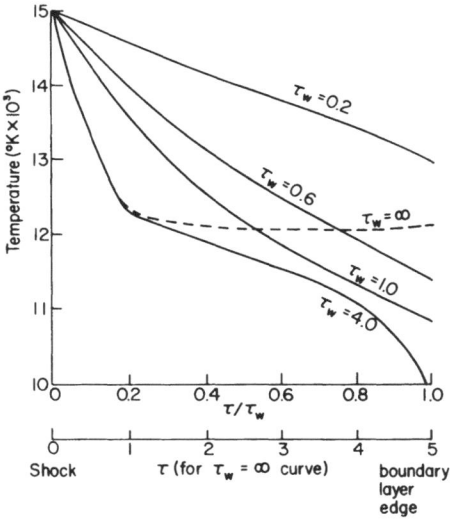

Fig. 14. Temperature distributions in inviscid shock layer near a stagnation point. From Yo-shikawa and Chapman ([13]), courtesy of NASA.

free path is much larger than the mean free path of the gas. Region III is the boundary layer flow near the stagnation point. When the Reynolds number is small, or the Knudsen number is large, these three regions may merge into one. We shall discuss only the case when Reynolds number is large, so that these three regions can be treated separately as follows:

We assume that the temperature in the free stream is not high enough that the thermal radiation effects in the free stream can be completely neglected. Hence we may use the ordinary Rankine–Hugoniot relation to determine the shock strength and the flow conditions immediately behind the shock.

Second, we may determine the flow in the inviscid shock region II from the initial conditions immediately behind the shock. In region II, we assume that the temperature is so high that the radiative heat flux should be considered. Since we are interested in the region near the axis of a body of revolution and near a stagnation point, the one-dimensional approxima-tion of radiative heat flux may be used as a first approximation. The most important coordinate is the distance along the axis of symmetry. For a first approximation, we may even neglect the curvature effect. Some typical temperature distributions in the inviscid shock layer region ([13]) are shown in Fig. 14 where $p_2 = 10$ atm and $T_2 = 15,000\,°K$, which is the temperature immediately behind the shock. For a given T_2, the temperature of the gas at the outer boundary of the boundary layer, i.e., $\tau = \tau_w$, depends on the optical thickness of the gas, i.e., the value of τ_w. This gas temperature at $\tau = \tau_w$ is the temperature of the gas at the outer edge of the boundary layer.

Finally, we may solve the stagnation-point boundary-layer flow with the free-stream temperature determined in the solution of the inviscid shock layer. The thermal radiation effect is to increase the thickness of the thermal boundary layer and decrease the temperature in the boundary layer.

ACKNOWLEDGMENT

This chapter was based on work supported in part by the US Air Force through the Air Force Office of Scientific Research under grant No. AFOSR-10: 5-67A.

NOTATION

$a = (\gamma RT)^{1/2}$, sound speed

$a_R = 7.67 \times 10^{-15}$ erg cm^{-3} °K^{-4}, Stefan–Boltzmann constant

B   magnetic induction

B $(\sigma/\pi)T^4$, integrated Planck function, defined in Eq. (46)

B_ν Planck radiation function, defined in Eq. (20)

C Chapman–Rubesin constant, defined in Eq. (116)

C_R sound speed when radiation effect is included, defined in Eq. (81) or Eq. (93)

$c = 3 \times 10^8$ m/sec, velocity of light in free space

c_p specific heat at constant pressure

c_v specific heat at constant volume

D_R Rosseland diffusion coefficient of radiation, Eq. (29)

E   electric field strength

E radiation energy

E_R radiation energy density, defined in Eq. (10)

E_ν radiation energy at frequency ν to $\nu + d\nu$

$\bar{e}_m$ total energy of the gas, defined in Eq. (38)

e_ν emissivity coefficient of a surface

F   body force

$f(\xi)$ Blasius function of a boundary layer flow

g   gravitational acceleration

H   magnetic field strength

$h = 6.62 \times 10^{-27}$ erg-sec, Planck's constant

I integrated intensity of thermal radiation, defined in Eq. (6)

I_ν specific intensity of radiation, defined in Eq. (4)

i, j, k   unit vectors in x, y, and z directions, respectively

$i = \sqrt{-1}$

J   electric current density

J_ν source function of radiation, defined in Eq. (19)

j_ν emission coefficient of radiation

K^* effective coefficient of heat conductivity, defined in Eq. (76)

K_p Planck mean absorption coefficient, defined in Eq. (32)

K_R Rosseland mean absorption coefficient, defined in Eq. (30)

Kn   Knudsen number, defined in Eq. (65)

Kn_r Knudsen number of radiation, defined in Eq. (62)

$k = 1.379 \times 10^{-16}$ erg/°K, Boltzmann constant

k_ν absorption coefficient of radiation

k_ν' reduced absorption coefficient, defined in Eq. (25)

L_f mean free path of a gas, defined in Eq. (66)

L_p Planck mean free path of radiation

L_R Rosseland mean free path of radiation

$L_{R\nu}$ mean free path of radiation, defined in Eq. (15)

l, m, n direction cosines of a ray with respect to x, y, and z axes, respectively

M   Mach number

m mass

n index of refraction

n^i ith component of direction cosine

Pe   Peclet number, defined in Eq. (64)

Pr   Prandtl number, defined in Eq. (60)

p pressure

p_R radiation pressure, defined in Eq. (12)

Q heat source

Q_R divergence of radiative heat flux, defined in Eq. (8)

q   velocity vector

$\mathbf{q}_R$ radiative heat flux

$\mathbf{q}_V$ convective heat flux

R gas constant

R_E	electric field number, defined in Eq. (105)
R_F	radiative flux number, defined in Eq. (67)
$R_F{'}$	radiative flux number, defined in Eq. (70)
R_f	radiative flux number, defined in Eq. (152)
R_H	magnetic pressure number, defined in Eq. (103)
R_h	Hartmann number, defined in Eq. (108)
R_p	radiation pressure number, defined in Eq. (63)
R_r	relativistic parameter, defined in Eq. (61)
R_t	time parameter, defined in Eq. (57)
Re	Reynolds number, defined in Eq. (59)
r_ν	reflection coefficient of a surface
s	distance along a radiation ray
T	temperature
t	time
tr_ν	transparency coefficient of a surface
U	typical velocity
U_m	internal energy of a gas per unit mass
U_λ	radiation energy density in wavelength interval λ to $\lambda + d\lambda$
U_ν	radiation energy density in frequency interval ν to $\nu + d\nu$
u, v	x and y components of velocity vector, respectively
u^i	ith component of a velocity vector
w	modified velocity component defined in Eq. (146)
x, y, z	Cartesian coordinates
x^i	ith component of Cartesian coordinates
Y	modified y coordinate, defined in Eq. (123)

Greek Letters

α, β	constants
γ	ratio of specific heats
γ_e	effective value of γ, defined in Eq. (90)
γ_0	injection parameter, defined in Eq. (129)
$\delta^{ij} = 0$ if $i \neq j$; $= 1$ if $i = j$	
ε_n	exponential integral, defined in Eq. (45)
ζ	dimensionless enthalpy, defined in Eq. (131)
η	similarity coordinate, defined in Eq. (126)
θ	angle
$\varkappa$	coefficient of heat conductivity

$\varkappa_R$	effective coefficient of heat conductivity, defined in Eq. (43)
λ	wavelength
$\lambda = \lambda_R + i\lambda_i$, complex wave number	
μ	coefficient of viscosity
ν	frequency
ν_g	coefficient of kinematic viscosity
ξ	similarity coordinate, defined in Eq. (147)
ϱ	density of a gas
ϱ_e	excess electric charge
$\sigma = 5.75 \times 10^{-5}$ erg cm^{-2} sec^{-1} °K^4 Stefan–Boltzmann constant for radiative flux	
σ_0	elementary area
τ_R	radiation stress tensor with ij component τ_R^{ij} defined in Eq. (11)
τ_s	viscous stress tensor
τ_ν or τ	optical thickness, defined in Eq. (14) or (41)
ϕ	potential energy
ψ	stream function, defined in Eq. (148)
ω	solid angle
ω	frequency of a wave in Section 6.

Subscripts

i	ith component of a vector
0	some reference value
∞	free-stream value
*	dimensionless value

REFERENCES

1. S. I. Pai, *Radiation Gasdynamics*, Springer-Verlag, Vienna and New York (1966).
2. S. Rosseland, *Theoretical Astrophysics*, (Oxford University Press, London) (1936).
3. M. Planck, *The Theory of Heat Radiation*, Dover Publications, New York (1959).
4. S. I. Pai, *Magnetogasdynamics and Plasma Dynamics*, Springer-Verlag, Vienna and New York (1962).
5. S. I. Pai and A. I. Speth, "The Wave Motions of Small Amplitude in Radiation Electromagnetogasdynamics," in: *Proc. 6th Midwest Conf. on Fluid Mech.*, Univ. of Texas Press (1959), p. 446.
6. V. A. Prokof'ev, "Propagation of Forced Plane Compression Waves of Small Amplitude in a Viscous Gas when Radiation is Taken into Account," *Am. Rocket Soc. J.* **31**, 988 (1961).

7. S. I. Pai and A. I. Speth, "Shock Waves in Radiation Magnetogasdynamics," *Phys. Fluids* **4**, 1232 (1961).

8. S. C. Traugott, "Shock Structure in a Radiating, Heat-Conducting and Viscous Gas," Martin Co. Research Report RR 57 (1964).

9. S. I. Pai, "Plane Couette Flow in Radiation Magnetogasdynamics," in: *Proc. 6th Intern. Symp. on Ionization Phen. of Gases*, S.E.R.M.A., Paris (1963), p. 431., vol. IV.

10. S. I. Pai and A. P. Scaglione, "Unsteady Laminar Boundary Layers on an Infinite Plate in Radiation Gasdynamics," Report SID 66-353, North American Aviation Inc. Space & Inf. Div. (1966).

11. S. I. Pai and C. K. Tsao, "A Uniform Flow of a Radiating Gas over a Flat Plate," in: *Proc. 3rd Intern. Conf. of Heat Transfer*, ASME, Chicago, American Institute of Chemical Engineers, New York (1966), p. 129.

12. S. I. Pai, *Viscous Flow Theory. I. Laminar Flow*, D. Van Nostrand Co., Princeton, New Jersey (1956).

13. K. K. Yoshikawa and D. R. Chapman, "Radiative Heat Transfer and Absorption behind a Hypersonic Normal Shock Wave," NASA TN D-1424 (1962).

14. S. M. Scala and D. H. Sampson, "Heat Transfer in Hypersonic Flow with Radiation and Chemical Reaction," in: *Supersonic Flow, Chemical Processes and Radiative Transfer*, Pergamon Press, New York (1964), p. 319.

Chapter 8

SOME PROBLEMS OF RADIATIVE TRANSFER

F. K. Moore

Professor and Chairman
Department of Thermal Engineering
Cornell University

INTRODUCTION *

This chapter is intended to describe certain current areas of research in the field of radiative transfer that seem significant and timely. Another view of the subject and appraisal of current research is provided in Chapter 7 by Pai.

A perennial problem in the field of radiative transfer is to assess the validity of the gray-gas approximation, whereby the medium absorbs and emits radiation with, it is assumed, an absorption coefficient independent of frequency. Although this approximation is physically disreputable, it has been, and no doubt will continue to be, extremely useful for estimating the role of radiative transfer in a complex problem. For this reason, the last year has seen a continuing effort to establish ranges of practical validity of the gray-gas approximation, and, where this proves impossible, to propose simple spectral models for absorption coefficients which retain at least some of the computational advantages of the gray gas. Certain theoretical studies of mean absorption coefficients will be described, and brief mention will be made of the engineering problem of calculating radiant heat transfer from rocket exhaust gases.

The so-called differential approximation to the equation of radiative transfer replaces a difficult integral equation with a simple and appealing second-order partial differential equation for the connection between radiative heat flux and emissive power of the medium. Like the gray-gas approximation, the differential approximation is severely limited in its

---

* Symbols used in this chapter are defined on pp. 328–329.

theoretical justification, and the last year has seen, on the one hand, greater use of this quick calculation method, and, on the other, the discovery of an increasing number of circumstances in which it fails rather badly. We will see that quite simple methods of improvement may assure the future respectability of this technique for three-dimensional and nongray problems.

Following discussion of the foregoing questions of methodology in radiative transfer, a specific problem of coupled radiative transfer and gas dynamics will be discussed, namely, the "thermal choking" of a channel flow by radiative transfer from upstream. In effect, this is a Rayleigh process with heat addition by radiation. The emphasis will be on an initially subsonic flow that is already very hot so that the coupling between radiation and gas dynamics is strong. The specific application then will be made to the generation of shock waves by this method. This problem corresponds to a certain interest in the generation of shock waves by radiation and to the acceleration of propulsive flows by the same method. The problem will make use of the gray gas and differential approximation ideas described below.

1. ABSORPTION COEFFICIENTS FOR RADIATIVE TRANSFER CALCULATION

1.1. Gray-Gas Approximations

The basic equation of radiative transfer is

$$-\partial I_\nu/\partial s = \varrho \varkappa_\nu (I_\nu - B_\nu) \tag{1}$$

which specifies the loss of spectral intensity I_ν along paths in the direction of the radiation. Due to absorption which is proportional to the intensity of the radiation and to a mass absorption coefficient $\varkappa_\nu$, the gain of intensity in the case of "local thermodynamic equilibrium" is, by Kirchhoff's law, the result of blackbody emission with a mass emission coefficient also equal to $\varkappa_\nu$. Then B_ν is the Planck function

$$B_\nu \equiv \frac{2h\nu^3}{c^2} \frac{1}{e^{h\nu/kT} - 1} \tag{2}$$

For a plane-parallel atmosphere or slab problem, it is appropriate to replace the variable s by a coordinate normal to the boundaries. Then if μ is the cosine of the angle between the intensity vector and the normal coordinate, Eq. (1) becomes

$$-\mu \, \partial I_\nu/\partial x = \varrho\varkappa_\nu(I_\nu - B_\nu) \tag{3}$$

The intensity I_ν is in units of power per unit area per steradian.

Our problem concerns the quantity $\varkappa_\nu$. It is clear that if $\varkappa_\nu$ were independent of frequency, then Eq. (3) could be integrated over all frequencies, so that the total intensity (integrated over the entire spectrum) upon which the radiative heat flux depends could be calculated directly in terms of the integrated Planck function $(\sigma T^4/\pi)$,

$$-\mu \, \partial I/\partial x = \varrho\bar{\varkappa}(I - B); \quad I \equiv \int_0^\infty I_\nu \, d\nu; \quad B \equiv \sigma T^4/\pi \tag{4}$$

If $\varkappa_\nu$ is not a constant, then this integration cannot be performed directly. In fact, $\varkappa_\nu$ is a widely-varying function of frequency and temperature for most substances. For gases at rather low temperatures, the absorption coefficient would have a discrete band structure. At higher temperatures, where dissociation and recombination processes dominate, the absorption would be continuum, but still very dependent on frequency. Figure 1 (based on information supplied the author by Dr. Treanor of CAL) shows the band and continuum absorption coefficients for a nearly transparent sample of oxygen at two different temperatures.

It is clear that there is no way in principle of saying that $\varkappa_\nu$ is constant. There remains, however, the possibility of justifying the use of a suitable average of $\varkappa_\nu$ over frequency. For high-temperature gases, perhaps the most useful of these is the "Planck mean":

$$\bar{\varkappa}_P \equiv \int_0^\infty \varkappa_\nu B_\nu \, d\nu \left/ \int_0^\infty B_\nu \, d\nu \right. \tag{5}$$

This definition can be used in Eq. (3), provided the intensity term on the right is neglected; i.e.,

$$-\mu \, \frac{\partial I}{\partial s} = -\varrho \, \frac{\displaystyle\int_0^\infty \varkappa_\nu B_\nu \, d\nu}{B} \, B \tag{6}$$

Now I_ν can be neglected relative to B_ν for an "emission-dominated" problem, such as the case of a transparent sample of gas which is radiating strongly as a black body. The typical high-energy shock-tube experiment furnishes an example. This "Planck mean" can be calculated once and for all from physical properties of the material, however rapidly they may vary with frequency.

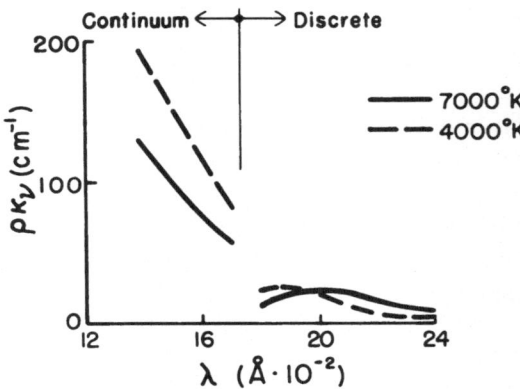

Fig. 1. Spectral absorption coefficient for oxygen.

If Eq. (3) is first divided by x_ν before integration over ν, then the "Rosseland mean"

$$(\bar{x}_R)^{-1} \equiv \frac{\int_0^\infty (x_\nu)^{-1}(dB_\nu/dT)\, d\nu}{\int_0^\infty (aB_\nu/dT)\, d\nu} \tag{7}$$

results, on the additional assumption that I is very nearly equal to B, i.e., the radiative situation is nearly opaque. In addition, one takes the variable x to be a unique function of temperature.

In a recent report, Traugott [1] has carefully considered the various possibilities for defining a satisfactory effective mean absorption coefficient. He considered various moments of the transfer equation, i.e., directional averages which would be chosen according to the quantity of interest. For example, heat flux is $2\pi \int_0^1 I\mu\, d\mu$. He distinquishes between a truly gray gas, in which x_ν is taken constant, a semigray gas, in which two coefficients are chosen, and a "quasigray" gas, in which all nongray effects are absorbed into the relation between optical depth and physical distance.

If Eq. (3) is integrated with no assumption about I_ν, the result may be written

$$-\mu\, \partial I/\partial x = \varrho(\bar{x}I - \bar{x}_P B) \tag{8}$$

where the "mean absorption coefficient" $\bar{x}$ is dependent on the unknown intensity,

$$\bar{x} \equiv \int_0^\infty x_\nu I_\nu\, d\nu \Big/ \int_0^\infty I_\nu\, d\nu \tag{9}$$

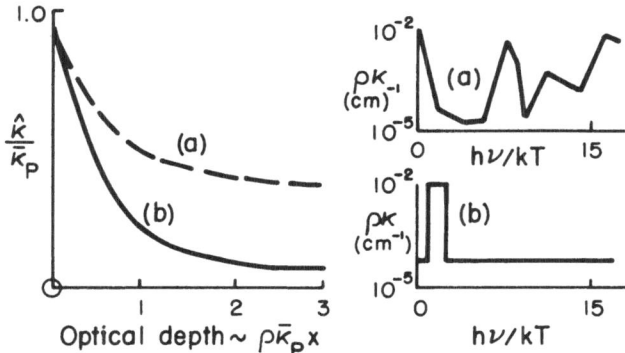

Fig. 2. Quasigray absorption coefficients $\varkappa$ for two nongray
models. From Traugott ([1]).

This approach leads to the semigray model, provided some means of estimating $\bar{\varkappa}$ is found. A quasigray model would involve a single mean absorption coefficient

$$(\hat{\varkappa})^{-1} \equiv \frac{\int_0^\infty (\varkappa_\nu)^{-1}(\partial I_\nu/\partial x)\, d\nu}{\int_0^\infty (\partial I_\nu/\partial x)\, d\nu} \tag{10}$$

of which the Rosseland mean [Eq. (7)] is a special case.

Obviously, both $\bar{\varkappa}$ and $\hat{\varkappa}$ depend on the unknown intensity I_ν, and, perhaps less seriously, on the unknown temperature upon which the intensity I_ν depends. Therefore Traugott considered as a test problem the heat conduction across a finite slab which is at constant temperature. One might imagine this to be possible if some mode of heat exchange other than radiation also operated in such a way as to keep the temperature constant. He then calculated how these various coefficients vary with the optical depth of the slab, assuming a variety of spectral functions $\varkappa_\nu$. Two such examples are shown in Fig. 2, which shows the result for a combined absorption coefficient $\hat{\varkappa}$. Figure 3 shows his result for the effective coefficient $\bar{\varkappa}$, which is seen to approach the Planck mean at the two extremes of opacity.

If a temperature is allowed to vary linearly through the slab, a different comparison is developed. Figure 4, based on Traugott's example, shows how doubtful it is that any gray coefficient could be chosen to give the dependence of heat flux on optical depth for a spectral function with a "window".

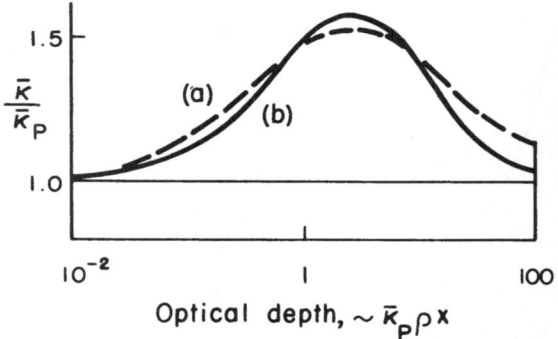

Fig. 3. Mean absorption coefficients $\varkappa$. From Traugott ([1]).

Traugott's conclusion from this study was that, of the proposals for mean absorption coefficient that he had examined in a logical manner, "none were able to survive comparison with nongray calculations." A similar study to Traugott's was provided by Patch ([2]).

1.2. Piecewise Gray Models

Especially for high-temperature gases, for which continuum radiation is very important, as in problems of stellar structure or reentry stagnation-point heating, it is often proposed to assume the absorption coefficient to be constant at two different levels in two ranges of frequency. I believe this idea was first used in a calculation for an astrophysical problem by Carrier and Avrett ([3]). Liu and Clarke ([4]) have renewed this proposal in the context of the differential approximation, of which more will be said later. A recent application of this idea to the reentry stagnation-point heating problem

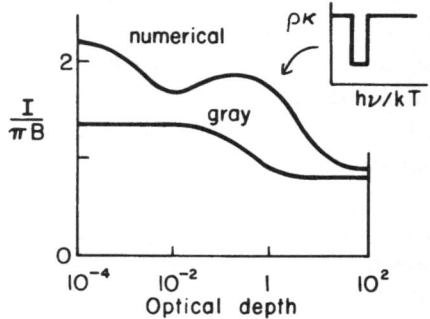

Fig. 4. Heat flux for slab; variable temperature. From Traugott ([1]).

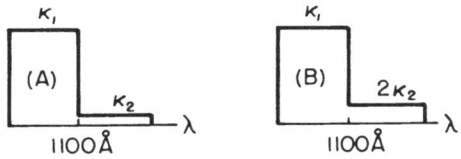

Fig. 5. Two nongray models for air. From
Anderson ([5]).

has been given by Anderson ([5]). We will briefly summarize Anderson's
results.

For continuum radiation only, a model of high-temperature air was
chosen as in Fig. 5, where $\varkappa_1$ represents vacuum ultraviolet absorption and
$\varkappa_2$ is determined from a survey of air data. Anderson carried out heat
transfer calculations based on this model, using the flow model of Howe
and Viegas ([6]). Figure 6 shows some Anderson's results for radiative heat
transfer plotted as a function of nose radius. Here, he distinguishes in the
solution between the radiative heat flux at wavelengths less than, and that
at wavelengths greater than, the wavelength which divides the range for his
model. Evidently, doubling the long-wavelength absorption coefficient
would make substantial changes for larger nose radii (because for small
nose radii, the gas cap is practically transparent to the longer wavelengths).
Comparisons provided in ([5]) with the exact calculations of Hoshizaki and
Wilson ([7]) suggest that the step model would be quite satisfactory for the
reentry problem.

A few comments are perhaps in order about this general line of en-
deavor. One should perhaps not take too seriously the problem of gray or
semigray or step-wise gray approximations for radiative transfer at a time
when there are an increasing number of calculations using high-speed
computing equipment, in which the spectral absorption coefficient is used

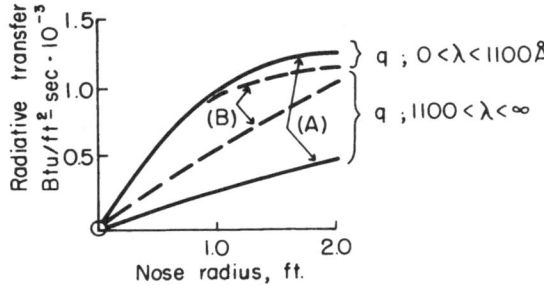

Fig. 6. Reentry radiative heat flux for two nongray
models. From Anderson ([5]).

"exactly," i.e., information about the spectrum of $\varkappa_\nu$ can be simply plugged into the program. The work reported in ([7]) furnishes an example of this approach. The merit of the model calculations, of course, must be in their utility for exploring various effects or making design choices where one may not want to incur the expense of a large machine calculation or may perhaps be in ignorance about the spectral distribution of $\varkappa_\nu$. This situation reminds me of the "Lighthill ideal-dissociating-gas" model, which no one really needs, in view of the information available about the physical chemistry of high-temperature gases, but which one finds very useful for exploratory work. In view of the foregoing argument, it would seem important to keep such models simple—two steps, or perhaps "windows," are probably the most one should tolerate in terms of complexity.

1.3. Models for Band Radiation

There is considerable current interest in finding the proper description of band radiation coefficients for heat transfer from gases at temperatures that are high enough to provide strong radiative transfer, but not so high that gasdynamic coupling is important. An example is the heating loads imposed by radiation from rocket exhaust plumes on the base structure of rockets ([10]). A recent NASA conference on this subject was held at Marshall Space Flight Center on October 5–6, 1967. Here, one is especially interested in gases that are far out of radiative equilibrium, and one is chiefly concerned with the estimation of changes of intensity through a highly nonuniform gas field. An important question is whether, along nonuniform paths, an average spectral distribution of emission coefficient may be used. This is referred to as the Curtis–Godson approximation, discussed at length in Goody's book ([8]). Apparently, this approximation is adequate and may be used with confidence for most problems. Particular band models used in conjunction with such calculations are the Elsasser and random Elsasser models also discussed in ([8]).

2. THE DIFFERENTIAL APPROXIMATION

2.1. For a Nongray Gas

We turn now to a brief introduction of the "differential approximation" for radiative transfer, showing first how a two-step model doubles the order of the appropriate equation [see ([4])]. The differential approximation for radiative transfer is very commonly used, and may be found in standard textbooks such as those by Goody ([8]) or by Vincenti and Kruger ([9]). For a gray gas in "local thermodynamic equilibrium," the one-dimensional equation of transfer may be written as follows:

$$\lambda^2 q_{xx} - 4\lambda\pi B_x - 3q = 0 \tag{11}$$

where a certain arbitrariness attaches to the numerical coefficients of the terms. Here, q is the one-dimensional heat flux and λ is the reciprocal absorption coefficient or photon-free path. This equation can be derived in a number of ways. One way, by inspection, is to notice that this is the simplest differential equation that will give the Rosseland heat conduction formula if the highest-order derivative is discarded, and the Newtonian or transparent formula if the lowest-order term is discarded. Now, suppose that instead of a fixed value of λ, we choose a step model involving two parameters λ_1 and λ_2 corresponding, e.g., to Fig. 5 without any long-wavelength cut off. Then Eq. (11) would be replaced by the pair of equations

$$\lambda_1^2 q_{1_{xx}} - 4\lambda_1\pi B_{1_x} - 3q_1 = 0 \tag{12}$$

$$\lambda_2^2 q_{2_{xx}} - 4\lambda_2\pi B_{2_x} - 3q_2 = 0 \tag{13}$$

applicable in the two frequency ranges. Rather than these equations, we would prefer to have an equation for the total flux $q = q_1 + q_2$. The result is

$$\lambda_1^2\lambda_2^2 q_{xxxx} + (\lambda_1^2 + \lambda_2^2)q_{xx} + 9q$$
$$+ 12\pi[\lambda_1 B_{1_x} + \lambda_2 B_{2_x} + \tfrac{4}{3}\lambda_1\lambda_2(\lambda_2 B_{1_{xxx}} + \lambda_1 B_{2_{xxx}})] = 0 \tag{14}$$

This equation is of fourth order, and is thus more complicated than the equation based on a gray-gas approximation. The higher order of the equation reflects the fact that transparent and opaque classifications are relative to two absorption coefficients, so that there are, so to speak, four choices instead of two. Liu and Clarke ([4]) consider the three-dimensional version of this development.

Thus we see that the simple differential equation (11) cannot be expected to succeed when nongray effects are important. To give this a physical illustration, consider the slab problem solved exactly by Heaslet and Warming for a gray gas ([11]). Their result for the temperature distribution in a slab for an optical depth of 1 is shown in Fig. 7, and shows a slightly S-shaped curve which has temperature slip at both walls and a constant heat flux across the entire slab. The differential approximation gives a very good approximation to this result—the straight line profile shown in the

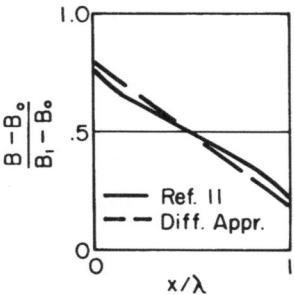

Fig. 7. Distribution of temperature in slab (unit optical thickness) with radiative heat transfer.

figure. Figure 8 shows the case of uniform heat addition within the slab when the wall temperatures are held equal. Here, a temperature profile may again be found by both methods, and again agreement is rather good. When the differential approximation is used, a parabolic temperature profile results.

Now consider the problem illustrated in Fig. 7 in an extremely non-gray form. We imagine that the walls are at the same temperature, but an intense source of monochromatic radiation shines across the slab from the left, as shown in Fig. 9. If we suppose that the absorption coefficient for that specific frequency is very much smaller than that for the thermal radiation in a gas, the situation is clearly one to which the solution shown in Fig. 7 is not applicable even though a constant flux is maintained across the slab. In fact, under the circumstances described, the imposed radiation would provide on a transparent basis just the uniform heat addition contemplated in the solution shown in Fig. 8. Thus a parabolic temperature

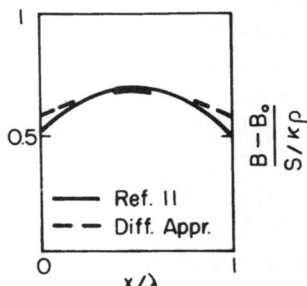

Fig. 8. Distribution of temperature in slab (unit optical thickness) with volume heat addition S.

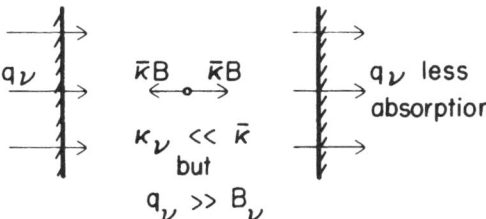

Fig. 9. Sketch of radiative heat transfer problem
with monochromatic source.

distribution would be anticipated. Notice that the walls of the slab in this
case must be regarded as transparent to the incident radiation but opaque
to thermal radiation at the temperature of the gas.

The foregoing is an extreme example, showing how the straightforward
application of the differential approximation cannot cope with a problem
in which nongray aspects are important. The fact that we were actually
able to give the extreme solution in terms of the differential approximation of
Eq. (8) is just lucky. We were able in that case to split the two frequency
ranges completely and apply the differential approximation to each. Thus
we have, in effect, done a nongray calculation.

2.2. Difficulties near Surfaces

We see from Fig. 7 that even if the gas is exactly gray, the differential
approximation is at least somewhat inaccurate near the surface, failing to
produce the slight S-shape that the Heaslet and Warming calculations
provide, leading to a somewhat larger error in heat transfer rate. Olfe [12]
has studied this matter, and has proposed a modification of the differential
approximation that seems to correct the difficulty. Independently, Landram
and Greif [13] made a similar proposal.

It can be shown that the source of the error lies in improper account
being taken of external radiation, e.g., from the surfaces of the slab. Therefore Olfe divided the intensity or energy flux into two parts:

$$q \equiv q_{\text{ext}} + q_g \tag{15}$$

where Q_{ext} is due to the walls, but is attenuated by absorption within the gas.
Treating this part exactly,

$$q_{\text{ext}} = 2\sigma T_1{}^4 E_3\left(\frac{x}{\lambda}\right) - 2\sigma T_2{}^4 E_3\left(\frac{L}{\lambda} - \frac{x}{\lambda}\right) \tag{16}$$

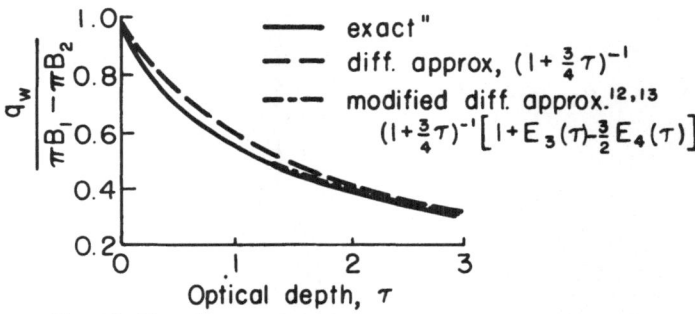

Fig. 10. Heat flux across slab. From Landram and Greif ([13]).

where E_3 is the third-order exponential integral and L is the width of the slab. This function in itself is not a solution of the differential equation (11); however, the heat flux contributed by emission from the gas is made subject to Eq. (11). The boundary condition which determines the solution is that the total heat flux is constant. Applying this condition at the two walls, and taking into account the flux toward the two walls due to q_g, completes the solution. Figure 10 shows Olfe's results for dimensionless heat flux as a function of optical depth of the slab. It is clear that the desired improvement has been accomplished, at least insofar as heat transfer is concerned. However, it appears that the temperature profile, by Olfe's method, would remain a straight line. Presumably, the next step in this development would be to recalculate the temperature distribution within the layer, beginning with corrected values of heat flux.

2.3. The Three-Dimensional Difficulty

The three-dimensional version of the differential approximation may be derived by various methods ([14,15]):

$$\lambda^2 \nabla (\nabla \cdot \mathbf{q}) - 4\pi\lambda \, \nabla B - 3\mathbf{q} = 0 \qquad (17)$$

This is a vector equation, and again by inspection, we see that the proper three-dimensional opaque and transparent limits are special cases. Again one sees that this equation has deficiencies in the vicinity of boundaries; not only the difficulties described by Olfe, but peculiarly three-dimensional problems occur as well. Notice that taking the curl of Eq. (17) yields

$$\nabla \times \mathbf{q} = 0 \qquad (18)$$

i.e., any solution of Eq. (17) must have curl-free heat flux. This is a most disappointing feature of the equation, because it precludes consideration

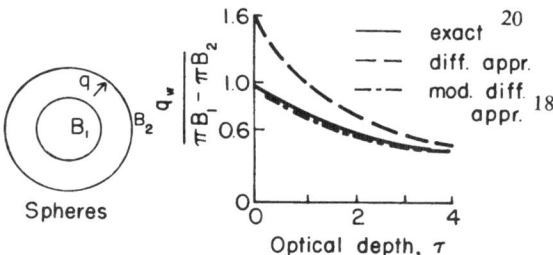

Fig. 11. Heat flux across spherical annulus; $r_2/r_1 = \frac{1}{2}$.
From Olfe ([18]).

of incident radiation fields having edges. For example, a searchlight beam, before being dispersed, would have a "shear" of heat flux; curl **q**, in fact, would be an infinite curl at the beam boundaries. It seems likely that this difficulty can systematically be avoided by application of Olfe's ideas, whereby we would simply consider the incident searchlight beam separately from Eq. (17).

However, attention at the moment appears focused on the shortcomings of Eq. (17) for solving the spherical or cylindrical slab problem—analogs to the problems described in Figs. 7 and 8. That is, one treats the problem shown in Fig. 11, where Eq. (18) is indeed satisfied, by symmetry. Heaslet and Warming ([16],[17]) have used various methods to predict radiative transfer in a homogeneous cylindrical or spherical medium. In those problems, heat is generated within the spherical surface and there is no inner wall. Perlmutter and Howell ([19]) have numerically calculated the heat transfer within a cylinder. The differential approximation appears to be quite successful for homogeneous heat addition, but not for the concentric sphere problem. Viskanta and Crosbie ([20]) have analyzed this spherical problem by solving the appropriate integral equation by a method of successive approximation. The result for heat flux is shown in Fig. 11 as a function of optical depth of the annulus. Heat flux is compared with the value it would have in a completely-transparent situation. Olfe ([18]) has shown that the differential approximation fails, except in the optically-thick limit, but that a differential approximation following the scheme of his earlier paper ([12]) will repair the situation very satisfactorily.

In summary, it appears that the differential approximation will continue to be useful for a wide variety of problems, since Olfe's and Landram and Greif's work shows that a special treatment of the boundary conditions can correct the most serious deficiencies of the method. Again, one must view this question in the light of our ability to solve these problems exactly by high speed computing methods: serious complications of the differential approximation ought not to be tolerated. Either the method should be simple to apply, or else it should be discarded. Olfe's corrected method would seem to be within reasonable limits of complication from this point of view.

3. A PROBLEM OF THERMAL CHOKING BY RADIATION*

Interest in problems of gas dynamics coupled with radiative energy transfer continues to be high. The study of shock waves as affected by radiative transfer has been carried out in great detail by Zeldovich [21], by Raizer [22], by Heaslet and Baldwin [23], and others. One is also interested in heat exchange by radiative means for propulsion in such proposed devices as the gas-core nuclear rocket [see the review by Cooper [24]]. A related line of inquiry is the generation of shock waves by neutron absorption [25], a matter akin to radiative transfer. One would like to understand the gas-dynamic aspects of such energy addition. For reasons that we have already discussed, the one-dimensional radiative transfer problem is by far the simplest. Thus let us consider the "Rayleigh" flow, which is a one-dimensional channel flow subject to energy addition, in our case by radiative transfer.

We suppose, as in Fig. 12, that a gas enters the channel at $x = 0$ with temperature T_0 and a subsonic velocity u_0. Then, as this gas flows downstream, it receives radiant energy from a black surface located at $x = 0$, at a temperature T_w considerably higher than T_0. We may for a moment think of the plane $x = 0$ as a hot, porous plug from which a cooler gas emerges. Then, as the gas heats up under the influence of the hot surface, its Mach number increases toward 1, and, finally, Mach 1 is achieved and thermal choking is accomplished. The temperature during this process must go through a maximum as dictated by the rules of Rayleigh flow [26]. Simply from gas dynamics, a graph of temperature vs Mach number shows a maximum at a Mach number of $1/\sqrt{\gamma}$ (Fig. 13). Thus it is necessary for the flow to negotiate this temperature maximum before reaching the sonic point.

Under the differential approximation of Eq. (11), the following differential equation results:

$$\left(\frac{\lambda^3}{3}\frac{\partial^2}{\partial x^2} - 1\right)\left[\left(\frac{u}{u_0} - \frac{u^*}{u_0}\right)^2\right] - K_0\frac{\lambda}{\sqrt{3}}\left[\left(\frac{u}{u_0}\right)^4\left(\frac{\gamma+1}{\gamma}\frac{u^*}{u_0} - \frac{u}{u_0}\right)^4\right]_x = 0$$

$$(19)$$

* The study summarized here was performed under NASA sponsorship.

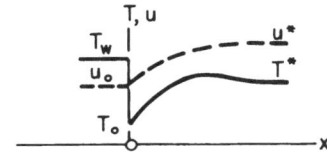

Fig. 12. Sketch of radiative "Rayleigh flow."

where u^* is the velocity when Mach number reaches 1, i.e., infinitely far downstream. The term in the first parentheses is, in effect, the heat flux which must vanish at the sonic point by the laws of gas dynamics. The square bracket is the gasdynamic expression for temperature. Thus we see that Eq. (11) is embodied in Eq. (19). Here K_0 is a parameter which compares the radiant flux associated with the incoming gas temperature to its kinetic energy flux:

$$K_0 = 4 \frac{\gamma - 1}{\gamma + 1} \frac{(\sigma M_0{}^2 T_0)^4}{\varrho_0 u_0{}^3} \tag{20}$$

Equation (19) is closely related to the equation solved by Heaslet and Baldwin ([23]). In fact, the equation

$$(y + K_0 F) y'(\theta) - \theta = 0 \tag{21}$$

to which Eq. (19) may be reduced appears in ([23]). This equation is Abel's equation of the second kind, and is obtained under the definitions

$$\theta \equiv \left(\frac{u^*}{u_0} - \frac{u}{u_0} \right)^2 \tag{22a}$$

$$F(\theta) \equiv \left(\frac{u}{u_0} \right)^4 \left(\frac{\gamma + 1}{\gamma} \frac{u^*}{u_0} - \frac{u}{u_0} \right)^4 \tag{22b}$$

$$y(\theta) \equiv \frac{\lambda}{\sqrt{3}} \frac{d\theta}{dx} - K_0 F \tag{22c}$$

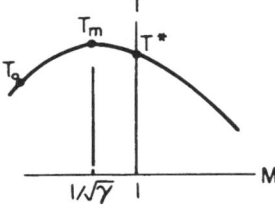

Fig. 13. "Rayleigh curve" of temperature vs Mach number.

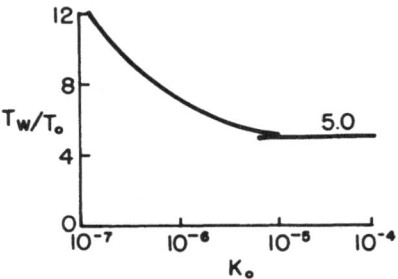

Fig. 14. Temperature to produce choking,
$M_0 = 0.2$.

The boundary condition for Eq. (21) is

$$y(0) = -K_0 F(0) \qquad (23)$$

i.e., the solution goes smoothly to Mach 1, and therefore heat flux θ must vanish.

The boundary value problem may be solved by isoclines. One presentation of the results of such a calculation appears in Fig. 14, which displays the temperature of the hot surface necessary to produce choking. Obviously, if the gas is initially very cool, a relatively high wall temperature would be required to produce the necessary heat flux. On the other hand, as the parameter K_0 increases, the wall temperature ratio reaches a finite value equal to the maximum temperature on the Rayleigh curve, denoted by T_m. Figure 14 may, in fact, be quite accurately reproduced by assuming the heat flux gasdynamically necessary to produce choking is a balance between blackbody flux from the wall at T_w and an equivalent black, i.e., optically-thick, region downstream at temperature T_m.

Figure 15, which shows how velocity increases downstream for various values of K_0, makes clear why the downstream region may be regarded as opaque. We see, in fact, that as K_0 increases, the part of the process prior to achieving the maximum temperature on the Rayleigh curve tends to be more and more sudden (i.e., transparent). At the same time, the process downstream from the top of the Rayleigh curve becomes optically very thick. We note also that for purposes of this discussion, large K_0 is not necessarily large numerically. A value of 10^{-5} would seem to be large enough to display the behavior described.

We see in this result the tendency of a Rayleigh process governed by radiation to sort itself out into a transparent part and an opaque part when radiative flux is intense. The reason this happens is easily understood.

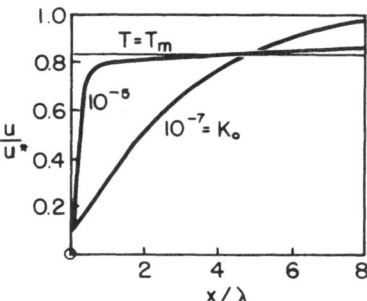

Fig. 15. Velocity for $M_0 = 2$.

One wishes to convey a finite amount of heat in this problem, fixed by the gas dynamics. As the radiative flux becomes very large, one finds that the only way to accommodate the limited heat with a very large temperature difference is either to have the absorbing layer very thin (transparent) or else to have the layer extremely thick, so that the Rosseland conduction of heat is small despite the large temperature difference. One finds, in fact, that the Rosseland region scales with K_0, whereas the transparent region scales inversely with K_0.

In view of the foregoing considerations, it seems sensible, and, in fact, proves to be easy, to solve the governing differential equation (21) in a transparent limit and again in the Rosseland limit. One finds for θ (the heat flux, or, alternatively, the velocity) the solutions shown in Fig. 16. It turns out that these solutions do not merge smoothly at maximum temperature. On the transparent scale, the velocity gradient is zero, but on the Rosseland scale, the velocity gradient is infinite. These are consistent as a first approximation when K_0 is large, but it is interesting to obtain by matched expansions the proper solution crossing through the maximum temperature. This proves to be feasible and quite simple analytically, with the

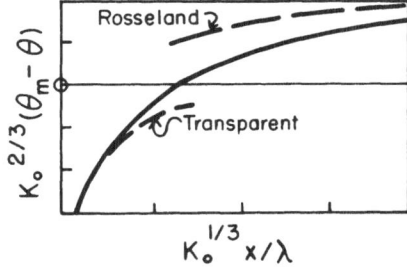

Fig. 16. Velocity near the temperature maximum.

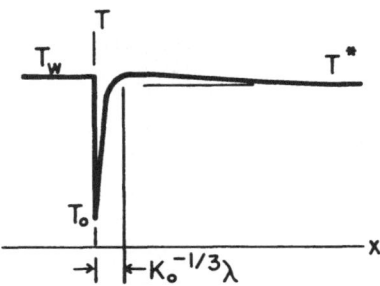

Fig. 17. Temperature for large K_0.

result appearing in Fig. 16, which shows that for K_0, the scaling is such that an essentially transparent process carries the gas through the maximum temperature point. The point is located at a distance inversely proportional to $K_0^{1/3}$. The temperature history of the flow is somewhat as shown in Fig. 17 for large K_0: The temperature suddenly rises by transparent absorption until the maximum temperature is reached at a level effectively equal to the wall temperature. Now, however, the gas is faced with the problem of continuing to gain energy so as to approach the sonic point, but with a falling temperature. This is possible only in a Rosseland or opaque process, and this is the way the flow proceeds.

Finally, let us return to the physical description of the problem, and ask if any physical circumstances can be visualized in which the flow described could actually occur. It seems clear that this solution would relate directly to the addition of heat behind a strong shock wave by radiation coming from upstream due to some external source of radiant energy (perhaps another shock wave). In this interpretation, the shock front is the origin of coordinates ($x = 0$). The gas to the left of the shock wave is assumed transparent to the approaching radiation. We see, too, that if the shock is strong enough (i.e., K_0 is large), we could quickly calculate the shock strength produced in the steady state by strong incoming radiation under the limiting condition of sonic flow far downstream. Perhaps a problem such as this would be important in certain astrophysical applications, where shock waves are known to occur under the influence of intense radiative energy flux.

NOTATION

B integrated Planck function, Eq. (4)

B_ν Planck function

c speed of light

$E_n(\xi) \equiv \int_1^\infty e^{-\xi z} z^{-n} \, dz$, exponential integral

F function of θ, Eq. (22b)

h Planck's constant

I integrated intensity

I_ν spectral intensity

K_0 a measure of gas temperature, Eq. (20)

k Boltzmann constant

M Mach number

q heat flux vector

s distance in I direction

T temperature

x distance normal to some surface

y function of θ, Eq. (22c)

Greek Letters

γ ratio of specific heats

θ related to velocity, Eq. (22a)

$\bar{\varkappa}$ effective absorption coefficient, Eq. (9)

$\hat{\varkappa}$ combined absorption coefficient, Eq. (10)

$\varkappa_P$ Planck mean absorption coefficient, Eq. (5)

$\varkappa_R$ Rosseland mean absorption coefficient, Eq. (6)

$\varkappa_\nu$ spectral mass absorption coefficient

λ wavelength of radiation

μ cosine of angle between s and x

ν frequency of radiation

ϱ density

σ Stefan–Boltzmann constant

Superscript

* evaluation at the sonic point ($M = 1$)

Subscripts

0 evaluation at origin

1 evaluation at surface 1

2 evaluation at surface 2

w temperature of surface

m evaluation at maximum of Rayleigh curve

REFERENCES

1. S. C. Traugott, "On Gray Absorption Coefficients in Radiative Transfer," RIAS Technical Report 67-9c, Martin Marietta Corp. (August 1967).
2. R. W. Patch, "Approximation for Radiant Energy Transport in Nongray, Nonscattering Gases," NASA TN-D-4001 (June 1967).
3. G. F. Carrier and E. H. Avrett, *Astrophys. J.* **134**, 469 (1961).
4. J. T. C. Liu and J. H. Clarke, "Differential Formulations for Radiative Transfer in Nongray Gases," *Phys. Fluids* **10**, 2088 (1967).
5. J. D. Anderson, Jr., "A Simplified Analysis for Re-Entry Stagnation Point Heat Transfer from a Viscous Nongray Radiating Shock Layer," AIAA Paper No. 68–164 (1968).
6. J. T. Howe and J. R. Viegas, "Solution of the Ionized Radiating Shock Layer, Including Re-Absorption and Foreign Species Effects and Stagnation Region Heat Transfer," NASA TR-159 (1963).
7. H. Hoshizaki and K. H. Wilson, "Convective and Radiative Heat Transfer during Superorbital Entry," *AIAA J.* **5**, 25 (1967).
8. R. M. Goody, *Atmospheric Radiation. I. Theoretical Basis*, Oxford University Press, London (1964).
9. W. G. Vincenti and C. H. Kruger, Jr., *Introduction to Physical Gas Dynamics*, J. Wiley and Sons, New York (1965).
10. S. J. Morizumi and H. J. Carpenter, "Thermal Radiation from the Exhaust Plume of an Aluminized Composite Propellant Rocket," *J. Spacecraft Rockets* **1**, 501 (1964).
11. M. A. Heaslet and R. F. Warming, "Radiative Transport and Wall Temperature Slip in an Absorbing Planar Medium," *Int. J. Heat Mass Transfer* **8**, 979 (1965).
12. D. B. Olfe, "A Modification of the Differential approximation for Radiative Transfer," *AIAA J.* **5**, 638 (1967).
13. C. S. Landram and R. Greif, "Semi-Isotropic Model for Radiation Heat Transfer," *AIAA J.* **5**, 1971 (1967).
14. P. Cheng, "Two-Dimensional Radiating Gas Flow by a Moment Method," *AIAA J.* **2**, 1662 (1964).
15. S. C. Traugott and K. C. Wang, "On Differential Methods for Radiant Heat Transfer," *Int. J. Heat Mass Transfer* **7**, 269 (1964).
16. M. A. Heaslet and R. F. Warming, "Theoretical Predictions of Radiative Transfer in a Homogeneous Cylindrical Medium," *J. Quant. Spectr. Radiative Transfer* **6**, 751 (1966).
17. M. A. Heaslet and R. F. Warming, "Application of Invariance Principles to a Radiative Transfer Problem in a Homogeneous Spherical Medium," *J. Quant. Spectr. Radiative Transfer* **5**, 669 (1965).
18. D. B. Olfe, "Application of a Modified Differential Approximation to Radiative Transfer in a Gray Medium between Concentric Spheres and Cylinders," to appear in *J. Quant. Spectr. Radiative Transfer*.
19. M. Perlmutter and J. R. Howell, *J. Heat Transfer* **C86**, 169 (1964).
20. R. Viskanta and A. L. Crosbie, "Radiative Transfer through a Spherical Shell of an Absorbing–Emitting Gray Medium," *J. Quant. Spectr. Radiative Transfer* **7**, 871 (1967).
21. Ya. B. Zeldovich, "Shock Waves of Large Amplitude in Air," *Soviet Phys.—JETP* **5**, 919 (1957).

22. Yu. P. Raizer, "On the Structure of the Front of Strong Shock Waves in Gases," *Soviet Phys.—JETP* **5**, 1242 (1957).

23. M. A. Heaslet and B. S. Baldwin, "Predictions of the Structure of Radiation-Resisted Shock Waves," *Phys. Fluids* **6**, 781 (1963).

24. R. S. Cooper, "Prospects for Advanced High-Thrust Nuclear Propulsion," *Astronaut. Aeronaut.* **4**, 60 (1966).

25. H. P. Smith, Jr., C. W. Busch, and A. K. Oppenheim, "Pressure Wave Generated in a Fissionable Gas by Neutron Irradiation," *Phys. Fluids* **7**, 676 (1964).

26. A. H. Shapiro, *The Dynamics and Thermodynamics of Compressible Fluid Flow*, Vol. I, Ronald Press, New York (1953).

Chapter 9

PLASMA DYNAMICS

S. I. Pai

Research Professor
Institute for Fluid Dynamics and Applied Mathematics
University of Maryland

1. PLASMAS AND PLASMA DYNAMICS [1]*

At very high temperatures, above 10,000°K, a gas will be ionized. The properties of an ionized gas, or *plasma*, differ considerably from those of a neutral gas. Hence we may consider the plasma as a fourth state of matter. The main difference between a plasma and a neutral gas is that electromagnetic forces play important roles in the dynamics of the plasma. Aside from this, the plasma behaves in a manner very similar to a gas in many flow problems. In plasma dynamics, we need to study simultaneously the electromagnetic fields and the gasdynamic field. Many new phenomena occur due to the interaction of the gasdynamic and electromagnetic forces.

The science which deals with the flow problems of a plasma is called "Plasma dynamics." The scope of plasma dynamics is very broad. It contains problems ranging from electrical discharge in a rarefied gas, to propagation of electromagnetic waves in ionized media, to problems in so-called "Magnetofluid dynamics." In this chapter, we consider only those problems of plasma dynamics in which the plasma may be considered as a continuum, and the electromagnetic forces are of the same order of magnitude as gasdynamic forces and their interaction is important, particularly for cases in magnetofluid dynamics.

The plasma may be considered as a mixture of N species which consist of ions, electrons, and neutral particles. In the flow field, ionization and recombination of ions and electrons may occur. Hence in plasma dynamics, we also have the effects of chemical reactions if we consider the ionization process as a chemical reaction. However, since the mass of an electron is

---

* Symbols used in this chapter are defined on pp. 375–377.

much smaller than those of ions or neutral particles, the diffusion velocity of electrons is very large, and the treatment using the diffusion coefficient approximation will not be satisfactory. A better treatment, known as the multifluid theory of plasma dynamics, will be discussed in Section 15. However, for a first approximation, we may consider the plasma as a single fluid. We use the single fluid concept in the major portion of this chapter.

In Section 2, we discuss the fundamental equations of the dynamics of an ionized gas, in which both the gasdynamic equations including the interaction of electromagnetic variables, and the electromagnetic field equations are given. In Section 3, we discuss further the electromagnetic equations and the boundary conditions of electromagnetic fields. In general, the electric field and the magnetic field are of equal importance. However, in many problems, the magnetic field is more important than the electric field. By means of the magnetogasdynamic approximations, we may simplify the general equations discussed in Section 2 to the equations of magnetofluid dynamics given in Section 4. On the other hand, in many other problems, the electric field is more important than the magnetic field. By means of electrogasdynamic approximations, we may simplify the general equations of Section 2 to the equations of electrogasdynamics given in Section 5. The most important dimensionless parameters are discussed in Section 6. In Sections 7–14, we discuss various flow problems of magnetofluid dynamics and electrogasdynamics, including one-dimensional flows, channel flows, waves and shocks, boundary layer flow, and turbulent flows. Two improvements in the ordinary theory of magnetofluid dynamics, the tensor electrical conductivity and multifluid theory, are discussed in Sections 13 and 15, respectively.

2. FUNDAMENTAL EQUATIONS OF THE DYNAMICS OF AN ELECTRICALLY-CONDUCTING FLUID [1]

We consider the plasma as a single fluid of definite composition. To describe the motion of such an electrically-conducting fluid, we have to know the six gasdynamic variables and ten electromagnetic variables as follows:

1. The velocity vector of the plasma $\mathbf{q}$ which has three components in general, u_i, $i = 1$, 2, or 3.

2. The temperature of the plasma T.

3. The pressure of the plasma p.

4. The density of the plasma ϱ.

5. The electric field strength **E** which has three components $\mathbf{E}_i$.

6. The magnetic field strength **H** which has three components H_i.

7. The excess electric charge ϱ_e.

8. The electrical current density **J**, which has three components J_i.

The sixteen relations which govern these variables are the fundamental equations of plasma dynamics, and are as follows:

a. Equation of State of a Plasma. A Plasma may be considered as an ideal gas, and to satisfy the ideal-gas law:

$$p = \varrho RT \tag{1}$$

where R is the gas constant of the plasma.

b. Equation of Continuity. The conservation of mass of the plasma gives

$$\frac{\partial \varrho}{\partial t} + \nabla \cdot (\varrho \mathbf{q}) = 0 \tag{2}$$

where $\nabla = \mathbf{i}(\partial/\partial x) + \mathbf{j}(\partial/\partial y) + \mathbf{k}(\partial/\partial z)$ is the gradient operator and **i**, **j**, and **k** are, respectively, the unit vectors in the x, y, and z directions. Equation (2) is the same as that in ordinary gas dynamics.

c. Equations of Motion. The conservation of momentum of a plasma in vector form is

$$\varrho \, D\mathbf{q}/Dt = -\nabla p + \nabla \cdot \tau + \mathbf{F}_e \tag{3}$$

The viscous stress tensor τ has as its ij component

$$\tau_{ij} = \mu\left(\frac{\partial u_i}{\partial x_j} + \frac{\partial u_j}{\partial x_i}\right) - \frac{2}{3}\mu\frac{\partial u_k}{\partial x_k}\delta_{ij} \tag{4}$$

where $\delta_{ij} = 0$ if $i \neq j$ and $\delta_{ij} = 1$ if $i = j$, and μ is the coefficient of viscosity.

The electromagnetic force $\mathbf{F}_e$ is

$$\mathbf{F}_e = \varrho_e\mathbf{E} + \mathbf{J} \times \mathbf{B} \tag{5}$$

where $\mathbf{B} = \mu_e\mathbf{H}$ is the magnetic induction and μ_e is the magnetic permeability. In free space, we have, in the MKS unit system,

$$\mu_e = 4\pi \times 10^{-7} \quad \text{kg-m/C}^2 \tag{6}$$

We neglect the gravitational force in Eq. (3).

d. Energy Equation. The conservation of energy gives

$$\varrho \frac{DU_m}{Dt} = -p(\nabla \cdot \mathbf{q}) + \Phi + \nabla \cdot (\varkappa \nabla T) + (I^2/\sigma) \tag{7}$$

where $\mathbf{I}$ is the electrical conduction current, $\mathbf{I} = \mathbf{J} - \varrho_e\mathbf{q}$, σ is the electrical conductivity of the plasma, $\varkappa$ is the coefficient of heat conductivity, and Φ is the viscous dissipation:

$$\Phi = \tau_{ij} \, \partial u_i/\partial x_j \tag{8}$$

where the summation convention is used. In (7), U_m is the internal energy of the plasma.

e. Maxwell's Equations of the Electromagnetic Fields. We have

$$\nabla \times \mathbf{H} = \mathbf{J} + \partial \mathbf{D}/\partial t \tag{9}$$

$$\nabla \times \mathbf{E} = -\partial \mathbf{B}/\partial t \tag{10}$$

where $\mathbf{D} = \varepsilon\mathbf{E}$ is the dielectric displacement and ε is the inductive capacity, which has the following value in free space in the MKS system:

$$\varepsilon = 8.854 \times 10^{-12} \quad \text{C}^2\text{-sec}^2\text{-kg}^{-1}\text{-m}^{-3} \tag{11}$$

f. Equation of Electrical Current Density J. The electrical current density $\mathbf{J}$ consists of two parts: one is the electrical conduction current $\mathbf{I}$ and the other is electrical convection current $\varrho_e\mathbf{q}$, i.e.,

$$\mathbf{J} = \mathbf{I} + \varrho_e\mathbf{q} \tag{12}$$

The electrical conduction current is due to the diffusion phenomena of charged particles. For a first approximation, we may use the generalized Ohm's law for the equation of electrical conduction current:

$$\mathbf{I} = \sigma(\mathbf{E} + \mathbf{q} \times \mathbf{B}) = \sigma\mathbf{E}_u \tag{13}$$

where $\mathbf{E}_u$ is the electric field in moving coordinates. In Eq. (13), we assume that the electrical conductivity σ is a scalar quantity. However, if the magnetic field strength is large and the density of the plasma is low, we should consider the electrical conductivity as a tensor quantity; this will be discussed in Section 13.

g. Equation of Conservation of Electric Charge

$$\frac{\partial \varrho_e}{\partial t} + \nabla \cdot \mathbf{J} = 0 \tag{14}$$

The fundamental equations of plasma dynamics are Eqs. (1)–(3), (7), (9), (10), (13), and (14) for the variables $\mathbf{q}$, p, ϱ, T, $\mathbf{E}$, $\mathbf{H}$, $\mathbf{J}$, and ϱ_e. These equations should be solved for given initial and boundary conditions. The boundary conditions of the gas dynamic fields are the same as those in ordinary gas dynamics, i.e., the no-slip conditions may be used. The boundary conditions of electromagnetic fields are given in next section.

3. EQUATIONS AND BOUNDARY CONDITIONS OF ELECTROMAGNETIC FIELDS [2]

In electromagnetic theory, it is sometimes convenient to use vector and scalar potentials instead of the electromagnetic fields. These potentials are defined as follows:

The divergence of Eq. (10) gives

$$\frac{\partial}{\partial t} (\nabla \cdot \mathbf{B}) = 0 \tag{15}$$

or $\nabla \cdot \mathbf{B} = \text{const}$ at every point in the field. This constant is zero if, at any point in its past or future history, the magnetic induction $\mathbf{B}$ may vanish. Hence we have

$$\nabla \cdot \mathbf{B} = \nabla \cdot (\mu_e \mathbf{H}) = 0 \tag{16}$$

From Eq. (16), we may introduce a vector potential $\mathbf{A}$ such that

$$\mathbf{B} = \nabla \times \mathbf{A} \tag{17}$$

with the condition

$$\nabla \cdot \mathbf{A} = 0 \tag{18}$$

Similarly, the divergence of Eq. (9), with the help of Eq. (14), gives

$$\frac{\partial}{\partial t} (\nabla \cdot \mathbf{D} - \varrho_e) = 0 \tag{19}$$

If in its past or future history, the electric field $\mathbf{E}$ and the excess electric charge ϱ_e may vanish simultaneously, we have

$$\nabla \cdot \mathbf{D} = \nabla \cdot (\varepsilon \mathbf{E}) = \varrho_e \tag{20}$$

Equation (20) is known as Poisson's equation.

Substituting Eq. (17) into Eq. (10), we have

$$\nabla \times \left(\mathbf{E} + \frac{\partial \mathbf{A}}{\partial t} \right) = 0 \tag{21}$$

From Eq. (21), we may introduce a scalar potential ϕ such that

$$-\nabla \phi = \mathbf{E} + \frac{\partial \mathbf{A}}{\partial t} \tag{22}$$

Substituting the above relations for these vector and scalar potentials into Eq. (9), we have

$$\frac{1}{c^2} \frac{\partial^2 \mathbf{A}}{\partial t^2} - \nabla^2 \mathbf{A} + \frac{1}{\nu_H} \left[\frac{\partial \mathbf{A}}{\partial t} - \mathbf{q} \times (\nabla \times \mathbf{A}) \right]$$
$$= -\frac{1}{c^2} \left(\frac{\partial \nabla \phi}{\partial t} - \mathbf{q} \nabla^2 \phi \right) - \frac{1}{\nu_H} \nabla \phi \tag{23}$$

where

$$c = 1/(\varepsilon \mu_e)^{1/2} \tag{24}$$

is the velocity of light in free space and

$$\nu_H = 1/\mu_e \sigma \tag{25}$$

is the magnetic diffusivity.

The Maxwell equations (9) and (10) and their equivalent equations are valid only for those points in whose neighborhood the physical properties of the medium vary continuously. On the boundary of the flow field, the physical properties of the medium may exhibit discontinuities. For instance, at a solid boundary, the electromagnetic properties of the plasma will change to those of the solid. Across such a surface of discontinuity of electromagnetic properties, the following four conditions hold:

1. The transition of the normal component of the magnetic induction $\mathbf{B} = \mu_e \mathbf{H}$ is continuous, i.e.,

$$(\mathbf{B}_2 - \mathbf{B}_1) \cdot \mathbf{n} = 0 \tag{26}$$

where $\mathbf{n}$ is the unit normal to the discontinuity surface. The subscripts 1 and 2 refer to the values immediately on each side of the discontinuity surface.

2. The behavior of the magnetic field $\mathbf{H}$ at the boundary is

$$\mathbf{n} \times (\mathbf{H}_2 - \mathbf{H}_1) = \mathbf{J}_s \tag{27}$$

where $\mathbf{J}_s$ is the surface current density. For finite electrical conductivity, $\mathbf{J}_s$ is zero, but for the case of infinite electrical conductivity, $\mathbf{J}_s$ may be different from zero.

3. The transition of the tangential component of the electric field $\mathbf{E}$ is continuous, i.e.,

$$\mathbf{n} \times (\mathbf{E}_2 - \mathbf{E}_1) = 0 \tag{28}$$

4. The behavior of the dielectric displacement $\mathbf{D}$ at the boundary is

$$\mathbf{n} \cdot (\mathbf{D}_2 - \mathbf{D}_1) = \varrho_{eS} \tag{29}$$

where ϱ_{eS} is the surface free-charge density.

For most problems of magnetogasdynamics, we may neglect the surface current density $\mathbf{J}_s$ and the surface free-charge density ϱ_{eS}. Hence our boundary conditions for the electromagnetic fields are that the tangential components of $\mathbf{H}$ and $\mathbf{E}$ and the normal components of $\mathbf{B}$ and $\mathbf{D}$ are all continuous across a surface of discontinuity which separates a solid body and a fluid or two different fluids. The distinctions between $\mathbf{H}$ and $\mathbf{B}$ and between $\mathbf{E}$ and $\mathbf{D}$ should be noticed, because the values of μ_e and ε may be different on the two sides of the surface.

4. MAGNETOGASDYNAMIC APPROXIMATIONS [1]

For many important flow problems, the following conditions are satisfied:

1. The time scale t_0 of our problem is of the same order of magnitude as L/U, where L is the characteristic length and U is the characteristic velo-

city of the flow field. In other words, the time or frequency parameter

$$R_t = \frac{t_0}{L/U} \tag{30}$$

is of the order of unity. This means that we shall not consider phenomena at very high frequencies where the time scale t_0 is very small.

2. The electric field, which may be characterized by a value E_0, is of the same order of magnitude as the induced electric field $\mathbf{q} \times \mathbf{B}$. In other words, the electric field parameter

$$R_E = E_0/UB_0 \tag{31}$$

is of the order of unity or smaller, where B_0 is a characteristic magnetic induction.

3. The velocity of the flow of the plasma is much smaller than the velocity of light c, i.e., the relativistic parameter

$$R_r = U/c \tag{32}$$

is much smaller than unity.

Under the above conditions, the displacement current $\partial \mathbf{D}/\partial t$ and the terms with the excess electric charge ϱ_e in the fundamental equations of Section 2 are negligible in comparison with those terms containing the magnetic field. For instance,

$$\varrho_e \mathbf{E} \cong (U^2/c^2)\mathbf{J} \times \mathbf{B} \tag{33}$$

Hence we may neglect all the terms containing ϱ_e and the displacement current. We may express the electric current density and the electric field strength in terms of magnetic field strength and the velocity as follows:

$$\mathbf{J} \approx \mathbf{I} \approx \nabla \times \mathbf{H} \tag{34}$$

and

$$\mathbf{E} = (\mathbf{J}/\sigma) - \mathbf{q} \times \mathbf{B} \approx (1/\sigma)(\nabla \times \mathbf{H}) - \mathbf{q} \times \mathbf{B} \tag{35}$$

As a result, we need to consider the interaction between the magnetic field strength $\mathbf{H}$ and the gasdynamic variables only. It is for this reason that we call the resulting field of study "magnetofluid dynamics" or magneto-hydrodynamics or magnetogasdynamics. Substituting Eqs. (34) and (35) into Eq. (10), we have a single vector equation for the magnetic field $\mathbf{H}$

which replaces the ten electromagnetic equations in Section 2:

$$\partial \mathbf{H}/\partial t = \nabla \times (\mathbf{q} \times \mathbf{H}) - \nabla \times [\nu_H(\nabla \times \mathbf{H})] \tag{36}$$

With the help of Eqs. (34) and (35) and $\varrho_e = 0$, the equations of motion (3) and the equation of energy (7) become, respectively,

$$\frac{\partial q}{\partial t} + (\mathbf{q} \cdot \nabla)\mathbf{q} - \frac{1}{\varrho}(\mathbf{B} \cdot \nabla)\mathbf{H} = -\frac{1}{\varrho}\nabla\left(p + \frac{1}{2}\mathbf{B} \cdot \mathbf{H}\right) + \frac{1}{\varrho}(\nabla \cdot \tau) \tag{37}$$

$$\varrho\frac{Dh_0}{Dt} = \frac{\partial p}{\partial t} + \nabla \cdot (\mathbf{q} \cdot \tau) + \nabla \cdot (\varkappa \nabla T) + (\nabla \times \mathbf{H}) \cdot (\nu_H \nabla \times \mathbf{B} - \mathbf{q} \times \mathbf{B}) \tag{38}$$

where $h_0 = U_m + RT + \frac{1}{2}q^2$ is the stagnation enthalpy of the gas. Equations (36)–(38) with Eqs. (1) and (2) are the fundamental equations of magneto-gasdynamics, in which there are nine unknowns $\mathbf{H}$, p, ϱ, T, and $\mathbf{q}$.

5. ELECTROGASDYNAMIC APPROXIMATIONS

There is another limiting case, at the other extreme from the magneto-gasdynamic approximation, which is known as the electrogasdynamic approximation. In this case, the plasma consists of essentially one kind of charged particle, e.g., electrons alone. In this case, the electric field is more important than the magnetic field. Hence the term *electrogasdynamics* ([3]).

In the present case, the electric field is very large and the excess electric charge is far from zero. The electrical conduction current is negligible in comparison with the electrical convection current. For a gas with a single kind of charged particle, there is no electrical conduction current. The order of magnitude of the electric field is

$$E \approx \varrho_e L/\varepsilon \tag{39a}$$

where L is a reference length. The order of magnitude of the convection current is

$$J \approx \varrho_e U \tag{39b}$$

where U is a reference velocity. Finally, the order of magnitude of the magnetic field for the ordinary flow problem without high-frequency phenomena is

$$H \approx \varrho_e UL \tag{39c}$$

Equations (39a–c) may be called the electrogasdynamic approximations.

With the help of these equations, we may show that the terms containing the magnetic field in the fundamental equations of electromagnetogasdynamics given in Section 2 are negligible in comparison with those containing the excess electric charge ϱ_e. For instance,

$$\varrho_e \mathbf{E} = (c^2/U^2)\mathbf{J} \times \mathbf{B} \tag{40}$$

The fundamental equations of electrogasdynamics are then:

$$p = \varrho RT \tag{41a}$$

$$(\partial \varrho/\partial t) + \nabla \cdot (\varrho \mathbf{q}) = 0 \tag{41b}$$

$$\varrho \, D\mathbf{q}/Dt = -\nabla p + \nabla \cdot \tau + \varrho_e \mathbf{E} \tag{41c}$$

$$\varrho \, DU_m/Dt = -p(\nabla \cdot \mathbf{q}) + \Phi + \nabla \cdot (\varkappa \, \nabla T) \tag{41d}$$

$$\nabla \times \mathbf{E} = 0 \tag{41e}$$

$$\nabla \cdot \mathbf{E} = \varrho_e/\varepsilon \tag{41f}$$

6. IMPORTANT PARAMETERS OF ELECTROMAG-NETOFLUID DYNAMICS

Before we discuss some flow problem of an electrically-conducting fluid under the influence of applied electromagnetic fields, we would like to discuss the important dimensionless parameters used to characterize such a flow field. These dimensionless parameters may be obtained from the fundamental equations of Section 2 if we express them in dimensionless form ([1]). Since the fundamental equations of electromagnetofluid dynamics include all the terms of the fundamental equations of fluid dynamics, all the dimensionless parameters of ordinary fluid dynamics, such as Mach number, Reynolds number, Prandtl number, etc., remain as important parameters in electromagnetofluid dynamics. The following are some new dimensionless parameters due to the electromagnetic properties:

1. *Magnetic pressure number R_H and magnetic Mach number $\mathbf{M}_m$:*

$$R_H = \frac{\text{Magnetic pressure}}{\text{Dynamic pressure}} = \frac{\frac{1}{2}\mu_e H^2}{\frac{1}{2}\varrho_0 U^2} = \frac{V_H^2}{U^2} = \frac{1}{\mathbf{M}_m^2} \tag{42}$$

where ϱ_0 and H are, respectively, reference values of density and magnetic field. The velocity V_H is known as the velocity of Alfvén's wave, which is the characteristic velocity of magnetofluid dynamics and is defined as

$$V_H = H(\mu_e/\varrho_0)^{1/2} \tag{43}$$

Alfvén ([16]) first predicted a wave in an incompressible and inviscid fluid of density ϱ_0 and of infinite electrical conductivity under a homogeneous magnetic field strength H propagated at a speed V_H.

When R_H is of the order of unity or higher, the fluid will be affected noticeably by the magnetic field. The Alfvén wave is one of the basic waves in magnetogasdynamics, and it will affect the effective sound speed in a plasma. The flow pattern in an electrically-conducting fluid will differ when M_m is greater and when it is less than unity. When M_m is greater than unity, the flow is said to be super-Alfvén flow, and when M_m is smaller than unity, the flow is said to be sub-Alfvén flow.

2. *Magnetic Reynolds number* Re_σ:

$$Re_\sigma = \mu_e \sigma UL = UL/\nu_H \tag{44}$$

This is another important parameter in magnetogasdynamics. The magnetic Reynolds number Re_σ determines the diffusion phenomena of the magnetic field along streamlines in a similar manner as the ordinary Reynolds number determines the diffusion phenomena of vorticity along streamlines. Hence the magnetic Reynolds number determines the effect of the flow field on the magnetic field. If Re_σ is negligibly small, the magnetic field is practically unaffected by the flow field; if Re_σ is very large, the magnetic field will stay with the flow—the so-called frozen-in field—and it is greatly influenced by the motion of the fluid. When Re_σ is very large, we have the magnetic boundary layer, the thickness of which is proportional to the square root of the magnetic diffusivity.

3. *Electric field parameter*:

$$R_E = E/UB \tag{45}$$

In magnetogasdynamics, the electric field parameter R_E is of the order of unity or less. It may be used as an extra parameter in the study of the flow problems of magnetogasdynamics. However in electrogasdynamics, this parameter R_E is very large.

It is possible to combine the above electromagnetic parameters with some other gasdynamic parameters to get new parameters which may be useful in special problems. We shall discuss such new parameters later.

7. ONE-DIMENSIONAL FLOW IN MAGNETOGASDY-NAMICS ([4])

The simplest flow in magnetogasdynamics is the one-dimensional flow problem, in which only one component of the velocity u is different from zero and all the variables are functions of one spatial coordinate x only. This corresponds to the flow of an electrically-conducting fluid in a channel of variable cross section $A(x)$, where x is the axial coordinate under the influence of applied electromagnetic fields. Even such a flow is very complicated if we consider a general case. For simplicity, we consider two limiting cases as follows:

1. *Approximately one-dimensional flow in transverse electromagnetic fields.* We consider the steady inviscid flow of a plasma in a nozzle of slowly-varying cross section $A(x)$. By approximately one-dimensional flow, we mean that we may consider the externally-applied magnetic field $B_y = B(x)$ and electric field $E_z = E(x)$ (where the subscripts y and z refer to the y and z components, respectively) as given functions of spatial coordinate x. Our fundamental equations for this problem are

$$A\varrho u = \text{const} \tag{46a}$$

$$\varrho u \frac{du}{dx} = - \frac{dp}{dx} - \sigma B(E + uB) \tag{46b}$$

$$\varrho u \frac{dU_m}{dx} + \frac{p}{A} \frac{d(uA)}{dx} = \sigma(E + uB)^2 \tag{46c}$$

$$p = \varrho RT \tag{46d}$$

Equations (46a–d) may be considered as a system of two first-order total differential equations as follows:

$$\frac{du}{dx} = \frac{1}{M^2 - 1} \left[\frac{u}{A} \frac{dA}{dx} - \frac{\sigma B^2}{p} (u - u_1)(u - u_3) \right] = \frac{F_1(u, M, x)}{M^2 - 1} \tag{47a}$$

$$\frac{dM}{dx} = \frac{1}{M^2 - 1} \left[\left(1 + \frac{\gamma - 1}{2} M^2 \right) \frac{M}{A} \frac{dA}{dx} \right.$$

$$\left. - \left(1 + \frac{\gamma - 1}{2} M^2 \right) \frac{\sigma B^2 M}{up} (u - u_2)(u - u_3) \right] = \frac{F_2(u, M, x)}{M^2 - 1} \tag{47b}$$

where

$$u_1 = - \frac{\gamma - 1}{\gamma} \frac{E}{B}; \quad u_2 = \frac{(1 + \gamma M^2)u_1}{2 + (\gamma - 1)M^2}; \quad u_3 = \frac{\gamma u_1}{\gamma - 1}; \quad M = \frac{u}{a} \tag{47c}$$

The velocities u_1, u_2 and u_3 are characteristic velocities of our problem. For instance, if we consider a nozzle of constant cross-sectional area $A = $ const, Eqs. (47a,b) become

$$\frac{du}{dx} = -\frac{\sigma B^2}{(M^2 - 1)p}(u - u_1)(u - u_3) \tag{48a}$$

$$\frac{dM}{dx} = -\left(1 + \frac{\gamma - 1}{2}M^2\right)\frac{\sigma B^2 M}{up}(u - u_2)(u - u_3) \tag{48b}$$

It is easy to show that the characteristics of the flow field depend on the initial value of velocity, u_0, relative to the characteristic velocities u_1, u_2, and u_3. For instance, if the initial Mach number $M = M_0$ is less than one, and $u_0 < u_2 < u_1 < u_3$, both du/dx and dM/dx are initially positive. This is the case of magnetogasdynamic acceleration. If the initial Mach number is sufficiently close to unity, the acceleration will soon become infinite, and the nozzle is choked. For other initial conditions, the velocity u may increase monotonically toward u_1, while the Mach number M first increases and then decreases $(u > u_2)$, and finally approaches an asympototic value $M_a < 1$. The third possibility is a case of a particular set of initial conditions for which M reaches unity and u reaches u_1 simultaneously. We have a smooth acceleration from a subsonic to a supersonic flow in a nozzle of constant cross-sectional area, which is not possible in ordinary gasdynamics.

For a nozzle of variable cross-sectional area $A(x)$, we have to solve Eqs. (47a) and (47b) simultaneously. The well-known method of nonlinear mechanics may be used in solving Eqs. (47a,b). For instance, in the phase plane M–u, Eqs. (47a,b) give

$$dM/du = F_2(u, M, x)/F_1(u, M, x) \tag{49}$$

Since both functions F_1 and F_2 are functions of x as well as of u and M, the trajectory through any point (u, M) is not unique. However, for any given shape of the nozzle, $A(x)$, and given electromagnetic fields $B(x)$ and $E(x)$, we can draw a definite set of trajectories for each station x in the nozzle. Furthermore, the most interesting features of Eq. (49) are the behaviors near the singular points of this equation. If we restrict our attention to the neighborhood of these singular points, we may consider the equation at the singular point $x = x_c$ and neglect the x-dependence of F_1 and F_2. To illustrate this point, let us consider the case of zero electric field, $E = 0$. In this case, Eq. (49) becomes

$$\frac{dM}{du} = \frac{M}{u}\left(1 + \frac{\gamma - 1}{2}M^2\right) \tag{50}$$

Integration of Eq. (50) gives

$$1 + \frac{\gamma - 1}{2} \frac{u^2}{a^2} = \frac{a_0^2}{a^2} \tag{51}$$

where a_0 is the stagnation sound speed, which is a constant. The relations (50) and (51) are the same as those in ordinary gas dynamics. We need to consider Eq. (47a) alone which for $E = 0$ becomes

$$\frac{du}{dx} = \frac{u}{M^2 - 1} \left(\frac{1}{A} \frac{dA}{dx} - \frac{\sigma B^2 u}{p} \right) \tag{52}$$

Equation (52) shows that the pondermotive force in the present case retards the velocity of the flow in the nozzle if the flow is supersonic, $M > 1$, and accelerates the flow if it is subsonic, $M < 1$. If either $\sigma = 0$ or $B = 0$, Eq. (52) reduces to the form as in ordinary gas dynamics, and at the singular point $M = 1$ and $dA/dx = 0$, the singular point of Eq. (52) is a saddle point. When both σ and B are different from zero, the flow behaves quite differently from that in ordinary gas dynamics. For pure subsonic flow, the velocity in the nozzle for the same pressure drop in the MGD case is always higher than the corresponding velocity in ordinary gas dynamics, and the maximum velocity occurs at a station downstream of the minimum section in the MGD case instead of at minimum section as in the ordinary gasdynamic case. The location of the critical station where maximum velocity occurs is given by the formula

$$\frac{L}{A} \frac{dA}{dx} = \gamma \mathrm{Re}_\sigma \frac{M^2}{M_m^2} \tag{53}$$

where L is a reference length, $\mathrm{Re}_\sigma = \sigma \mu_e U L$ is the local magnetic Reynolds number, $M = u/a$ is the local Mach number, $M_m = u/V_H$ is the local magnetic Mach number, and $V_H = H_y(\mu_e/\varrho)^{1/2}$ is the local Alfvén wave speed.

If the condition (53) and $M = 1$ occur simultaneously, we have the critical case $x = x_c$, where the singular point of Eq. (52) occurs. For ordinary gas dynamics, this singular point of Eq. (52) is a saddle point. For the MGD case, depending on the variation of $A(x)$ and $B(x)$, we may have a saddle point, a nodal point, or a spiral point. Hence the MGD flow has much more variations than occur in ordinary gas dynamics.

2. *Strictly one-dimensional flow in transverse electromagnetic fields.* In this type of analysis, we assume that all variables are strictly independent of the transverse coordinates y and z and Maxwell's equations must be obeyed. In quasione-dimensional, or approximately one-dimensional, flow,

we consider the average value over a section, and for local values of variables, these may be functions of x, y, and z.

The fundamental equations for one-dimensional steady flow in a nozzle of cross-sectional area $A(x)$ are

$$A\varrho u = \text{const} \tag{54a}$$

$$\varrho u \frac{du}{dx} = - \frac{dp}{dx} - B \frac{dH}{dx} \tag{54b}$$

$$\varrho u c_p \frac{dT}{dx} + \varrho u^2 \frac{du}{dx} = \frac{dB}{dx} \left(v_H \frac{dH}{dx} - uH \right) \tag{54c}$$

$$p = \varrho RT \tag{54d}$$

$$v_H \frac{dH}{dx} = uH + E_0 \tag{54e}$$

The main difference between Eqs. (46) and Eqs. (54) is that the magnetic field H in Eqs. (54) is not a given function of x, but should be determined by solving this set of equations. For the case of infinitely large electrical conductivity $\sigma = \infty$, or $v_H = 0$, we have, from Eqs. (54),

$$\frac{du}{dx} = \frac{ua^2}{(u^2 - a_e^2)} \frac{1}{A} \frac{dA}{dx} \tag{55}$$

where a_e is the effective sound speed given by the relation

$$a_e^2 = a^2 + V_H^2 \tag{56}$$

In this case, Eq. (55) is identical to the expression of ordinary gas dynamics except that the effective sound speed a_e is used instead of ordinary sound speed a in the denominator. Hence at the throat of the nozzle, the choke speed will be a_e instead of a. The general conclusion of ordinary gas dynamics may be applied here with a_e for a as the critical speed, or effective Mach number $M_e = u/a_e$ should be considered as the important dimensionless parameter here.

8. ONE-DIMENSIONAL FLOW IN ELECTROGASDY-NAMICS [5]

We consider the one-dimensional flow of an inviscid, nonheat-conducting, electrically-charged fluid in a nozzle of cross-sectional area $A(x)$. We may use the electrogasdynamic approximations and consider four unknowns: p, n, u, and $E_x = E$, where the temperature T may be expressed in terms of pressure p and number density n of the gas. The fundamental equations of electrogasdynamics are:

$$nAu = \text{const} = A_0 N \tag{57a}$$

$$\frac{dp}{dx} + mnu\frac{du}{dx} = ZeE \tag{57b}$$

$$\varepsilon\frac{dE}{dx} = Zen \tag{57c}$$

$$p = p(n) \tag{57d}$$

where we replace the energy equation by the barotropic relation (57d), i.e., the pressure p is a function of number density n only. The energy equation (41d) can be reduced to

$$dS/dx = 0 \tag{58}$$

if the flow is adiabatic. We may consider the isentropic flow with $p = \text{const } (n)^\gamma$. The other simple case is the isothermal process, $T = \text{const}$ and $p = (\text{const})n$. In these equations, Ze is the charge on a particle of the gas. For singly-charged ion, $Z = 1$, and for the electron, $Z = -1$.

Equations (57a–d) are a set of nonlinear differential equations. We may use the well-known method of nonlinear mechanics to study this set of equations. For simplicity, let us consider the case of isothermal flow, $T = \text{const}$ in a channel of constant cross-sectional area $A(x) = A_0 = \text{const}$. For a constant-area channel, Eqs. (57a–d) give

$$p + (mN^2/n) - \tfrac{1}{2}\varepsilon E^2 = \text{const} \tag{59}$$

where m is the mass of a particle of the gas. Since $p = p(n)$, Eq. (59) gives the phase plane relation between n and E. Clauser [5] discussed extensively the case of isothermal flow, i.e., $p = knT$. He introduced the following dimensionless quantities for this case:

$$n^* = n/n_a; \qquad E^* = \frac{E}{(2n_a kT/\varepsilon)^{1/2}} \tag{60}$$

where $n_a = (mN^2/kT)^{1/2}$ is the value of n at $u = a = (kT/m)^{1/2}$. Equation (59) becomes

$$n^* + (1/n^*) - E^{*2} = \text{const} \tag{61}$$

Equation (61) is plotted in Fig. 1. There is a singular point at $E^* = 0$ and

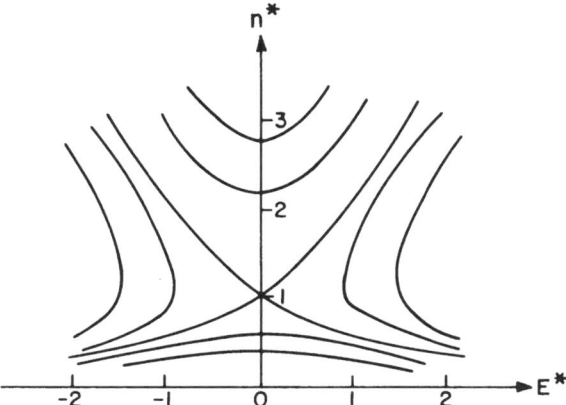

Fig. 1. One-dimensional flow in an isothermal, constant-area channel in electrogasdynamics.

$n^* = 1$, and it is a saddle point. If we substitute the relation $E^* = E^*(n^*)$ obtained in Eq. (61) into Eq. (57c), we have a first-order total differential equation for $E^*(x)$ which can be numerically integrated. Some typical results are shown in Fig. 2. The most interesting point is that the integral curve of E^* has a cusp at the sonic line $n^* = 1$ or $M = u/a = 1$. This means that the results cannot represent the real flow condition at the sonic point. The only curves which are reasonable are those integral curves which pass through the saddle point.

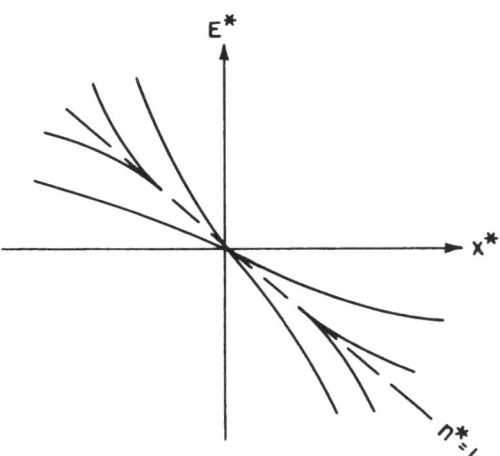

Fig. 2. Electric field along x direction in a constant-area channel with an isothermal flow in electrogasdynamics.

9. CHANNEL FLOW IN MAGNETOHYDRODYNAMICS [6]

In the last two sections, we assumed that the fluid is inviscid. At low Reynolds number, the viscous effect may not be neglected. In this section, we consider the viscous flow in a two-dimensional channel of an electrically-conducting liquid such as mercury. Such a flow was first studied by Hartmann and Lazarus [17] in the 1930's.

We consider the fully developed flow between two parallel plates under transverse electromagnetic fields of a steady, two-dimensional, laminar flow of an incompressible, electrically-conducting fluid of constant viscosity and constant electrical conductivity. We assume that the two walls are situated at $y = \pm L$. One of the walls, at $y = -L$, is at rest and is an insulated plate, while the other wall is moving with a speed U_w in the x direction or is at rest (Fig. 3). There is a constant magnetic field in the y direction, H_0, and a constant electric field E_0 in the z direction. All the variables are functions of y only, except for the pressure p, which is a function of both x and y.

We define the following variables for our problem (dimensionless variables denoted by an asterisk):

$$u = U_0 u^*(y^*); \qquad v = w = 0; \qquad E_x = E_y = 0;$$

$$E_z = E_0(E_z^* = 1); \qquad y = Ly^*$$

$$H_x = H_0 H_x^*(y^*); \qquad H_y = H_0(H_y^* = 1); \qquad H_z = 0; \qquad (62)$$

$$x = Lx^*; \qquad p_x = \partial p/\partial x$$

$$p = \varrho U_0^2 p^*(x^*, y^*) = \varrho U_0^2[x^* P_x^* + p_1^*(y^*)]$$

where U_0 and L are, respectively, the reference velocity and length. In dimensionless form, our fundamental equations are

$$\frac{1}{R_h^2} \frac{d^2 u^*}{dy^{*2}} - u^* = \frac{\mathrm{Re}}{R_h^2} p_x^* + R_E = C = \text{const} \qquad (63)$$

$$\frac{dp_1^*}{dy^*} = \mathrm{Re}_\sigma R_H H_X^*(u^* + R_E) \qquad (64)$$

$$\frac{du^*}{dy^*} + \frac{1}{\mathrm{Re}_\sigma} \frac{d^2 H_x^*}{dy^{*2}} = 0 \qquad (65)$$

where $\mathrm{Re} = \varrho U_0 L/\mu$ is the Reynolds number, $R_H = B_0 H_0/\varrho U_0^2$ is the magnetic pressure number, $\mathrm{Re}_\sigma = \sigma \mu_e U_0 L$ is the magnetic Reynolds number,

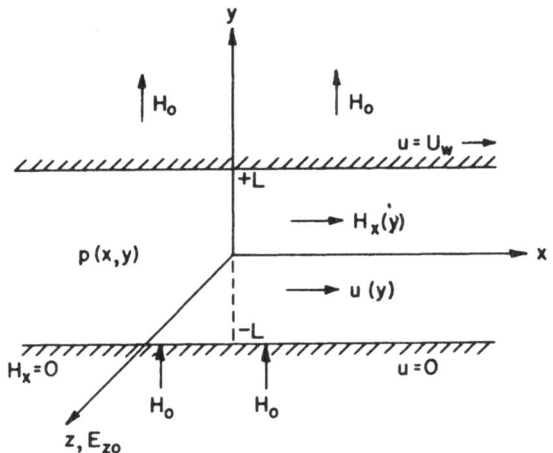

Fig. 3. Generalized Hartmann flow in magnetohydrodynamics.

$R_h = (R_e R_H R_\sigma)^{1/2}$ is the Hartmann number, $R_E = E_0/U_0 B_0$ is the electric field number, and $B_0 = \mu_e H_0$ is the magnetic induction of the applied field. The boundary conditions are

$$u^* = 0 \qquad \text{at } y^* = -1 \qquad (66a)$$

$$u^* = U_w/U_0 = u_w{}^* \qquad \text{at } y^* = +1 \qquad (66b)$$

$$H_x{}^* = 0 \qquad \text{at } y^* = -1 \qquad (66c)$$

The general solution of Eq. (63) for u^* is

$$u^* = A\cosh(R_h y^*) + B\sinh(R_h y^*) + C \qquad (67)$$

where A and B are arbitrary constants to be determined by the boundary conditions. The constant C is the effective electric field parameter, or the effective x-wise pressure gradient, which is given in our problem and defined in Eq. (63). From the boundary conditions (66), we have

$$u^* = \tfrac{1}{2}u_w{}^*\left(\frac{\cosh(R_h y^*)}{\cosh R_h} + \frac{\sinh(R_h y^*)}{\sinh R_h}\right) + C\left(1 - \frac{\cosh(R_h y^*)}{\cosh R_h}\right) \qquad (68)$$

Equation (68) shows that the dimensionless velocity distribution u^* depends on the Hartmann number R_h, effective x-wise pressure gradient C, which consists of both the actual x-wise pressure gradient and the electric field number, and the velocity of the upper wall $u_w{}^*$. When we plot u^* vs

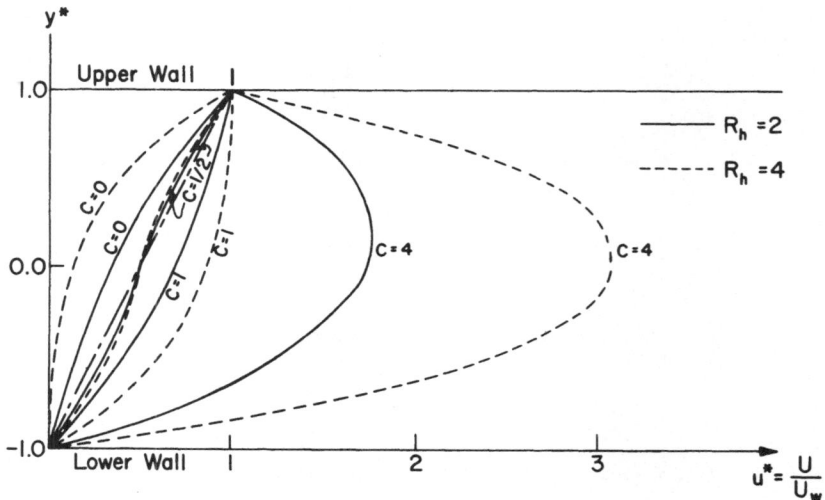

Fig. 4. Velocity distributions for MHD Couette flow.

y^*, we have to specify these parameters. Furthermore, the shape of the u^* vs y^* curves depend on the choice of the reference velocity U_0.

If we use the velocity of the upper wall as the refrence velocity, we have $u_w^* = 1$ and $u^*(y^*)$ is a function of R_h and C. Some typical velocity distributions for this case are shown in Fig. 4.

If we take the spatial mean velocity V flowing through the channel as a reference velocity, we have

$$\int_{-1}^{1} u^* \, dy^* = 2 \tag{69}$$

Substituting Eq. (68) into (69), we have

$$C = \frac{R_h - \tfrac{1}{2}u^* \tanh R_h}{R_h - \tanh R_h} \tag{70}$$

In this case, the velocity distribution u^* depends on R_h and u_w^*. Some of the typical velocity distributions for the case $u_w^* = 0$ are shown in Fig. 5. Even though the velocity distribution u^* of Fig. 5 is independent of C explicitly, the actual mass flow rate depends on both R_h and C because the characteristic velocity U_0 is

$$U_0 = \frac{(E_0/B_0) + (p_x/\sigma B_0^2)}{(R_h - \tfrac{1}{2}u_w^* \tanh R_h)/(R_h - \tanh R_h)} \tag{71}$$

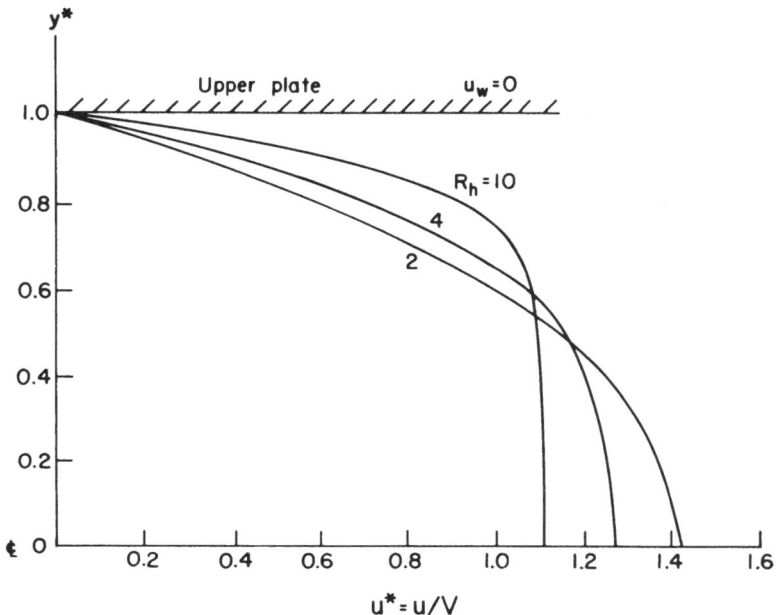

Fig. 5. Velocity distributions for MHD Hartmann flow.

For given values of p_x and E_z, the mass flow Q decreases as the strength of the magnetic field H_0 increases.

The z-wise current density J_z is

$$J_z = \sigma(E_z + uB_y) = \sigma U_0 B_0(R_E + u^*) \qquad (72)$$

The x-wise magnetic field may be calculated from the equation

$$dH_x/dy = -J_z = -\sigma U_0 B_0(R_E + u^*) \qquad (73)$$

With the boundary condition Eq. (66c), we have

$$H_x^* = -R_o\left\{\frac{1}{2}\,u_w^*\left[\frac{\sinh(R_h y^*) + \sinh R_h}{R_h \cosh R_h} + \frac{\cosh(R_h y^*) - \cosh R_h}{R_h \sinh R_h}\right]\right.$$

$$\left. + C\left[(y^* + 1) - \frac{\sinh(R_h y^*) + \sinh R_h}{R_h \cosh R_h}\right] + R_E(y^* + 1)\right\} \qquad (74)$$

It is interesting to notice that the electric field E_0 may be considered as a free parameter which may be chosen by the investigator. Some values of R_E have special significance:

1. Without electric field, we have $R_E = 0$. In this case, the total current through the channel is not zero, and the x-wise magnetic field on the upper wall is also not zero.

2. If the total current flowing through the channel is zero, we have

$$I_t = \int_{-1}^{1} J_z \, dy^* = 0 \tag{75}$$

or

$$(u_w{}^* - 2C) \frac{\sinh R_h}{R_h \cosh R_h} + (C + R_E) = 0 \tag{75'}$$

If we consider the case $u_w{}^* = 0$ and take the mean flow velocity as the reference velocity $U_0 = V$, Eq. (75') gives $R_E = -1$. It should be noticed that the critical value of R_E depends on the choice of U_0. For the MHD Couette flow of Fig. 4, the critical value of R_E is $\frac{1}{2}$ when there is no total current, i.e., when both walls are insulated.

After u^* and $H_x{}^*$ are obtained, we may calculate $p_1{}^*$ from Eq. (64) by simple quadrature. In ordinary hydrodynamics, p is independent of y, but in magnetohydrodynamics, p is a function of y as well as x.

10. WAVES AND SHOCKS IN MAGNETOGASDYNAMICS [1]

The study of wave propagation in an electrically-conducting fluid has both academic interest and practical application. The wave motion will bring out many special features in magnetofluid dynamics which may differ significantly from those in ordinary fluid dynamics. The practical applications of wave propagation are numerous, including space communication systems, wave propagation in the ionosphere, and MHD power generation for ac current, as well as in many astrophysical and geophysical problems.

The properties of a wave in an electrically-conducting fluid depend on the amplitude of the wave. The simplest type of wave is the wave of infinitesimal amplitude. Ordinary sound and radio waves belong to this group. They may be considered as special cases of magnetogasdynamic waves. Magnetogasdynamic waves are resultant waves due to the interaction of gasdynamic waves and electromagnetic waves by means of an externally-applied magnetic field. Such an interaction will give us many new phenomena which occur neither in ordinary gasdynamics nor in ordinary electrodynamics. Mathematically speaking, we may linearize the equations which govern the wave of infinitesimal amplitude, and the superposition principle is applicable to such waves. We shall study this type of wave first.

For waves of finite amplitude, the shape of the wave will distort as the wave propagates, while the wave of infinitesimal amplitude will maintain its shape when it propagates. When the distortion is large, an ordinary wave will develop into a shock wave, in which a large change of physical quantities occurs in a very narrow region.

We assume that ordinarily the plasma is at rest with a pressure p_0, a temperature T_0, and a density ϱ_0, and that it is subjected to an externally-applied uniform magnetic field $\mathbf{H}_0 = \mathbf{i}H_x + \mathbf{j}H_y + \mathbf{k}0$, where we choose the coordinate system such that $H_z = 0$ and H_x and H_y are constants but may be zero. There is no electric current, excess electric charge, or applied electric field. The plasma is perturbed by a small disturbance, so that in the resultant motion of the plasma, we have

$$u = u(x, t); \qquad v = v(x, t); \qquad w = w(x, t);$$
$$p = p_0 + p'(x, t); \qquad T = T_0 + T'(x, t)$$
$$\varrho = \varrho_0 + \varrho'(x, t); \qquad \mathbf{E} = \mathbf{E}(x, t); \qquad \mathbf{J} = \mathbf{J}(x, t);$$
$$\varrho_e = \varrho_2(x, t); \qquad \mathbf{H} = \mathbf{H}_0 + \mathbf{h}(x, t)$$

$$(76)$$

For simplicity, we assume that all the perturbed quantities are functions of only one spatial coordinate x and of time t. Thus we consider the wave propagation in the x direction. We have 16 perturbed quantities. If we substitute Eqs. (76) into the fundamental equations of electromagneto-gasdynamics of Section 2, we have 16 linear equations for these variables if we neglect the higher-order terms of the perturbed quantities. These linearized equations may be divided into three independent groups:

1. $h_x = 0$. The x component of $\mathbf{h}$ is independent of all the other 15 variables and may be set equal to zero without loss of generality.

2. The equations governing the variables w, h_z, J_x, J_y, E_x, E_y, and ϱ_e are coupled, where the subscripts x, y, and z refer, respectively, to the corresponding components of a vector quantity. These equations may be called the equations of transverse waves because they deal with the velocity and magnetic field components perpendicular to the applied magnetic field H_0.

3. The rest of the equations, which govern the variables u, v, p', ϱ', T', h_y, J_z, and E_z are known as the equations of longitudinal waves, of which the ordinary sound wave is a special case.

We are looking for a periodic solution in which all the perturbed quantities are proportional to

$$\exp[i(\omega t - \lambda x)] = \exp[-i\lambda_R(x - Vt)]\exp(\lambda_i x) \tag{77}$$

where ω is a given real angular frequency, $\lambda = \lambda_R + i\lambda$ is the complex wave number, $i = \sqrt{-1}$ and

$$V = \omega/\lambda_R \tag{78}$$

is the speed of wave propagation. Substituting an expression of the form of (77) into the linearized equations of electromagnetogasdynamics, we obtain the dispersion relations $\lambda(\omega)$ for both the transverse and longitudinal waves.

Transverse Waves

The dispersion relation for the transverse waves is

$$\left(i\omega - \nu_H \frac{\omega^2}{c^2}\right)\left[\left(i\omega + \nu_g\lambda^2\right)\left(i\omega + \nu_H\lambda^2 - \nu_H \frac{\omega^2}{c^2}\right) + V_x^2\left(\lambda^2 - \frac{\omega^2}{c^2}\right)\right]$$
$$- \frac{\omega^2}{c^2} V_y^2\left(i\omega + \nu_H\lambda^2 - \nu_H \frac{\omega^2}{c^2}\right) = 0 \tag{79}$$

where ν_g is the kinematic viscosity, ν_H is the magnetic diffusivity of the plasma, $V_x = H_x(\mu_e/\varrho_0)^{1/2}$ is the x component of the velocity of the Alfvén $V_y = H_y(\mu_e/\varrho_0)^{1/2}$ is the component of the velocity of the Alfvén wave. Equation (79) is a quadratic equation in λ^2, and hence we have two transverse waves. Under MFD approximations, the terms including the speed of light c may be neglected, and Eq. (79) becomes

$$\nu_g\nu_H\lambda^4 + [V_x^2 + i(\nu_g + \nu_H)^\omega]\lambda^2 - \omega^2 = 0 \tag{80}$$

If there is no external magnetic field ($H_x = 0$), Eq. (80) gives two simple modes: one is the viscous wave depending on ν_g, and the other is the magnetic wave depending on ν_H. These are the two basic transverse waves. If H_x is different from zero, there are couplings between these two basic waves, and we have two new transverse MFD waves. For an ideal plasma with $\nu_g = \nu_H = 0$, Eq. (80) gives

$$V = \omega/\lambda = V_x \tag{81}$$

This is known as the Alfvén wave, which has a speed of propagation V_x. It was Alfvén [16] who first showed that if there is a homogeneous magnetic field H_x in an incompressible, inviscid fluid of density ϱ_0 and of infinite electrical conductivity $\sigma = \infty$, the disturbance in this fluid will propagate as a wave in the direction of H_x with a speed V_x. The Alfvén wave and the corresponding velocity of propagation play an important role in magneto-fluid dynamics.

Longitudinal Waves

The dispersion relation for the longitudinal waves is

$$\left\{ \varkappa \left(\frac{1}{\varrho_0} + \frac{4i\omega v_q}{3p_0} \right) \lambda^4 - \left[\frac{\omega^2 \varkappa}{p_0} + \frac{4v_q \omega^2}{3T_0(\gamma - 1)} - ic_p\omega \right] \lambda^2 - \frac{i\omega^3}{T_0(\gamma - 1)} \right\}$$

$$\times \left[(v_H \lambda^2 + i\omega)(v_g \lambda^2 + i\omega) + V_x^2 \lambda^2 \right]$$

$$- \lambda^2 V_y^2 (i\omega + v_g \lambda^2) \left[\frac{\omega^2}{T_0(\gamma - 1)} - \frac{i\omega \lambda^2 \varkappa}{p_0} \right] = 0 \qquad (82)$$

The first line of Eq. (82) represents the sound waves in a viscous, heat-conducting medium, and the second line of Eq. (82) represents the transverse MFD wave. If there is no transverse magnetic field ($H_y = 0$), there is no coupling between the sound waves and the magnetic waves. If H_y is different from zero, i.e., $V_y \neq 0$, there are couplings between the gasdynamic and magnetic waves and Eq. (82) gives in general four coupled waves from the four basic waves: the basic sound, heat, viscous, and magnetic waves.

The effect of H_y on the sound waves may be seen clearly by considering an ideal plasma, in which $\varkappa = v_g = v_H = 0$; Eq. (82) gives

$$(a_0^2 - V^2)(V_x^2 - V^2) - V^2 V_y^2 = 0 \qquad (83)$$

In this case, there are two longitudinal waves because Eq. (83) gives two solutions for the velocity of wave propagation $V = \omega/\lambda_R$. We may call one wave a fast wave, with $V = V_f$, and the other a slow wave, with $V = V_s$. Equation (83) gives

$$V_s \leq a_0 = (\gamma R T_0)^{1/2} \leq V_f \qquad (84a)$$

and

$$V_s \leq V_x = H_x(\mu_e/\varrho_0)^{1/2} \leq V_f \qquad (84b)$$

For an ideal plasma, we have three characteristic velocities of wave propagation, V_x, V_f, and V_s, where V_x is for the transverse wave and the other two are for the longitudinal waves. The wave patterns in magnetofluid dynamics are considerably different from those in ordinary gas dynamics.

The wave patterns in magnetogasdynamics for an ideal plasma may be shown by the Friedrich diagram ([7,8]) of Fig. 6, in which the shape of the

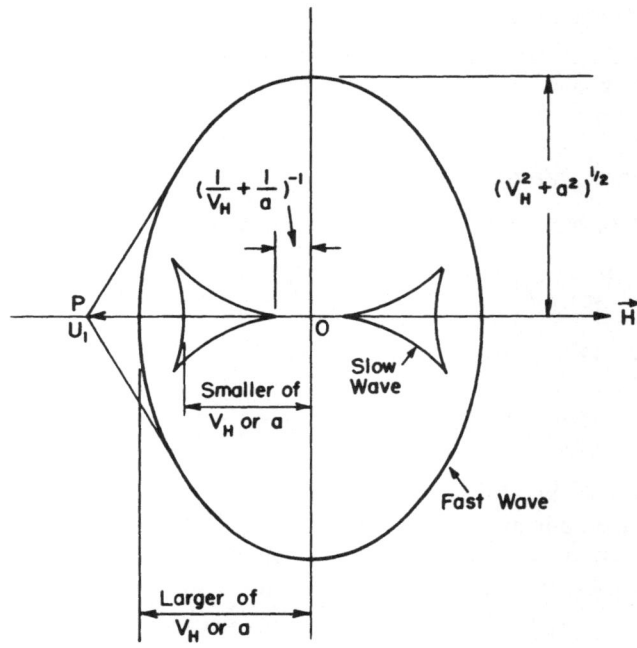

Fig. 6. Friedrich diagram of magnetogasdynamic wave in an ideal plasma.

disturbance that propagates from a point disturbance at the origin is shown. In Fig. 6, the abscissa is in the direction of the magnetic field and the flow direction may be in any arbitrary direction with respect to the magnetic field. Basically, we have two characteristic speeds: one is the ordinary sound speed $a = (\gamma RT)^{1/2}$, and the other is the Alfvén wave speed $V_H = H(\mu_e/\varrho)^{1/2}$, where H is the magnitude of the magnetic field and ϱ is the local density of the plasma. The speed of wave propagation in magnetogasdynamics V may be expressed in terms of a and V_H as follows [cf. Eq. (83)]:

$$V = \{\tfrac{1}{2}[(a^2 + V_H^2) \pm (a^2 + V_H^2)^2 - 4a^2 V_H^2 \cos^2 \theta]\}^{1/2} \qquad (85)$$

where θ is the angle between the magnetic field $\mathbf{H}$ and the velocity of the point source $OP = \mathbf{u}$. The plus sign in Eq. (85) is for the fast wave, while the minus sign is for the slow wave. Figure 7 shows some typical wave patterns for various velocities of the point source. When the velocity of the point source is $\mathbf{u} = OP_1$, which is larger than both a and V_H and is parallel to $\mathbf{H}$, we have the wave pattern shown in Fig. 7a, which is known as super-Alfvén flow, and is a set of backward-inclined waves as in ordinary

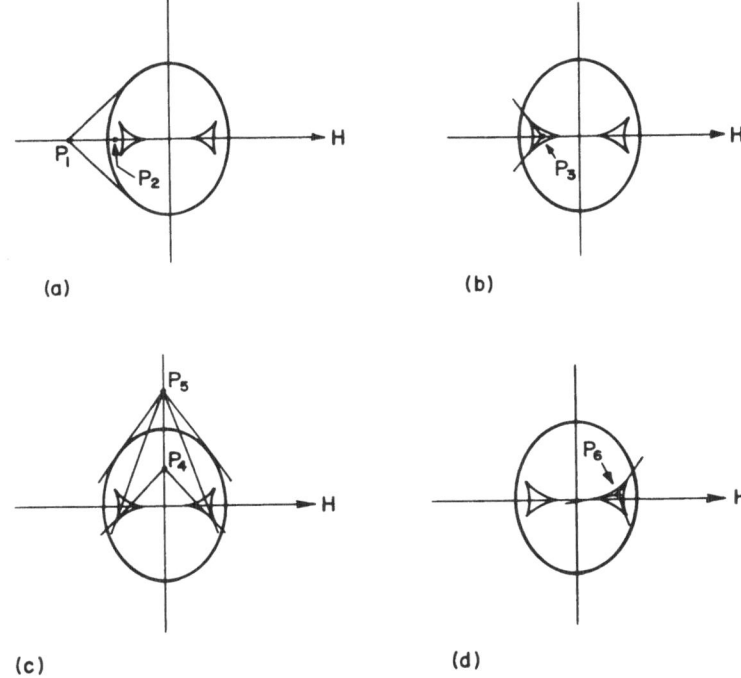

Fig. 7. Wave patterns in magnetogasdynamics of an ideal plasma. (a) Super-Alfvén wave. (b) Sub-Alfvén wave. (c) Transverse magnetic field. (d) Two wave patterns.

supersonic flow. If $\mathbf{u} = OP_2$, there will be no standing wave, which is similar to the case for subsonic flow in ordinary gas dynamics. If $\mathbf{U} = OP_3$ in the curved triangle of the slow wave, we have a forward-inclined wave pattern which is not possible in ordinary gas dynamics (Fig. 7b). When the velocity $\mathbf{u}$ is perpendicular to $\mathbf{H}$, we have one set of waves if $\mathbf{u} = OP_4 < (a^2 + V_H^2)^{1/2}$ and two sets of waves if $\mathbf{u} = OP_5 > (a^2 + V_H^2)^{1/2}$. The case of two sets of waves is not possible in ordinary gas dynamics (Fig. 7c). Finally, if $\mathbf{u} = OP_6$, which is in the curved triangle but not parallel to $\mathbf{H}$, we have two sets of waves: one set is inclined forward, and the other, backward (Fig. 7d). This is also a case which cannot occur in ordinary gas dynamics.

It should be noticed that not only do we have slow and fast waves of infinitesimal amplitude, but we also have slow and fast waves of finite amplitude and the corresponding shock waves, too.

If we study the three-dimensional unsteady flow of an ideal plasma, we will easily find that the characteristic velocities are V_x, V_s, and V_f, which are given in Eq. (83), except that the local values of a, V_x, and V_y should be used. Similarly, we have three types of magnetogasdynamic shock.

For an oblique magnetogasdynamic shock, we have to consider both the direction of the magnetic field and that of the flow velocity with respect to the shock front. In ordinary gas dynamics, by proper choice of the coordinate system, it is always possible to reduce the oblique shock to a corresponding normal-shock case. However, it is not in general possible to do so in magnetogasdynamics. However, it is possible to choose the coordinate system so that the velocity and the magnetic field are parallel both in front of and behind the shock front. In this way, we may classify the shock waves.

Let us consider the case where the velocity and magnetic field are parallel on both sides of a shock front (Fig. 8). Let the subscript 1 refer to a value in front of the shock, and the subscript 2 to a value behind the shock. Let θ be the turning angle of the magnetic field across the shock and ω the shock angle. Figure 8 shows the orientation of the fast and slow shocks [9]:

1. If the shock angle is less than 90°, we have the fast shock, which is similar to that in ordinary gas dynamics.

2. When the shock angle is between 90 and 180°, we have the slow shock, which is different entirely from those in ordinary gas dynamics. We have here an upstream-inclined shock.

3. When the shock angle is 90°, the tangential component of the magnetic field is zero in front of the shock, but different from zero behind the shock. This is the "switch-on" shock S_n.

4. When the shock angle is equal to $90° + \theta$, the tangential magnetic

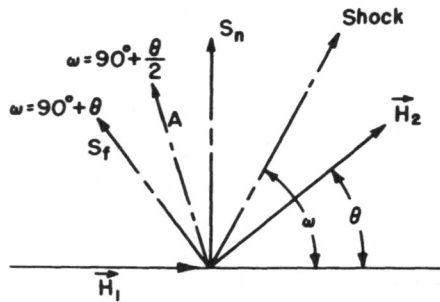

Fig. 8. Magnetogasdynamic oblique shocks
with $\mathbf{u} \parallel \mathbf{H}$.

field is different from zero in front of the shock, but vanishes behind the shock. This is the "switch-off" shock S_f.

5. When the shock angle is equal to $90° + \frac{1}{2}\theta$, we have Alfvén's shock, or a shock corresponding to the transverse wave.

The Rankine–Hugoniot relations across a magnetogasdynamic shock are different from those for an ordinary shock in gas dynamics. To illustrate this difference, let us consider a normal shock under a transverse magnetic field, which corresponds for the special case of a fast shock with flow velocity u normal to the shock front and with magnetic field parallel to the shock front. In this case, the ratio of the velocity behind the shock u_2 to that in front of the shock u_1 is

$$
\frac{u_2}{u_1} = \frac{1}{2}\left[\frac{\gamma - 1}{\gamma + 1} + \frac{2}{(\gamma + 1)M_1^2} + \frac{2\gamma h_1^2}{\gamma + 1}\right]
$$
$$
+ \frac{1}{2}\left\{\left[\frac{\gamma - 1}{\gamma + 1} + \frac{2}{(\gamma + 1)M_1^2} + \frac{2\gamma h_1^2}{\gamma + 1}\right]^2 + \frac{8(2 - \gamma)h_1^2}{\gamma + 1}\right\}^{1/2} \quad (86)
$$

where M_1 is the Mach number (u_1/a_1) in front of the shock and $h_1 = H_1/(2\varrho_1 u_1/\mu_e)^{1/2}$ is the dimensionless magnetic field in front of the shock. When there is no magnetic field, Eq. (86) reduces to the normal-shock relation in ordinary gas dynamics. For a given value of M_1, the ratio u_2/u_1 increases and the strength of the shock decreases as the magnetic field h_1 increases. This occurs because the shock strength depends on the effective Mach number $M_{e,1} = u_1/a_{e,1}$ and

$$
a_{e,1} = (a_1^2 + V_y^2)^{1/2} \quad (87)
$$

If $h_1^2 > h_c^2 = \frac{1}{2}[1 - (1/M_1^2)]$ and $M_{e,1} < 1$, there will be no shock.

11. BOUNDARY LAYER FLOW IN MAGNETOFLUID DYNAMICS [10]

In this section, we discuss some flow problems for a viscous, electrically-conducting fluid. We shall assume that the Reynolds number is large and the Prandtl number of the fluid is of the order of unity, but the magnetic Prandtl number may be of the order of unity or very small. We are going to treat these two magnetic Prandtl number cases separately, because when the magnetic Reynolds number is small, we do not have the magnetic boundary layer, and when the magnetic Reynolds number is large, we shall have the magnetic boundary layer as well as the ordinary boundary layer.

Case of Very Small Magnetic Reynolds Number [10]

First we consider the case of large Reynolds number but small magnetic Reynolds number. We have boundary layer flow of velocity and temperature, but not for the magnetic field. Hence in the boundary layer flow we may assume that the applied electromagnetic fields E_0 and H_0 or B_0 are unaffected by the flow field. For simplicity, let us consider the case where there is an externally-applied transverse magnetic field $H_{y,0}$ in the direction normal to the main flow direction u. The boundary layer direction is y, with $u \ll v$. The pondermotive force has a component in the x direction only, i.e.,

$$\mathbf{F}_e = \mathbf{J} \times \mathbf{B} = -\mathbf{i}\sigma u B_{y,0}^2 \tag{88}$$

where $\mathbf{J} = \mathbf{k}\sigma u B_{y,0}$ is the z direction only and $\mathbf{i}$ is the unit vector in the x direction. The Joule heat in the present case is

$$J^2/\sigma = \sigma u^2 B_{y,0}^2 \tag{89}$$

The two-dimensional boundary layer equations for the present case are

$$\frac{\partial \varrho}{\partial t} + \frac{\partial \varrho u}{\partial x} + \frac{\partial \varrho v}{\partial y} = 0 \tag{90a}$$

$$\frac{\partial u}{\partial t} + u \frac{\partial u}{\partial x} + v \frac{\partial u}{\partial y} = -\frac{1}{\varrho} \frac{\partial p}{\partial x} + \frac{1}{\varrho} \frac{\partial}{\partial y} \left(\mu \frac{\partial u}{\partial y} \right) - \frac{\sigma u}{\varrho} B_{y,0}^2 \tag{90b}$$

$$\varrho \left(\frac{\partial h}{\partial t} + u \frac{\partial h}{\partial x} + v \frac{\partial h}{\partial y} \right)$$
$$= \frac{\partial p}{\partial t} + u \frac{\partial p}{\partial x} + \frac{\partial}{\partial y} \left(\varkappa \frac{\partial T}{\partial y} \right) + \mu \left(\frac{\partial u}{\partial y} \right)^2 + \sigma u^2 B_{y,0}^2 \tag{90c}$$

$$p = \varrho RT \tag{90d}$$

where $h = c_p T$ is the enthalpy of the plasma, and the pressure is assumed to be a given function of x.

For steady flow, Eqs. (90a–d) give the relation

$$\varrho \left(u \frac{\partial h_0}{\partial x} + v \frac{\partial h_0}{\partial y} \right) = \frac{\partial}{\partial y} \left(\mu \frac{\partial h_0}{\partial y} \right) + \left(\frac{1}{\text{Pr}} - 1 \right) \frac{\partial}{\partial y} \left(\mu \frac{\partial u}{\partial y} \right) \tag{91}$$

If the Prandtl number $\text{Pr} = c_p \mu / \varkappa$ is unity, we have

$$h_0 = h + \tfrac{1}{2} u^2 = \text{const} = \text{stagnation enthalpy} \tag{92}$$

as a particular integral of Eq. (91). This is known as the Busemann relation in ordinary gas dynamics, and holds here too; however, the corresponding Crocco relation does not hold true here.

Equations (90a–d) may be solved by the usual method of solving boundary layer equations. For instance, if we consider the case of boundary layer flow over a semiinfinite plate in an incompressible fluid, we may use the power series method in x for the stream function ψ as follows:

$$\psi = (Uv_gx)^{1/2}[f_0(\eta) + mxf_2(\eta) + (mx)^2f_4(\eta) + \cdots] \tag{93}$$

where $m = (\sigma/\varrho U)B_{y,0}^2$, $\eta = y(U/v_gx)^{1/2}$, and U is the free-stream velocity. Substituting Eq. (93) into Eq. (90b), we have a set of total differential equations for f_n as follows:

$$2f_0''' + f_0''f_0 = 0 \tag{94a}$$

$$2f_2''' = 2f_0'f_2' - f_0f_2'' - 3f_2f_0'' + 2f_0' \tag{94b}$$

$$2f_4''' = 4f_0'f_4' + 2f_2'f_2' - f_0f_4'' - 3f_2f_2'' - 5f_4f_0'' + 2f_2' \tag{94c}$$

with the boundary conditions

$$
\begin{aligned}
f_0 = f_2 = f_4 = \cdots = 0 && \text{at} \quad \eta = 0 \\
f_0' = f_2' = f_4' = \cdots = 0 && \text{at} \quad \eta = 0 \\
f_0' = 1; \quad f_2' = -1; \quad f_4' = \cdots = 0 && \text{at} \quad \eta = \infty
\end{aligned}
\tag{95}
$$

Equation (94a) is the well-known Blasius equation of ordinary boundary layer flow. Equations (94b), (94c), etc., are linear equations which can be numerically integrated from the boundary conditions (95). Up to the term f_2, the skin friction coefficient is

$$c_f = \frac{1}{\frac{1}{2}\varrho U^2} \mu\left(\frac{\partial u}{\partial y}\right)_{y=0} = \frac{0.664 - 1.788mx + \cdots}{(R_{ex})^{1/2}} \tag{96}$$

where $R_{ex} = Ux/v_g$.

After $u(x, y)$ is obtained, we may either use the relation (92) to find the temperature if the wall is insulated, or to solve Eq. (90c) by means of series expansion with the known values of velocity and stream function.

Case of Large Magnetic Reynolds Number

If the magnetic Reynolds number and the ordinary Reynolds number are both large, we can derive the boundary layer equations for velocity, magnetic field, and temperature from Eqs. (37) and (38). For two-dimensional flow, we have the following boundary layer equations for a plasma with a high magnetic Reynolds number:

$$p = \varrho RT \tag{97a}$$

$$\frac{\partial \varrho}{\partial t} + \frac{\partial \varrho u}{\partial x} + \frac{\partial \varrho v}{\partial y} = 0 \tag{97b}$$

$$\frac{\partial H_x}{\partial t} + u \frac{\partial H_x}{\partial x} + v \frac{\partial H_x}{\partial y} + H_x \frac{\partial v}{\partial y} - H_y \frac{\partial u}{\partial y} = \frac{\partial}{\partial y} \left(\nu_H \frac{\partial H_x}{\partial y} \right) \tag{97c}$$

$$\frac{\partial H_x}{\partial x} + \frac{\partial H_y}{\partial y} = 0 \tag{97d}$$

$$\varrho \left(\frac{\partial u}{\partial t} + u \frac{\partial u}{\partial x} + v \frac{\partial u}{\partial y} \right) - B_y \frac{\partial H_x}{\partial y} = - \frac{\partial p}{\partial x} + \frac{\partial}{\partial y} \left(\mu \frac{\partial u}{\partial y} \right) \tag{97e}$$

$$p + \tfrac{1}{2} B_x H_x = p_0(x) = \text{given function of } x \tag{97f}$$

$$\varrho \left(\frac{\partial h}{\partial t} + u \frac{\partial h}{\partial x} + v \frac{\partial h}{\partial y} \right)$$

$$= \frac{\partial p}{\partial t} + u \frac{\partial p}{\partial x} + \frac{\partial}{\partial y} \left(\varkappa \frac{\partial T}{\partial y} \right) + \mu \left(\frac{\partial u}{\partial y} \right)^2 + \nu_H \left(\frac{\partial H_x}{\partial y} \right)^2 \tag{97g}$$

In solving Eqs. (97), we have two new parameters: magnetic Reynolds number and magnetic pressure number, or the magnetic Mach number. There are many new phenomena due to these new parameters. For instance, in the sub-Alfvén flow, i.e., $M_m = u/V_x < 1$, we may have an upstream influence of the boundary layer flow. Without going into lengthy calculations, we shall consider the MHD wake problem in the next section to illustrate the upstream influence of a boundary layer flow.

12. WAKES IN MAGNETOHYDRODYNAMICS [11]

We consider the two-dimensional flow of a uniform stream U over a body under a uniform magnetic field $\mathbf{H}_0 = \mathbf{i}H_{x,0} + \mathbf{j}H_{y,0} + \mathbf{k}0$ in an incompressible, electrically-conducting liquid. We assume that the deviations of the velocity and of the magnetic fields from the values in the uniform state are small. Hence we write $u = U + u'$, $v = v'$, $H_x = H_{x,0} + h_x$, and $H_y = H_{y,0} + h_y$, such that u', $v' \ll U$ and h_x, $h_y \ll H_{x,0}$, $H_{y,0}$.

If we neglect the higher-order terms of the perturbed quantities u',

v', etc., the magnetohydrodynamic equations (36) and (37) give the following dimensionless linearized equation:

$$\nabla^2\left[\left(\frac{\partial}{\partial x} - \frac{1}{\mathrm{Re}}\nabla^2\right)\left(\frac{\partial}{\partial x} - \frac{1}{\mathrm{Re}_\sigma}\nabla^2\right) - \left(R_{H,x}\frac{\partial}{\partial x} - R_{H,y}\frac{\partial}{\partial y}\right)^2\right]u^* = 0 \tag{98}$$

where x and y are the dimensionless coordinates in terms of a reference length L, $u^* = u'/U$, $\nabla = (\partial^2/\partial x^2) + (\partial^2/\partial y^2)$, $\mathrm{Re} = UL/v_g$, $\mathrm{Re}_\sigma = UL/v_H$, $R_{H,x} = \mu_e H_{x,0}^2/\varrho U^2$, and $R_{H,y} = \mu_e H_{y,0}^2/\varrho U^2$.

The operator of the square bracket may be considered as a product of two Oseen-type operators. For instance, if we assume that $\mathrm{Re} = \mathrm{Re}_\sigma$, Eq. (98) becomes

$$\nabla^2\left[(1 + R_{H,x})\frac{\partial}{\partial x} + R_{H,y}\frac{\partial}{\partial y} - \frac{1}{\mathrm{Re}}\nabla^2\right]$$

$$\left[(1 - R_{H,x})\frac{\partial}{\partial x} - R_{H,y}\frac{\partial}{\partial y} - \frac{1}{\mathrm{Re}}\nabla^2\right]u^* = 0 \tag{99}$$

The two operators in the square brackets in Eq. (99) are Oseen-type operators, which may be written in the general form:

$$\left[A_1\frac{\partial}{\partial x} + B_1\frac{\partial}{\partial y} - \frac{1}{\mathrm{Re}}\nabla^2\right]u^* = 0 \tag{100}$$

The solution of Eq. (100) is

$$u^* = \mathrm{const}\,\exp\left[\frac{(A_1 x + B_1 y)\mathrm{Re}}{2}\right]K_0\left[\frac{(A_1^2 + B_1^2)\mathrm{Re}\,r}{2}\right] \tag{101}$$

where $r^2 = x^2 + y^2$ and K_0 is the Bessel function of zeroth order.

For a given radial distance from the origin, the maximum value of u^* occurs at

$$\tan\theta = y/x = B_1/A_1 \tag{102}$$

Thus we have a wake along the direction

$$\theta = \theta_m = \tan^{-1}(B_1/A_1) \tag{103}$$

Since there are two Oseen operators in Eq. (99), we have two wakes, which are, respectively, along the lines (Fig. 9)

$$\theta_{m1} = \tan^{-1}\left(\frac{R_{H,y}}{1 + R_{H,x}}\right); \qquad \theta_{m2} = \tan^{-1}\left(-\frac{R_{H,y}}{1 - R_{H,x}}\right) \tag{104}$$

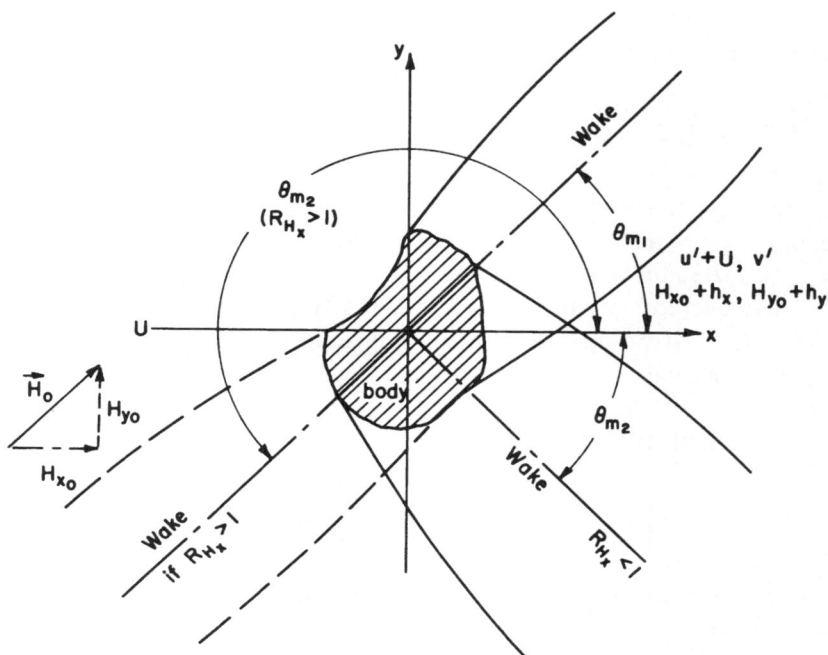

Fig. 9. MHD wakes around a body in an electrically-conducting liquid.

If $R_{H,x}$ is greater than unity, i.e., $M_m < 1$, we have an upstream-inclined wake, which is impossible in ordinary hydrodynamics. There are two wakes instead of one if $R_{H,y}$ is different from zero. This is also a new phenomenon appearing in magnetohydrodynamics.

From Eq. (101), we see that the rate of decrease of u^* with r increases with Reynolds number Re and we have boundary layer phenomena for large value of Re, i.e., the spread of the wake is within a narrow region from the axis which makes an angle θ_{m1} or θ_{m2} with respect to the main flow U. Thus if $R_{H,x} > 1$, we have the upstream influence of the boundary layer flow.

13. TENSOR ELECTRICAL CONDUCTIVITY [12]

In our analysis of electromagnetogasdynamics in previous sections, we assumed that the electrical conductivity σ is a scalar quantity, and Eq. (13) is used for the electrical conduction current. Hence the resultant conduction current is parallel to the electric field $\mathbf{E}_u$. For such an electric current in a magnetic field $\mathbf{H}$, there is an electromagnetic force $\mathbf{I} \times \mathbf{B}$ in the direction perpendicular to both $\mathbf{H}$ and $\mathbf{I}$ or $\mathbf{E}_u$. As a result, the charged particles will move in the direction perpendicular to $\mathbf{E}_u$ too. Thus the resultant electric current will not be parallel to $\mathbf{E}_u$. We have a Hall current. When we take the Hall current into account, we should consider the electrical conductivity σ in Eq. (13) as a tensor. We shall derive the electric current equation including Hall current in Section 15. In this section, we simply give the expression of the generalized Ohm's law with Hall current and ion slip included and discuss their effects on the flow problems.

The generalized Ohm's law with Hall current and ion slip included is

$$
\begin{aligned}
\mathbf{I} &= \frac{1 + \beta_i\beta_e + \beta_e{}^2}{(1+\beta_i\beta_e)^2+\beta_e{}^2}\,\sigma\mathbf{E}_u'' + \frac{(1 + \beta_i\beta_e)\sigma\mathbf{E}_u'}{(1+\beta_i\beta_e)^2+\beta_e{}^2} + \frac{1 + \beta_i\beta_e + \beta_e{}^2}{(1+\beta_i\beta_e)^2+\beta_e{}^2}\left(\frac{\mathbf{B}}{B}\times\mathbf{E}_u\right) \\
&= \sigma_T\mathbf{E}_u
\end{aligned}
\tag{105}
$$

where σ is the scalar electrical conductivity of Eq. (13); $\mathbf{E}_u = \mathbf{E}_u' + \mathbf{E}_u''$, with $\mathbf{E}_u'' = (\mathbf{E}_u \cdot \mathbf{B})\mathbf{B}/B^2$ the component of $\mathbf{E}_u$ parallel to $\mathbf{B}$ and $\mathbf{E}_u'$ $= [\mathbf{B}\times(\mathbf{E}_u\times\mathbf{B})]/B^2$ the component of $\mathbf{E}_u$ perpendicular to $\mathbf{B}$; $\beta_e = en_eB/K_{ei}$ $= \omega_c/f = WB(\varrho_0/\varrho)$, with $f = K_{ei}/m_e$ the collision frequency between ions and electrons, K_{ei} the friction coefficient between ions and atoms, m_e the mass of the electron, and n_e the number density of the electron; $\sigma = e^2n_e{}^2/K_{ei}$ is the scalar electrical conductivity; $\omega_c = en_eB/m_e$ is the cyclotron frequency of the electron, with B the magnitude of magnetic induction $\mathbf{B}$; and β_i $= en_eB/[1 + (n_e/n_a)]K_{ai}$, with K_{ai} is the friction coefficient between ions and atoms and n_a the number density of atoms in the plasma.

In Eq. (105), we neglect the effect of collisions between electrons and atoms. The factor β_e determines essentially the Hall effect, while the factor ϱ_i determines the ion-slip effect. If the magnetic field strength is small, we have both β_e and β_i negligibly small in comparison with unity, and Eq. (105) reduces to Eq. (13), i.e.,

$$
\mathbf{I} = \sigma\mathbf{E}_u
\tag{106}
$$

Let us consider the case where the ion-slip factor β_i is negligible, but the Hall factor β_e is not negligible. Equation (105) may be written as follows:

$$
\mathbf{I} = \sigma\mathbf{E}_u'' + \sigma'\mathbf{E}_u' + \sigma_h(\mathbf{E}_u' \times \mathbf{B}_1)
\tag{107}
$$

where $\mathbf{B}_1$ is the unit vector in the direction of $\mathbf{B}$. Equation (107) shows that in the direction of the magnetic field $\mathbf{B}$, the electrical conductivity of the plasma has the value of the scalar electrical conductivity σ, while in the other two directions perpendicular to **B**, the electrical conductivity has a value less than the scalar electrical conductivity σ. In the direction of $\mathbf{E}_u'$, the electrical conductivity is σ',

$$\sigma' = \frac{\sigma}{1 + (\omega_c/f)^2} \tag{108}$$

The value of σ' decreases with increase of magnetic induction **B**. Only when (ω_c/f) is negligible will σ' be equal to σ.

The last term on the right-hand side of Eq. (107) is the Hall current, which is in the direction perpendicular to both **B** and $\mathbf{E}_u$. The electrical conductivity of the Hall component of the conduction current may be written as

$$\sigma_h = (\omega_c/f)\sigma' \tag{109}$$

Hence only when (ω_c/f) is negligible may we neglect the Hall current.

To show the effects of tensor electrical conductivity on the flow pattern, we reexamine the channel flow of magnetohydrodynamics discussed in Section 9 by including the Hall current, i.e., we use Eq. (107) to replace Eq. (13). The general configuration of the channel is still the same as that shown in Fig. 3. With the Hall current, we cannot assume the z component of velocity, w, and the z component of the magnetic field, H_z, to be zero. The Hall current produces a force in the z direction which causes the motion in the z direction even though there is no net flow in the z direction. The pressure gradient in the z direction p_z will not be zero either.

With the Hall current, we have the following expressions for the dimensionless unknowns of the present problem [13]:

$$u = U_0 u^*(y^*), \qquad v = 0, \qquad w = U_0 w^*(y^*),$$

$$E_x = E_{x,0}, \qquad E_y = E_{y,0}, \qquad E_z = E_{z,0}$$

$$H_x = H_0 H_x^*(y^*), \qquad H_y = H_0, \qquad H_z = H_0 H_z^*(y^*) \tag{110}$$

$$p = \varrho U_0^2 p^*(x^*, y^*, z^*) = \varrho U_0^2 [x^* p_x^* + z^* p_z^* + p_1^*(y^*)]$$

The definition of all variables is the same as in Eq. (62). The electric field components $E_{x,0}$, $E_{y,0}$, and $E_{z,0}$ are constant or zero. The pressure gradients p_x^* and p_z^* are constant. Here we have five unknowns, u^*, w^*, H_x^*, H_z^*, and p_1^* which are governed by the following equations:

$$\frac{1}{\mathrm{Re}}\frac{d^2 u^*}{dy^{*2}} + R_H \frac{dH_x^*}{dy^*} = p_x^* \tag{111a}$$

$$\frac{1}{\text{Re}} \frac{d^2 w^*}{dy^{*2}} + R_H \frac{dH_z^*}{dy^*} = p_z^* \tag{111b}$$

$$\frac{dp_1^*}{dy^*} = -\frac{1}{2} R_H \frac{d}{dy^*} (H^{*2} + H^{*2}) \tag{111c}$$

$$(1 + \beta_e^2) \frac{d^2 H_x^*}{dy^{*2}} + \text{Re}_\sigma \frac{du^*}{dy^*} + \beta_e \text{Re}_\sigma \frac{dw^*}{dy^*} = 0 \tag{111d}$$

$$(1 + \beta_e^2) \frac{d^2 H_z^*}{dy^{*2}} + \text{Re}_\sigma \frac{dw^*}{dy^*} - \beta_e \text{Re}_\sigma \frac{du^*}{dy^*} = 0 \tag{111e}$$

where Re is the Reynolds number, Re_σ is the magnetic Reynolds number, and R_H is the magnetic pressure number. All these dimensionless parameters have the same definitions as given in Section 9. Strictly speaking, the Hall factor β_e depends on the resultant magnetic field B instead of the applied magnetic induction B_0. We shall assume that the induced magnetic induction is small and use B_0 in the definition of $\beta_e = en_e B_0/K_{ei} = w_{c,0}/f$.

We consider here only the Poiseuille flow with the boundary conditions:

$$u^* = w^* = 0, \qquad H_x^* = H_z^* = 0 \qquad \text{at} \quad y^* = \pm 1$$
$$p_1^* = p_0 = \text{const} \qquad\qquad\qquad \text{at} \quad y^* = 1 \tag{112}$$

In solving Eq. (111), it is convenient to introduce the following complex variables:

$$V^* = u^* - iw^*; \qquad H^* = H_x^* - iH_z^*; \qquad p^* = p_x^* - ip_z^* \tag{113}$$

where $i = \sqrt{-1}$. With the complex variables of Eq. (113), Eqs. (111a–e) may be reduced to the following equation:

$$\frac{d^3 V^*}{dy^{*3}} - R_h^* \frac{dV^*}{dy^*} = 0 \tag{114}$$

where

$$R_h^* = (\text{Re}\,\text{Re}_\sigma R_H)^{1/2}/(1 - i\beta_e)^{1/2} = R_h/(1 - i\beta_e)^{1/2}$$

is the complex Hartmann number.

Equation is the (114) has the same form as Eq. (63), and we have a similar solution but with complex variables, i.e.,

$$V^* = A_1 \cosh R_h^* y^* + B_1 \sinh R_h^* y^* + C_1 \tag{116}$$

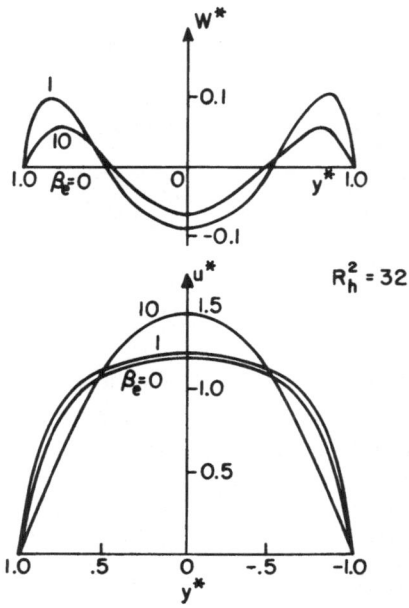

Fig. 10. Velocity distributions in a MHD
channel flow with Hall current.

where A_1, B_1, and C_1 are arbitrary complex constants to be determined by the boundary conditions. Some typical velocity distributions of MHD channel flow with Hall current included are shown in Fig. 10. Since R_h^* is a complex number, in the expression for V^*, we have sinusoidal terms such as $\sin a_1 y^*$ and $\cos a_1 y^*$, as well as the real hyperbolic terms which occurred in the case of scalar electrical conductivity (Section 9). Since we assume that there is no net flow in the z direction, the velocity distribution of w is of sinusoidal form, with the net flow equal to zero. The introduction of such a sinusoidal z velocity component is one of the new features of flow with Hall current.

Similarly, the Hall current, or the effect of tensor electrical conductivity, makes the flow pattern much more complicated than the corresponding case of scalar electrical conductivity. For instance, consider the force on a symmetrical body in a uniform stream without angle of attack under an applied transverse magnetic field. When β_e is small, the flow pattern will by symmetrical and there is drag force but no lift force; if β_e is not negligibly small, the flow pattern will be asymmetrical, and there will be both lift and drag [14].

14. TURBULENT FLOW IN MAGNETOHYDRODYNAMICS [14]

At high Reynolds number, flow in magnetofluid dynamics may be turbulent. In turbulent MFD flow, there are many new correlation terms between the magnetic field and the velocity field. In general, the problem is very complicated. In order to illustrate some essential features of turbulent flow in magnetofluid dynamics, we consider the case of incompressible fluid, i.e., turbulent MHD flow. First, we write

$$Q = \bar{Q} + Q' \tag{117}$$

where Q may represent any one of the MHD variables. The mean value of the fluctuating part Q' is zero, and $\bar{Q}$ is the mean value of Q. Substituting Eq. (117) into the fundamental equations of magnetohydrodynamics and taking the average of the resultant equations, we have the Reynolds equations of turbulent MHD flow as follows:

$$\frac{\partial \bar{u}_i}{\partial x_i} = 0 \tag{118a}$$

$$\varrho \frac{\partial \bar{u}_i}{\partial t} + \frac{\partial}{\partial x_k}(\varrho \overline{u_i u_k} - \mu_e \overline{H_i H_k}) = -\frac{\partial}{\partial x_i}\left(p + \frac{1}{2}\mu_e \bar{H}^2\right) + \mu \frac{\partial^2 \bar{u}_i}{\partial x_k^2} \tag{118b}$$

$$\frac{\partial \bar{H}_i}{\partial t} + \frac{\partial}{\partial x_k}(\overline{H_i u_k} - \overline{H_k u_i}) = \nu_H \frac{\partial^2 \bar{H}_i}{\partial x_k^2} \tag{118c}$$

where $\overline{u_i u_k} = \bar{u}_i \bar{u}_k + \overline{u_i' u_k'}$ and $\bar{H}^2 = \overline{H_i H_i} = \bar{H}_i \bar{H}_i + \overline{H_i' H_i'}$, etc., and the summation convention is used.

The seven equation represented by (118) may be used to determine the seven mean values of the pressure, three velocity components, and three magnetic field components. We still need some other relations for such turbulent correlation terms as $\overline{u_i' u_k'}$, $\overline{H_i' H_k'}$, and $\overline{u_i' H_k'}$. Similar to the case in ordinary gas dynamics, if we try to derive equations for these second-order correlation terms, we introduce third-order correlation terms. This is one of the main difficulties in trying to study turbulent flow based on the Reynolds equations. Usually, for engineering problems, we develop some empirical relations between these second-order correlation terms and the mean flow variables, using some empirical constants determined experimentally.

To illustrate this semiempirical theory of turbulent MHD flow, let us reconsider the two-dimensional channel flow of Fig. 3. We assume that the two plates are at rest and are insulated. There is no applied electric field. The boundary conditions for the mean and fluctuating quantities are as follows:

1. Both the mean and the fluctuating velocity components vanish on the plates.

2. We assume that the electrical conductivity is a scalar quantity; the boundary conditions for the mean magnetic field are $\bar{H}_x = 0$ and $\bar{H}_y = H_0$ at $|y| \geq L$.

3. The curl and divergence of the fluctuating magnetic field on the plates must be zero, and we have $\overline{H_x'H_y'} = 0$, $\overline{H_y'H_z'} = 0$, $\overline{H'^2} - \overline{H_x'^2} - \overline{H_z'^2} = 0$ at $y = L$.

Now we use the same type of dimensionless variables as those in Eq. (82), i.e., all the velocities are in terms of U_0 and all magnetic fields are in terms of H_0. The Reynolds equations (118) for the present case become:

$$\frac{1}{\mathrm{Re}}\frac{d^2\bar{u}^*}{dy^{*2}} + R_H\frac{d\bar{H}_x^*}{dy^*} - \frac{d}{dy^*}\left(\overline{u'^*v'^*} - R_H\overline{H_x'^*H_y'^*}\right) = \bar{p}_x^* \tag{119a}$$

$$\frac{1}{\mathrm{Re}_\sigma}\frac{d^2\bar{H}_x^*}{dy^{*2}} + \frac{d\bar{u}^*}{dy^*} - \frac{d}{dy^*}\left(\overline{H_x'^*v'^*} - \overline{H_y'^*u'^*}\right) = 0 \tag{119b}$$

$$\bar{p}_w^* - \bar{p}^* = \overline{v'^{*2}} + \tfrac{1}{2}R_H\left(\overline{H_x^{*2}} - \overline{H_y'^{*2}} - \overline{H_z'^{*2}}\right) \tag{119c}$$

Equations (119) have similar form as those for the laminar flow case in Section 9 except that we have three more Reynolds stresses. We may consider each of the terms in parentheses as a new variable, such as $(\overline{H_x'^*v'^*} - \overline{H_y'^*u'^*})$, etc. We have to make some reasonable assumptions to express these Reynolds stresses in terms of the mean flow variables $\bar{u}^*$ and $\bar{H}_x^*$ before we can solve for these variables. A preliminary attempt has been made in ([15]). However, since we do not have the necessary experimental results about the velocity distribution of MHD channel flow, it is not possible to decide how good a given assumption might be. We need some critical experimental investigations for turbulent MHD flow.

15. MULTIFLUID THEORY OF MAGNETOFLUID DYNAMICS ([12])

An ionized gas, or plasma, may be considered as a mixture of N species consisting of ions, electrons, and neutral particles. From a macroscopic point of view, a complete description of the flow field of a plasma should consist of the gasdynamic variables of all species, i.e., the velocity vectors $\mathbf{q}_s$, pressure p_s, density ϱ_s, and temperature T_s of each species s and the electromagnetic fields $\mathbf{E}$ and $\mathbf{H}$. Such an analysis is known as the multifluid theory of magnetogasdynamics. It is not feasible to discuss the multifluid theory in detail in this chapter because of limited space. Those interested in details should refer to (¹).

Since we have $6N + 6$ variables, we should use $6N$ gasdynamic equations for N species and 6 electromagnetic field equations as the fundamental equations of the multifluid theory of magnetogasdynamics. From the partial variables p_s, T_s, etc., we may define the gross variables of the plasma as a whole as follows: Pressure of the plasma

$$p = \sum_{s=1}^{N} p_s \tag{120}$$

number density of the plasma

$$n = \sum_{s=1}^{N} n_s \tag{121}$$

density of the plasma

$$\varrho = mn = \sum_{s=1}^{N} m_s n_s \tag{122}$$

(where m_s is the mass of a particle of the sth species and m is the mean mass of the plasma, which is a function of the composition of the plasma); temperature of the plasma

$$T = (1/n) \sum_{s=1}^{N} n_s T_s \tag{123}$$

ith velocity component of the plasma

$$u^i = (1/\varrho) \sum_{s=1}^{N} m_s n_s u_s{}^i \tag{124}$$

ith component of diffusion velocity of the plasma

$$w_s{}^i = u_s{}^i - u^i \tag{125}$$

excess electric charge

$$\varrho_e = \sum_{s=1}^{N} n_s e_s \tag{126}$$

(where e_s is the electric charge on a particle of the sth species); and ith component of electric current density

$$J^i = \sum_{s=1}^{N} n_s e_s u_s{}^i = \sum_{s=1}^{N} n_s e_s w_s{}^i + u^i \varrho_e = I^i + \varrho_e u^i \tag{127}$$

This shows that the electrical conduction current I^i is due to the diffusion velocities of all the charged particles. Hence the phenomenon is a very complicated one. Since the electrical conduction current depends on the movement of the charged particles, it must depend on both the electromagnetic forces and the gasdynamic forces. The electrical conduction current, or the diffusion velocity of the charged particles, is governed by a complicated differential equation which is due to the balance of various kinds of forces. For instance, the differential equation for the diffusion velocity $w_r{}^i$ is

$$\frac{\partial w_r{}^i}{\partial t} + u_r{}^j \frac{\partial u_r{}^i}{\partial x^j} - u^j \frac{\partial u^i}{\partial x^j} = \frac{1}{\varrho_r} \frac{\partial \tau_r^{ij}}{\partial x^j} + \frac{1}{\varrho} \frac{\partial p}{\partial x^i} - \frac{1}{\varrho} \frac{\partial \tau^{ij}}{\partial x^j}$$

$$+ \frac{X_r{}^i}{\varrho_r} - \frac{F_e{}^i + F_q{}^i}{\varrho} + \frac{\sigma_r}{\varrho_r} \, (Z_r{}^i = u_r{}^i) \quad (128)$$

where τ^{ij} is the ijth component of the viscous stress and F^i is the body force and $Z_r{}^i$ is the momentum associated with the mass source σ_r.

If we solve the diffusion velocities from equations of the type of (128), we may easily calculate the electrical conduction current. However, Eq. (128) is very complicated, and is very difficult to solve. Usually, we have to make simplifying assumptions in order to obtain some simple relation for the electrical conduction current. The assumptions which are made are:

1. The electrical conduction current, or the diffusion velocity, is explicitly independent of the time t and spatial coordinates x^i.

2. The electromagnetic forces are the only dominant forces in determining the diffusion velocities, and then the electrical conduction current.

3. There is no source term in the process.

Under the above three assumptions, Eq. (128) becomes

$$\frac{F_{er}^i + F_{0r}^i}{\varrho_r} = \frac{F_e{}^i}{\varrho} \quad (129)$$

or

$$(\varrho\varrho_{er} - \varrho_r\varrho_e)\mathbf{E}_u + (\varrho\varrho_{er}\mathbf{w}_r - \varrho_r\mathbf{I}) \times \mathbf{B} = \varrho \sum_{s=1}^{N} K_{rs}(\mathbf{w}_r - \mathbf{w}_s) \quad (130)$$

Equation (130) gives a set of linear algebraic equations for ω_r if we assume that ϱ_r, ϱ_{er}, $\mathbf{E}_u$, $\mathbf{B}$, and K_{rs} are given. For instance, if we consider the plasma as a mixture of ions, electrons, and one type of neutral particle, we obtain Eq. (105) [see [12]].

Since the diffusion velocity also depends on gasdynamic forces, Eq.

(130) is only a first approximation. We should include other forces if more accurate results are required. If we include the pressure gradient of the electrons and neglect β_i but include β_e, the equation of electrical conduction current becomes

$$\mathbf{I} + \frac{1}{en_e}(\mathbf{I} \times \mathbf{B}) = \sigma\left(\mathbf{E}_u + \frac{1}{en_e}\nabla p_e\right) = \sigma\mathbf{E}_{uT} \qquad (131)$$

The effect of the pressure gradient of the electrons is to increase the effective electric field strength. If we include β_i, additional terms will be obtained to Eq. (131). Since such a piecewise improvement of electrical conduction current will not include all the gasdynamic effects, particularly the effects of different temperatures between species, it is better to use the multifluid theory for accurate analysis, in which all the gasdynamic effects on the electric current density will be included.

ACKNOWLEDGMENT

This chapter is based on work supported in part by the US Air Force through the Air Force Office of Scientific Research under grant No. AFOSR-1015-67A.

NOTATION

$\mathbf{A}$	vector potential, defined in Eq. (17)
$A(x)$	cross-sectional area of a channel as a function of x
$a = (\gamma RT)^{1/2}$, sound speed	
a_e	effective sound speed, defined in Eq. (56)
$\mathbf{B}$	magnetic induction
c	velocity of light
c_p	specific heat at constant pressure
c_v	specific heat at constant volume
$\mathbf{D}$	dielectric displacement, given in Eq. (9)
$\mathbf{E}$	electric field strength
$\mathbf{E}_u$	electric field strength in moving coordinates, gives in Eq. (13)
e_m	total energy of gas
$\mathbf{F}$	body force
f_n	stream functions, defined in Eq. (93)
g	gravitational acceleration

H	magnetic field
h	perturbed magnetic field
h	enthalpy
h_0	stagnation enthalpy, given in Eq. (38)
I	electrical conduction current density
$i = \sqrt{-1}$	
i, j, k	unit vectors in the x, y, and z directions, respectively
J	electrical current density
K_{rs}	friction coefficient between species r and s
k	Boltzmann constant
M	Mach number
M_m	magnetic Mach number, defined in Eq. (42)
m	mass of a particle
n	unit normal vector
n	number density
Pr	Prandtl number
p	pressure
p_e	pressure of electrons
p_x	x-wise pressure gradient
p_z	z-wise pressure gradient
q	velocity vector
R	gas constant
R_E	electric field number, defined in Eq. (31)
R_H	magnetic pressure number, defined in Eq. (42)
R_h	Hartmann number, defined in Eqs. (67) and (115)
$R_h{}^*$	complex Hartmann number, defined in Eq. (115)
R_r	relativistic parameter, defined in Eq. (32)
R_t	time parameter, defined in Eq. (30)
Re	Reynolds numbers
Re_σ	magnetic Reynolds number, defined in Eq. (44)
S	entropy
T	temperature
t	time
U_m	internal energy
u, v, w	x, y, and z components of velocity, respectively
V_H	speed of Alfvén wave (components: V_x, V_y), defined in Eq. (43)
$\mathbf{w}_s$	diffusion velocity of sth species
x, y, z	Cartesian coordinates
x^i	ith Cartesian coordinate

Greek Letters

β_e	Hall factor, defined in Eq. (105)
β_i	ion-slip factor, defined in Eq. (105)
γ	ratio of specific heats, c_p/c_v

$\delta^{ij} = 0$ if $i \neq j$, $\delta^{ij} = 1$ if $i = j$

ε	inductive capacity
η	similarity coordinate defined in Eq. (16.93)
$\varkappa$	coefficient of heat conductivity

$\lambda = \lambda_R + i\lambda_i$ complex wave number

μ	coefficient of viscosity
μ_e	magnetic permeability
ν_g	coefficient of kinematic viscosity
ν_H	magnetic diffusivity, defined in Eq. (44)
ϱ	density of a fluid
ϱ_e	excess electric charge
σ	scalar electric conductivity
σ_T	tensor electrical conductivity
τ	viscous tensor with component τ^{ij}
Φ	viscous dissipation, defined in Eq. (8)
ϕ	scalar potential, defined in Eq. (22)
ψ	stream function
ω	frequency
ω_c	cyclotron frequency, defined in Eq. (105)

Subscripts

i, j	ith, jth components of a vector quantity
s, r	values of sth, rth species

Superscripts

*	dimensionless quantity
i, j	ith, jth components of a vector quantity

REFERENCES

1. S. I. Pai, *Magnetogasdynamics and Plasma Dynamics*, Springer Verlag, Vienna and New York (1962).
2. J. A. Stratton, *Electromagnetic Theory*, McGraw-Hill Book Co., New York (1941).
3. H. Hasimoto and S. Kuwabara, "Electrogasdynamics," *J. Phys. Soc. Japan* **20** (5), 859 (1965).

4. S. I. Pai, "Quasi-One Dimensional Analysis of Magnetogasdynamics. Electricity from MHD," in: *Proc. International Symposium on MHD Electrical Power Generation*, Vol. I, Salzburg, Austria, International Atomic Energy Agency (1966), p. 283.

5 F. H. Clauser, *Plasma Dynamics*, Aero. & Astro. International Series of Aero. Sci. and Space Flight, Div. IX, Vol. 4, Pergamon Press, New York (1960), p. 305.

6. S. I. Pai, "Magnetohydrodynamics of Channel Flow," in: *Advances in Hydroscience*, Vol. 3, V. T. Chow, ed., Academic Press, New York (1966), p. 63.

7. K. O. Friedrichs and H. Kranzer, "Notes on MHD VIII, Nonlinear Wave Motion," NYU Report NYO 6486 (July 1958).

8. W. R. Sears and E. L. Resler, Jr., "Sub- and Super-Alfvénic Flows Past Bodies," in: *Advances in Aeronautical Science*, Vol. 4, Pergamon Press, New York (1961), p. 657.

9. J. Bazer and W. B. Ericson, "Oblique Shock Waves in a Steady Two-Dimensional Hydromagnetic Flow," in: *Proc. of Symp. on Electromagnetics and Fluid Dynamics of Gaseous Plasma*, Vol. IX, Interscience Publishers, New York (1961).

10. V. J. Roosow, "On the Flow of Electrically Conducting Fluids over a Flat Plate in the Presence of a Transverse Magnetic Field," NACA Report 1358 (1958).

11. H. Hasimoto, "Magnetohydrodynamic Wakes in a Viscous Conducting Fluid," *Rev. Mod. Phys.* **32**, 860 (1960).

12. S. I. Pai, "Modern Aspects of Magnetofluid Dynamics (Tensor Electrical Conductivity and Multifluid Theory)," ARL Report 66-0060, Aero. Res. Lab. DAR, USAF, Wright-Patterson AFB, Ohio (1966).

13. A. Sherman and G. W. Sutton, "The Combined Effect of Tensor Conductivity and Viscosity on MHD Generator with Segmented Electrodes," in: *Magnetohydrodynamics*, Proc. 4th Biennial Gasdynamics Symp., Northwestern Univ. Press (1962), Chapter 12.

14. F. Fishman, J. Lothrop, R. Patrick, and H. Petschek, "Supersonic Two-Dimensional MHD Flow," Res. Report 39, AVCO Res. Lab. (1959).

15. L. P. Harris, *Hydromagnetic Channel Flows*, John Wiley and Co., New York (1960).

16. H. Alfvén, "On the Existence of Electromagnetic–Hydrodynamic Waves," Arkiv Mat. Astron. Fysik, **29B**, (2) (1942).

17. J. Hartmann, "Hg-Dynamics I," Kgl. Danske Videnskab. Selskab, Mat. Fys. Medd., **15** (6) (1937); J. Hartmann and F. Lazarus, "Hg-Dynamics II," Kgl. Danske Videnskab. Selskab, Mat. Fys. Medd., **15** (7) (1937).

AUTHOR INDEX

SUBJECT INDEX

A

Ablation cooling, 77
Absorption coefficient, 311, 312, 313, 314, 315, 316, 317, 319, 320
Absorption coefficient of radiation, 259
Absorption coefficient of surface, 269
Accommodation coefficient, 240, 243, 244, 247
Alfvén's shock, 361
Alfvén's wave, 342, 356
Angle of attack on hypersonic transition, 104
Approximate one-dimensional flow of MGD, 344
Axisymmetrical hypersonic needle problem, 161
Axisymmetrical slender bodies, 162

B

Band radiation coefficients, 318
Base flow, 219, 224, 228
Black surface, 270
Blasius function, 300
Blasius laminar boundary layer solution, 93
Blast wave parameter, 86, 87
Bluntness-induced pressure, 83, 87, 90
Boundary conditions, 267, 337
Boundary layer flow, 292, 298, 361
Boundary-layer-induced pressure, 83–86
Boundary layer transition, 100–114

C

Channel flow, 286, 350, 368
Chapman–Rubesin constant, 292
Classification of flow regions, 167
Combined bluntness- and viscosity-induced pressures, 83, 87–90
Constitutive equations, 67
Couette flow, 289

Couette-type equation, 169
Critical layer, 101
Critical velocity, 13, 28
Cross-flow, 161
Cyclotron frequency, 367

D

Damped oscillation, 142
Dielectric displacement, 336
Differential approximation, 267
Diffusion thermo-effect, 67
Displacement thickness, laminar flow, 98
Dissociation, 189, 227, 243
Distribution density, 236, 237, 238, 243, 245, 250, 251
Dufour effect, 67

E

Eddy viscosity, 224
Electric field parameter, 343
Electrogasdynamic approximation, 341
Electromagnetic field, 337
Electromagnetic force, 263, 335
Emission coefficient of radiation, 259
Entropy and speed defects, 132
Entropy wake, 139
Equation of conservation of electric charge, 335
Equation of dynamics of electrically-conducting fluid, 334
Equation of electric current density, 336
Equation of radiation gas dynamics, 263
Equivalent body of revolution, 174

F

Falkner–Skan velocity gradient parameter, 95
Fast shock, 360
Fast wave, 365
Fick's diffusion, 67